How to Use an Astronomical Telescope

A BEGINNER'S GUIDE TO OBSERVING THE COSMOS

by James Muirden

A Fireside Book
Published by Simon & Schuster Inc.
New York London Toronto Sydney Tokyo

Copyright © 1985, 1988 by James Muirden
All rights reserved
including the right of reproduction
in whole or in part in any form
First Fireside Edition, 1988
Published by Simon & Schuster Inc.
Simon & Schuster Building
Rockefeller Center
1230 Avenue of the Americas
New York, New York 10020
FIRESIDE and colophon are registered trademarks of Simon & Schuster Inc.
Designed by Irving Perkins Associates
Manufactured in the United States of America
10 9 8 7 6 5 4 3 2 1
10 9 8 7 6 5 4 3 2 1 Pbk.
Library of Congress Cataloging in Publication Data
Muirden, James.
  How to use an astronomical telescope: a beginner's guide to observing the
cosmos/by James Muirden.—1st Fireside ed.
  p. cm.
  Reprint. Originally published: New York: Linden Press/Simon & Schuster, 1985.
  "A Fireside Book."
  Includes index.
  1. Telescope—Amateurs' manuals. I. Title.
QB88.M85 1988   88-1917
522'.2—dc19   CIP
ISBN 0-671-47744-7
ISBN 0-671-66404-2 Pbk.

To my wife
HELEN

Contents

Preface

Astronomy, as a pastime, has never been so popular as it is today —if we rate popularity by the number of books and telescopes that are being sold. Nor has the average amateur ever been so well-equipped as he is today. Twenty years ago, a telescope with an aperture greater than 150 mm was considered pretty large, but nowadays many amateurs start their observing careers with instruments bigger than this. The choice of instruments, once fairly straightforward, is now bewildering.

The choice is no less bewildering when the time comes to turn your telescope to the sky. What should you observe, and why, and how? It is not surprising that many of these fine telescopes soon find themselves lying idle, perhaps brought out to show a visitor the Moon's craters, but not used in any systematic way. And, without a system and a plan, your observational career will soon languish!

The idea behind this book is simple enough. It is to give you some guidelines when selecting an instrument. It then surveys some fields of observation that are open to you, and supplies sufficient information for you to make a start. It assumes minimum knowledge, but maximum enthusiasm!

The author makes no apology for emphasizing the vital importance of training the eye to use a telescope properly, and of observing systematically. If you don't practise, you will not get the best out of your instrument; if you are not systematic, you will not get the best out of the sky. Astronomy is not an easy hobby. No matter how expensive your telescope, you will suffer cloud and cold, unsteady air conditions that blur the view, unwanted bright lights that blind you, and at times sheer tiredness. This is the necessary route towards becoming a true amateur astronomer; and perhaps, when we consider the obstacles, it is not surprising that so many good instruments remain unused, or used only to a fraction of their potential. The author hopes that this book will help reduce that number; for, inside every astronomical telescope, you will find a whole universe waiting to be explored.

CHAPTER 1

Astronomical Telescopes

What is a telescope? It is an instrument that forms an image of a distant object, and it is thanks to a marvelous property of light rays —that they can be bent or "refracted" by a piece of glass, or reflected by a shiny surface—that telescopes are possible. With mirrors, or glass lenses, we can manipulate light rays in any way we wish, casting images of remote objects onto the eye's highly sensitive screen, the *retina*. Countless nerve endings then transmit the color and intensity responses from different parts of this image to the brain, which in turn decodes the information and presents the viewer with a mental image of the physical image produced by the telescope.

The observer's task The view produced by the telescope will be both larger (the *magnification* aspect) and brighter (the *light-collecting* aspect) than what is seen with the unaided eye. However, it must never be forgotten that the telescope's task is only to throw the view onto the retina; it does not, itself, "see" anything. Unscrambling the image is the observer's job. The most perfect image will be wasted if the observer does not put it to good use, and the act of observing is a highly personal one. Set two people down side by side to look at the same object and to draw what they see, and you will notice enormous differences between the results. Some of these differences will be due to fluency with the pencil and general artistic competence, but outside of these effects lie real differences in what is perceived. Some people are very sensitive to color differences, others to symmetry, and so on. Some may try hard to detect fine details while

9

others could find themselves more concerned with overall proportion.

It is true that some people have such defective eyesight that no reasonable comparison is possible. But the differences outlined above do not refer to the *clarity* of vision. It is the way in which the image—as far as we know, the identical image—is used by different people that is so intriguing. The reason for mentioning this individuality of vision so early is to emphasize what a telescope *cannot* do. It can produce an image, but it cannot see; unless it is being used as a camera—and astronomical photography is a very important branch of amateur astronomy—it is only as good as the observer who is looking through the eyepiece. Nobody would expect someone who has just passed the driving test to get the most out of a high-performance automobile; on the other hand, someone with particular aptitude for driving will rapidly overtake (in both senses of the word!) another individual who is merely competent. It isn't easy, however, to convince someone buying a fine new telescope that he or she is going to have to work hard and persistently in order to get the best out of it. Advertisements have a habit of making astronomy look easier than it really is!

However, this chapter is about how telescopes work, so let us defer discussion of the observer until the appropriate place, and take a look at the very important considerations of magnification and, to begin with, aperture.

Light-collecting power The eye's light-collecting power is controlled by the *pupil*, a variable aperture that opens to its widest amount (about 8 mm in diameter) in dim light, and closes down to about 2 mm in sunlight. At night, then, we are observing with an 8-mm aperture "telescope." Even this modest instrument is sufficient to reveal some 2000 stars at any one time if the air is very clear, there are no nearby artificial lights, the Moon is absent, and the eye focuses sharply. However, there are innumerable stars which are too faint to be seen with the keenest unaided eye, and to detect them we have to use an aperture larger than 8 mm, so that more light can be collected and focused, and fainter stars therefore have a chance of energizing the nerve endings.

The area of a circle is proportional to the square of its diameter. Logic suggests, therefore, that a telescope with a light-collecting aperture of 16 mm will collect four times as much light as will the naked eye, making the same stars appear four times as bright, and revealing stars that are only a quarter as bright as the dimmest naked-eye stars. The same reasoning suggests that a telescope with an aperture of 100 mm—which is modest by amateur standards—will show the same star looking $(100/8)^2$ or about 156 times as bright as when seen with the naked eye, or reveal stars 156 times fainter than those visible without any optical aid. This is an enormous increase in light-gathering power, and it is not surprising that even a relatively small telescope utterly transforms our view of the universe.

Does 156 times the light-gathering power mean that a 100-mm aperture telescope will reveal 156 times as many stars in the sky? To investigate this question, it is necessary to understand how star brightnesses are graded, and this is important enough to be worth examining straight away. The brightness of a star is called its *magnitude,* and the magnitude scale is based on an ancient system of measurement in which the brightest naked-eye stars were called "1st magnitude" and the faintest were called "6th magnitude"—so the higher the magnitude number, the fainter the star. This was, originally, a very approximate grading, but modern brightness-measuring devices known as *photometers* permit the brightness of a star to be measured to within a hundredth of a magnitude unit. One magnitude step now corresponds to a brightness ratio of 2.512 times. The reason for choosing this number is that five magnitudes correspond to a brightness difference of exactly 100 (or 2.512 to the power of 5).

Theoretically, a 100-mm aperture telescope will gain about 5.2 magnitudes over the unaided eye. Therefore, whereas the naked eye will normally see stars no fainter than the 6th magnitude (although some observers, under extraordinarily good conditions, have reported stars of magnitude 6.5 or even fainter), the eye and telescope combined should reach the 11th magnitude. On any one night, there are several *million* stars of the 11th magnitude and brighter above the horizon. The telescope will, therefore, reveal perhaps a thousand times as many stars as the naked eye, and not just 156 times as many: far more than you could hope to observe, individually, in a lifetime. Imagine a thousand separate skies full of stars, and that is the gift to your eye of a 100-mm telescope.

We have already stated that this aperture is a modest one by commercial standards. Most amateur-owned telescopes fall in the 75- to 250-mm range, but some are much larger. In any of these instruments, the view is at first bewildering: the crowds of stars cannot be related to anything seen with the naked eye. A small low-power telescope attached to the main instrument, known as a *finder,* is of great value in locating objects, and an instrument of any reasonable size needs one. The task of the finder is to negotiate between what the eye sees and what the telescope reveals. If it is too small, it won't show enough telescopic stars; if too large, it will defeat its own purpose and confuse the eye with too many. For most amateur instruments, an aperture of about 50 mm and a magnification of about eight times is ideal.

Magnification "How much does it magnify?" is the frequently heard inquiry when a telescope is mentioned. The answer to this is "It depends." With the exception of small hand telescopes and binoculars, which have a fixed magnification, the magnifying power of a telescope depends upon the eyepiece or *ocular* that is used with it. Most astronomical telescopes are equipped with several eyepieces, giving a range of magnification. A telescope requires several different magnifications or "powers" because the answer to the question "How much *should* it magnify?" is not always the same. At first sight, it might seem obvious that the highest possible power will give the most detailed view of an object. While this may be true if a finely marked planetary disk is being examined, it certainly is not true if we want to survey a scattered cluster of stars or a large, hazy nebula. A lower magnification shows a larger area of sky at one view than does a higher one, and for some purposes it is important to have a wide rather than a narrow view.

Field of view In astronomy, the diameter of the field of view (which is the width of sky that can be seen at one time, without moving the telescope) is reckoned in angular measure. An angle of 1° corresponds to twice the diameter of the Moon or Sun in the sky. Binocular fields of view are usually rated in terms of the number of meters that can be seen at a distance of 1000 meters. This is all right for terrestrial viewing, but astronomically the same field could show a few craters on the Moon or encompass a whole group of galaxies! This is why the astronomer normally uses angular units of distance —in other words, how far apart objects appear to be in terms of degrees, minutes of arc, and seconds of arc. These latter units are

more correctly styled *arcmin* and *arcsec,* expressions which the
writer finds particularly ugly, and the old-fashioned symbols ' arc and
" arc will be used in this book to signify angular minutes and seconds
respectively. One degree (1°) is equivalent to 60' arc, each one of
which is equivalent to 60" arc. Although 1" arc may seem a tiny angle,
it is one that can be divided or *resolved* by most good amateur telescopes. It corresponds, approximately, to the thickness of a human
hair viewed from a distance of 2½ meters.

As a rough guide, a magnification of about 50 times—which means
that the apparent width or height of an object is 50 times larger than
when viewed without the telescope—will reveal a circle of sky about
1° across. If it is doubled to 100 times (more conveniently written as
×100), the diameter of the field of view is halved, to ½° or 30' arc,
and the whole of the Moon's disk could just be included in one view.
At ×200, the diameter of the field of view is halved again, to about
15' arc, and so on. The diameter of the field of view depends only
upon the design of eyepiece and the magnification it provides—it has
nothing to do with the size or type of telescope. Some designs of
eyepiece give fields of view noticeably wider than those quoted here,
while others have a narrower inherent field of vision.

Eyepieces What, then, *is* an eyepiece? It is a powerful magnifying
glass, consisting usually of at least two small lenses arranged close
together inside a metal or plastic mount. (The lens closest to the eye
is called the *eye lens,* while the one facing the telescope's upper end
is known as the *field lens,* its function being to increase the useful
field of view.) An eyepiece works by permitting the eye to be brought
very close to the telescopic image formed by its large lens or mirror.
The closer we are to something, the larger that thing appears to be.
The normal human eye cannot focus—at least not without some dis-
comfort—upon any object that is less than about 25 cm away. It is
the function of a magnifying glass (the ordinary kind) to allow the eye
to be brought much closer to whatever is being examined, so that it
appears larger and more detailed. For instance, a "×5" magnifier
makes an object appear as if it were being viewed from a distance of
5 cm rather than 25.

This value of 5 cm corresponds pretty closely to the *focal length*
of the magnifying glass. If the focal length were only 10 mm, then
this would be the effective distance from which the object were being
viewed, equivalent to a magnification of ×25 compared with the

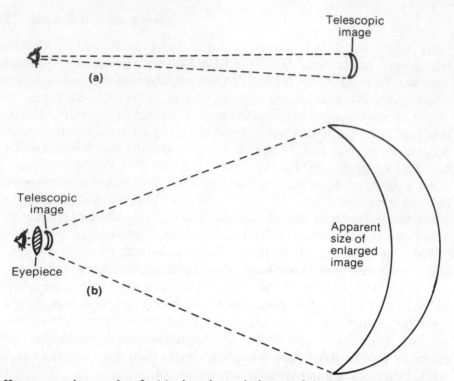

How an eyepiece works In (a), the telescopic image of the Moon is being viewed from the smallest distance of comfortable distinct vision—normally about 25 cm—and appears rather small. In (b), the use of a magnifying lens permits the eye to be brought very close to the image, which appears larger in direct proportion to the decrease of distance. For convenience, the drawing shows the enlarged image of the Moon apparently lying at the smallest distance of distinct vision; but the correct practise is to relax the eye, so that the image appears to be at infinity.

closest naked-eye view. Therefore, the shorter the focal length of the magnifying glass or eyepiece, the more powerful a magnifier it becomes.

Astronomical eyepieces are available in a considerable range of focal lengths. The longest are up to 50 mm or so, the shortest may be as little as 4 mm. There are many different optical designs, some optimizing critical viewing of fine detail, others offering a very wide field of view—some very expensive ones, such as the Plossl, claiming to achieve both simultaneously! Normally, a telescope comes ready equipped with standard eyepieces, which should be of reasonable quality. Designs such as the Achromatic Ramsden, Orthoscopic, Erfle, and Monocentric are all frequently encountered, but the design on its own means little if the eyepiece is poorly made, and a sky test by an expert is the only reliable guide if you are not happy with the performance of the telescope, because image faults can arise in the telecope's optical system as well as in the eyepiece. However, if you are using a reflecting telescope (page 24), and the image of a planet or bright star has a noticeable colored fringe around it, then the fault must lie in the eyepiece, which should be rejected.

Many smaller telescopes, that come with a standard set of eyepieces, have the magnifications given by these eyepieces already listed. If they are not, then it is necessary to know the focal length of the telescope because this determines the size of the image that the eyepiece is called upon to magnify. More will be said about focal length later in this chapter, but it can be pointed out here that the longer the focal length, the larger the image of a celestial object formed by the telescope; with a focal length of one meter, the image of the Sun or Moon is about 9 mm across, and telescopic focal lengths of from one to two meters are the most common. The magnification given by a particular eyepiece is determined by dividing its own focal length into that of the telescope.

High powers and the Barlow lens Suppose that you have a telescope of 1500 mm focal length, and wish to use a magnification of ×300. The focal length of the eyepiece will therefore need to be 1500/300 or 5 mm. The eye lens of such an eyepiece is so small that some difficulty may be experienced in seeing anything through it at all because the eye must be placed in exactly the right position behind it—a location not achieved without some practice. Even then, tiny eyepieces are not comfortable to use.

Many experts therefore recommend the use of a *Barlow* lens for high-power work. This is a device which is placed a little way inside the focal point of the telescope, having the effect of increasing (usually doubling) the effective focal length. With a Barlow in place, the magnifying powers of all existing eyepieces are doubled, and so the more comfortable, longer-focus oculars can be used on all occasions.

Resolving power Magnification cannot be increased without limit. The Earth's flickering atmosphere rarely allows powers higher than about ×350 or ×400 to be used with any telescope. There is, however, another and more basic limit to the useful range of magnification, set by the amount of detail which the telescope itself imprints into the image. This is known as the *resolving power* of the telescope. The telescopic image is effectively made up of a great number of tiny units, somewhat analogous to the dots in a half-tone photograph. Viewed from a normal distance, the picture appears to be continuous. Examined through a magnifying glass, however, it breaks down into a pattern of individually meaningless large and small dots, and the contours of the photograph are lost.

The effect of very high magnification is not to break up the image in this dramatic way, but to lose contrast. Overmagnifying does not produce new detail on a planet, but merely spreads and dims that which is already visible with a lower power. Having said all this, however, it is still a fact that the atmosphere almost always sets a magnifying limit below this image-degradation limit. In other words, a telescope set up in space could habitually profit more from high magnifications than those regularly used on the Earth's surface. Those rare nights of superb seeing will, frustratingly, tend to support this view!

To fix ideas, magnifications of about twice the aperture expressed in millimeters are necessary to give ready visibility to the resolved detail in a telescopic image. If the atmosphere normally sets a limit of about ×400 to what can usefully be employed, it is evident that large instruments are frequently being used below their true potential, while smaller instruments are more likely to give optimum performance. This does not, however, mean that a large instrument will reveal no more detail than a small one, for its extra light-grasp will always be a powerful factor, and even on the worst nights there are moments of relative steadiness when fine detail jumps into view.

Test double stars The night sky contains a great many convenient resolution tests. These are *binary stars:* pairs of suns revolving around each other, separated by distances from millions to thousands of millions of kilometers, and taking from years to centuries to complete a revolution. In the sky, some appear to be several seconds of arc apart, while others may be separated by only a fraction of one second. The stars themselves are so distant from the Earth that, despite their large size, they appear only as points of light far smaller than the resolving power of an ordinary telescope.

Point a telescope of small aperture towards a binary star whose components are separated by an angle smaller than the telescope's resolving limit. Then, no matter how high a magnification is used, the object will appear as a single star, though possibly of elliptical outline if it is almost resolved. Now examine the same double star with a telescope of much larger aperture, whose resolving power exceeds the angular separation of the binary. The object should now be seen as a close pair, even though the magnification may be the same!

Aperture and theoretical resolving power go hand in hand, and can be enshrined in a table as shown here:

Aperture of telescope (mm)	Resolving power (" arc)
50	2.3
75	1.5
100	1.2
150	0.77
200	0.58
250	0.46
300	0.39
400	0.29
500	0.23

Having done so, it should be added straight away that such a table can be highly misleading! This does not prevent manufacturers from stating that their particular model will resolve to the limit—often referred to as the *Dawes limit*. It may well do so in the laboratory, looking at an artificial double star. But there are plenty of double stars in the sky, wider than the theoretical resolution limit, that it has no hope of resolving. To begin with, the stars must be of approximately equal brightness, or else the fainter one will be obliterated by the brighter. Furthermore, even if they are of similar brightness, they may just be of the optimum magnitude for the particular aperture being used: too faint, and the eye will have difficulty in seeing them sufficiently clearly; too bright, and the individual stars will seem to enlarge and blur together. We shall have much more to say about observing double stars in Chapter 9. For the moment, the best advice is not to be beguiled by tables of "theoretical" performance because they can lead to quite unjustified disappointment.

Refracting telescopes The first telescopes, which appeared back in the 17th century, were of the *refracting* kind, and there are many people today who say that a good refractor is the best of all telescopes. The refractor uses an *object glass* or *objective,* which consists of a pair of lenses mounted in a cell at the upper end of the tube, to collect light and focus the image. The *focal point* of the objective (in other words, the place where the image is formed) is at the lower end of the tube. Here, the eyepiece fits into the *drawtube,* that can be moved in and out slowly with a knob to permit sharp focusing. The focal length of the object glass is, as would be expected, about equal to the length of the telescope tube.

Achromatism The performance of any telescope depends critically upon two things: the steadiness of the mounting that carries the tube, and the quality of its image-forming components. We shall have more to say about mountings in due course. In a refractor, then, the object glass must be singled out for particular attention. First of all, it must be *achromatic,* a word literally meaning "without color." A single

A refracting telescope   (a) The tube and fittings of a refractor are simple and fairly standard. The object glass is mounted in a cell which should, if the aperture is about 75 mm or more, have adjusting screws for bringing the lens accurately square-on to the axis of the tube. These are not shown. The drawtube, to which the eyepiece is attached, slides in and out when the focussing knob is turned. (b) The inner or flint lens of an achromatic object glass increases the focal length of the outer crown lens and effectively eliminates the color dispersal of the latter. It does this by superimposing its own, reversed, color dispersal on the optical system. The effect is greatly exaggerated in the diagram, and the focal length of an achromatic object glass is typically about fifteen times its aperture.

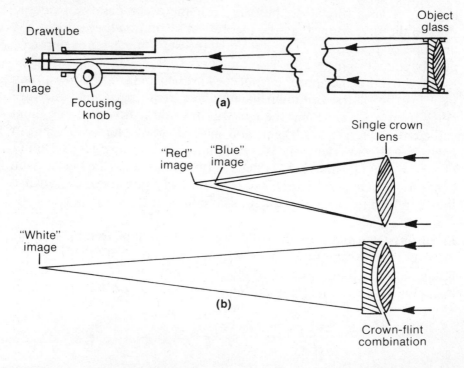

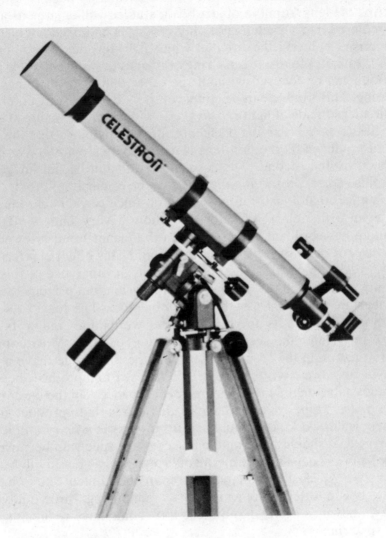

A refracting telescope   An 80-mm (3.1-inch) refractor made by Celéstron. It is mounted on a German equatorial stand, similar to the Newtonian telescope shown on page 24. Note the right-angled eyepiece attachment, which makes for a more comfortable viewing position at all altitudes of the telescope. A dewcap extends in front of the objective lens.

glass lens certainly can form an image if it is curved to a suitable cross-section, but this image will be of poor quality because the light entering the lens from the object being studied will emerge as a rainbow-colored blur. The lens not only produces an image but also acts as a prism, splitting the original white light up into its component colors. In fact, the result is a string of images of different color, none of which can be focused without the other, out-of-focus images, interfering. This trouble can be almost entirely overcome by combining the first lens (made of light *crown* glass) with a second (made of dense *flint* glass), the pair having been carefully computed so that the color spread produced by the first lens is matched by a color convergence supplied by the second. The result is, or should be, an image free from false color, preserving the tones of the original.

It is a fact that no pair of lenses can produce *perfect* achromatism, and even a first-class refractor will show a very faint violet halo around a very brilliant object, such as the planet Venus viewed in a dark sky, or perhaps the edge of the Moon. This effect is referred to as *secondary spectrum.* The obtrusiveness or otherwise of this halo depends partly upon the observer's color vision, but principally upon the aperture and focal length of the objective. The longer its focal length in relation to its aperture (in other words, the greater its *focal ratio*), the fainter the secondary spectrum becomes. Most commercial refractors, in the 75-mm to 150-mm aperture range, have a focal ratio of about 15, written f/15. This gives a compromise between unwieldy tube length and color-free performance. On the other hand, some modern types of glass have made it possible to produce object glasses which work acceptably at f/10: if you see a *fluorite* objective advertised, it should have minimal secondary spectrum because it is made from a special and rare type of crown glass—but it will be very expensive. At least one manufacturer in the United States has recently begun offering object glasses containing three individual lenses, known as *triplets,* and these too should have reduced secondary spectrum.

However, for most visual work with normal apertures, secondary spectrum is quite unimportant. It becomes serious with increasing aperture, out of the amateur range—but even the world's largest refractor, the 1-meter (40-inch) at Yerkes Observatory, has been in constant and successful use since 1897! It is certainly a nuisance in photography because photographic emulsions have more of an appetite for violet light than does the human eye—and in this connection it may be worth remarking that violet sensitivity does vary from one individual to another.

Virtues and drawbacks of the refractor Why, therefore, do so many people have such a deep regard for the refractor, if it has this inherent although certainly not insuperable drawback? Simply because a good refractor can most easily approach the most perfect clarity of definition possible with any telescope. The writer must choose his words very carefully here because the "refractor versus reflector" controversy is a notorious minefield; and he is certainly not saying that the refractor is the best kind of telescope, full stop! Such a judgment would be absurd. For example, not all astronomical work demands exquisite definition of image. But, for the work that does—which means that we wish to distinguish between those instruments defining adequately from those defining superbly—there is a good chance that the refractor will out-perform another type of telescope of similar aperture. Three reasons are offered:

1. Optical errors on the surface of a lens have less effect on the final image quality than would the same errors on the surface of a mirror: approximately a quarter as much.
2. The reflecting surfaces of mirrors, particularly as they age, tend to scatter more light around the image than do the glass surfaces of a lens. Therefore, the contrast of the image may be better.
3. A lens is much less sensitive to errors of adjustment than is a mirror.

What these points are saying is this: although it might be hard to distinguish the performance of typical refractors and reflectors under laboratory conditions, what matters is the way they work under the sky. In practice, the refractor may well prove more durable and consistent, able to suffer more mishandling and degradation without its optical performance being compromised.

However, the refractor has a serious drawback: its bulk. At f/15, even a 100-mm refractor can reach an adult's shoulder. Because the eyepiece is at the lower end of the tube, the telescope has to be mounted well above the ground if the eyepiece is to be brought to a comfortable height. Furthermore, for viewing objects at an altitude of more than about 60° above the horizon, an attachment known as a *zenith prism,* which turns the ocular at right-angles to the drawtube, is practically essential. A zenith prism can be a nuisance because it reverses the image top for bottom. It should be noted, in passing,

that most astronomical telescopes completely reverse the image anyway, so that what is up in the sky appears down, and what is right
appears left. But this can be overcome simply by turning the map or
chart, if one is being used, through 180°. A one-way reversal produced by a zenith prism cannot be accommodated in this way, and
the only course of action is to look through the back of the chart, a
not always tenable procedure.

Not only is a long tube awkward: it is also difficult to mount steadily. This is a most important consideration, for the excellence of a
telescope's mounting is at least as important as the quality of its
optics.* A refractor is, undeniably, difficult to mount. Whether this
tips the final judgment away from it must depend upon the individual
observer and the particular telescope; what must be said here is that
refractors of about 150 mm aperture and over should be mounted on
a solid pillar in an observatory rather than on a wooden or even steel
tripod. If both portability and large aperture are considered necessary, the refractor does not come into the reckoning.

Reflecting telescopes A mirror with a slightly concave surface can reflect light to form an image in a way analogous to the refractive
formation of a lens. In fact, it can do the job rather better because
mirrors do not disperse light into colors—they are perfectly achromatic. This is a great advantage enjoyed by reflecting telescopes.
Another benefit is that the mirror does not need to be made of super-
fine optical glass because the aluminum coating is applied to the front
rather than to the back of the glass, the reason being that passage
through a thick glass disk would introduce unwanted color into the
image.

Ordinary thick plate glass can be used for making telescope mirrors, but most are made of low-expansion glass, such as Pyrex, or
even from pure silica (fused quartz), which has practically zero expansion with temperature change. This can be important because the
definition of a slightly expanding or contracting mirror can be ruined.
A reflecting surface is also particularly sensitive to flexure, and mirror disks are commonly made very thick, with a diameter : thickness
ratio of 6 or 8. If the mirror is large, however, such a cross-section
results in an expensive and weighty disk, so the modern tendency is

---

* In assessing a telescope, optical quality is usually given precedence over everything else.
However, unless the telescope itself is mounted properly, it will be irritating to use, and
observational work will languish. The finest optics cannot perform well if poorly mounted.

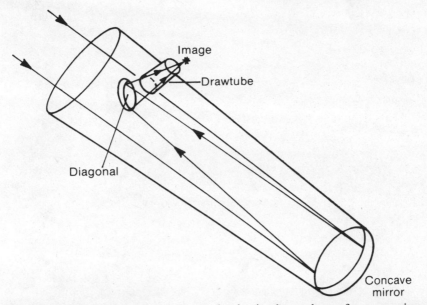

A Newtonian reflecting telecope A round tube is shown here, for convenience. Many Newtonians, particularly home-constructed ones, have square or framework tubes.

to use thinner glass and to take care to mount it in a cell where carefully designed supports preserve its optical form. Telescope advertisements often refer to 9-, 18-, or even 27-point flotation systems, but a typical thick 150-mm aperture mirror can be mounted perfectly well on just three locating points.

Other agreeable advantages accrue from using a mirror as the light-collector and image-former. It is, for instance, easy to obtain mirror disks up to 400 mm or 500 mm across. A number of amateurs have ground and polished their own mirrors of this order of aperture. Therefore, light-collecting power is there in plenty. Furthermore, because a mirror is perfectly achromatic, there is no need to make reflecting telescopes with the large focal ratios necessary for refractors. Many reflectors work in the range of f/5 to f/6. At f/5 you can have three times the aperture (hence, nine times the light-collecting power) of a typical refractor, for the same tube length. This is of enormous benefit when the telescope mounting is considered.

Newtonian reflectors All reflecting telescopes use a concave mirror, situated at the lower end of an open-mouthed tube, to reflect and

A Newtonian reflector This 200-mm (8-inch) aperture model by Meade Instruments possesses the standard features to be expected on a telescope of this type. The electric drive is enclosed in a box at the lower end of the polar axis (left of pillar); the declination axis has adjustable counterweights (right of pillar) to balance the weight of the telescope tube and accessories. Without this facility, too much strain would be put on the driving gears. The finder has small adjusting screws, which permit it to be aligned with the main telescope.

focus the light back up the tube. What happens then depends upon the design. In the common *Newtonian* form, a small, optically plane mirror of rectangular or elliptical outline intercepts the reflected light and reflects it through the side of the tube to where the eyepiece is situated. This mirror, which is referred to as the *diagonal* or *flat* (not as the "secondary," which is something else altogether), should preferably be elliptical, so that its outline, when viewed at an angle of 45°, is circular. For most amateur purposes the Newtonian is the most satisfactory, as well as the cheapest, reflecting telescope. Not the least of its advantages is that the eyepiece is near the upper end of the tube, and more or less equally convenient for the observer regardless of the altitude of the object being observed.

The cross-sectional shape of the concave mirror's surface is not part of a sphere, but part of a paraboloid; or at least it should be. A spherical concave mirror will not give a perfect image unless the focal ratio is unusually large—about f/12 with 150 mm aperture, and greater with larger apertures. Effectively, a parabolic mirror is a spherical mirror with the center deepened very slightly. In amateur sizes, this deepening can be measured only in terms of wavelengths of light: the difference in depth at the center of spherical and parabolic f/8 mirrors of 150 mm aperture is only 0.0003 mm, or 0.6 of the wavelength of yellow light! Yet this minute amount of glass makes a huge difference to the quality of the final image. Controlling such figuring in commercial production is a highly skilled business, and by no means all mirrors are as good as they should be. Typically, makers offer different grades of optical finish, depending upon how closely the surface approaches theoretical perfection.

Mirrors are usually graded in terms of wavelength error. A sample might be quoted as "⅛th-wave quality." This, by itself, does not mean anything very definite. It could mean that some parts of the surface are ⅛th of a wavelength of light deeper than the theoretical paraboloid, while others are up to ⅛ of a wave higher. In other words, the total "ripple" on the surface is up to ¼ of a wave. This would be a poor quality component. Alternatively, it could refer to a total ripple of ⅛th of a wave, or a maximum deviation of ¹⁄₁₆th of a wave from the paraboloid. Such a mirror should approach theoretical perfection of definition.

But a wavelength of what? Red, yellow, or blue light? Deep red light has almost twice the wavelength of violet light; therefore, a "red light" error will be physically much more objectionable than a "blue

light'' error. The eye is most sensitive to yellow light, roughly midway in wavelength between red and blue, and it is errors measured
at this wavelength (about 550 nanometers, or 0.00055 mm) that are
significant.

The optical quality of the flat is often ignored, or taken for granted.
But its optical tolerance is just as critical as that of the main mirror.
Therefore, this needs to be known too.

What we are saying is that there is plenty of scope for sharp practice in the manufacture and sale of Newtonian reflectors. Never judge
the optical worth of an instrument—any instrument—by a maker's
claims. A star test is the only way of establishing its excellence or
otherwise, and the section on page 30 is devoted to this important
procedure.

Cassegrain-type telescopes The basic Cassegrain design is rarely seen
in amateur sizes because very few manufacturers offer this type of
instrument for sale. Large professional telescopes, however, almost
always belong to the Cassegrain family. The Newtonian diagonal is
replaced by a small, convex, *secondary* mirror that reflects the con-

A Cassegrain-type reflecting telescope   Cassegrain telescopes can take many forms,
but the important principle is the interposition of a small convex mirror into the
converging beam from the primary mirror. The resultant effective focal length is
equal to the focal length of the primary mirror multiplied by $\dfrac{b}{a}$ (the amplification of
the secondary). Amplifications of 4 or 5 are common. Very low amplifications are
undesirable because the diameter of the secondary becomes objectionably large.

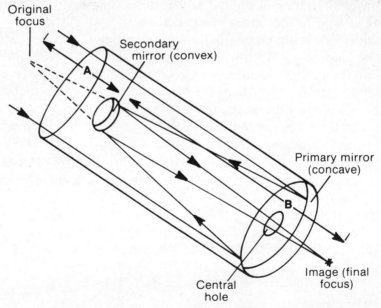

verging beam from the primary mirror back down the tube. This secondary effectively increases the focal length of the primary without necessitating a corresponding increase in tube length. The returned beam either passes through a small hole cut in the center of the primary mirror, or is reflected by a Newtonian-style plane mirror through a hole in the side of the tube, somewhere near the lower end.

The effective focal length of a typical Cassegrain may be between f/10 and f/30—separate secondaries can be supplied to give different amplifications. A long focal length, or large focal ratio, has some definite advantages. If the telescope is being used photographically, the image scale is increased: a focal length of 1000 mm produces a 9-mm lunar image, but this expands to 27 mm if the focal length is 3000 mm. A 3-meter tube would be extremely cumbersome, and the benefits of compressing this focal length into a much shorter tube are obvious. Visually, too, a long focal length means that high magnifications can be obtained without using very short-focus eyepieces, or a Barlow lens.

The larger the focal ratio of an object glass or mirror, the better is an eyepiece likely to perform. It must be remembered that eyepieces, like object glasses, have to be made achromatic by the introduction of separate lenses, and that their residual color is tested more severely when they are being used in conjunction with an object glass or mirror of short focal length. An f/5 Newtonian makes much greater demands upon eyepiece perfection than does an f/10 or f/15 Cassegrain or refractor.

The following table may be found useful. It sets out the eyepiece focal length necessary to produce a given magnification with telescopes of various focal lengths.

Magnification required	Focal length of telescope (mm)				
	1000	1500	2000	2500	3000
× 50	20	30	40	50	—
× 100	10	15	20	25	30
× 150	6.5	10	13	16.5	20
× 200	5	7.5	10	12.5	15
× 250	4	6	8	10	12
× 300	—	5	6.5	8.5	10
× 350	—	4.5	5.5	7	8.5

(All values have been rounded off to the nearest 0.5 mm.)

The advantage of a Cassegrain-type telescope for high-power work is obvious. On the other hand, an f/20 system would not be convenient for low-power, wide-field work. The diameter of the drawtube into which the eyepiece fits is rarely more than 30 mm; in the United States, one common ''standard'' eyepiece fitting takes ocular mounts that are exactly 1¼ inches across. Used at the focus of a telescope with a focal length of 3000 mm, this width is only a little greater than the diameter of the lunar image (½°), so that a view of the sky encompassing a greater diameter than this would be unobtainable.

Catadioptric or ''lens and mirror'' telescopes The handy compactness of the Cassegrain design has sent manufacturers seeking ways of combining a short, portable tube with an effective focal length that is not too long. The result has been the family of squat-tubed instruments, familiar from advertisements in astronomical and scientific magazines, that at first sight look utterly unlike a telescope at all. The

Catadioptric Cassegrain telescopes   The Schmidt (upper) and Maksutov (lower) systems both utilise an optical component at the front of the tube to correct the aberrations of the spherical primary mirror. Some Maksutovs, such as the one shown here, employ an aluminized spot on the rear surface of the shell as the convex secondary mirror. These instruments are relatively very compact, and have the benefit of a tube sealed from dust. The undulating curve on the Schmidt plate is greatly exaggerated in the diagram.

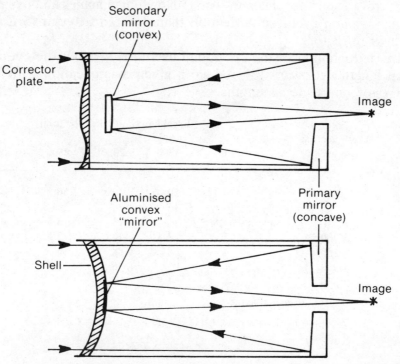

tube of a typical 200-mm aperture catadioptric telescope is less than 450 mm long, as opposed to a probable length of 800 mm for a true Cassegrain of the same aperture, or perhaps three meters for a refractor! How is this extreme compactness possible—and what are the benefits to the observer?

It would be possible, both in theory and in practice, to make an ordinary Cassegrain telescope with a tube length of only twice its aperture. It would, however, be very expensive to figure a mirror of such short focal length to the correct parabolic shape. Furthermore, the mirror, once finished, would be extremely sensitive to maladjustment. If it were even slightly shifted from its perfect position, the image quality would be ruined—parabolic mirrors of very small focal ratio need to be adjusted very critically indeed. It is true that the recent generation of huge observatory telescopes are Cassegrains with primaries of very small focal ratio—about f/2.5—but there is a world of difference between making a single expensive instrument and mass-producing telescopes for the amateur market.

Schmidt and Maksutov systems The seed of the catadioptric idea was planted in the mind of the Estonian astronomical optician Bernhard Schmidt half a century ago. He realized that a spherical rather than a parabolic mirror could give a perfect image if the light first passed through a very thin, almost flat plate of glass. This correcting plate has so slight a curve or "figure" on it that, to a casual glance, it looks just like an ordinary piece of glass. Schmidt was interested in the wide-field photographic possibilities of this device, and the so-called *Schmidt camera* was born. Much later, it was realized that the same principle could be used for correcting a visual telescope, using a spherical rather than a parabolic primary mirror. There are several advantages:

1. We have already seen that the surface of a lens does not have to be figured as accurately as that of a mirror. Because a spherical mirror can be mass-produced quite easily, it may be cheaper to "figure" it by making a corrector plate.
2. The corrector plate/spherical mirror combination is not so sensitive to misalignment as is a single, short-focus parabolic mirror.
3. The small secondary mirror of the Cassegrain-type system can be mounted directly on the corrector plate, obviating the need for a supporting arm.
4. By sealing the contents of the tube from the outside air, the corrector plate protects the optical surfaces from dust and tarnish.

5. It is possible to design a corrector plate/mirror combination that gives a wider field of good definition than could an ordinary Cassegrain. However, this is usually significant only with photographic work.

This was the genesis of the *Schmidt-Cassegrain* telescope, of which the best-known, at the present time, is the Celestron. There is also a rather similar system, in which a very strongly curved lens or *shell* replaces the almost flat corrector plate. This is the *Maksutov-Cassegrain*, of which the best-known example is the Questar. In some Maksutov designs, the center of the shell is converted directly into the secondary mirror by coating it with a circle of aluminum. Although there are considerable technical differences between them, both systems embody the same principle of a relatively long focal length compressed into a short tube. For example, the Celestron 8, of 200 mm aperture, has an effective focal ratio of 10, so that a 2000-mm focal length is compressed into a tube only 450 mm long.

Advantages of the catadioptric Even though it contains a refracting element, whether plate or shell, a well-made catadioptric telescope gives an image that is practically as achromatic as a pure reflector. Also, the use of the Cassegrain system adds another dimension to the possibilities of this telescope because, by altering the distance between the primary and secondary mirrors, it can be made to focus on objects near at hand. An ordinary refractor or reflector can also do this, but only by pulling the eyepiece a long way out beyond its normal setting for "infinity." In the case of the Cassegrain, however, the position of the final focus is very sensitive to changes in the inter-mirror distance, and an increase in this distance of only a centimeter or so will allow the observer to focus on a bird in a nearby tree instead of on the Moon! This is accomplished by turning a knob that moves the primary mirror back and forth inside the tube.

As achromatic as a reflector, fantastically compact, and highly versatile . . . is the catadioptric telescope the ultimate choice for the observer? Before the case can be fairly examined, we need to look at the ways in which the different types of telescopes can be mounted.

Testing a telescope Only someone with a certain amount of experience can make a really worthwhile judgment of a telescope's optical quality. You are, therefore, strongly advised to seek the judgment of

Plate 3

A Schmidt-Cassegrain catadioptric telescope Compare the length of this 200-mm (8-inch) telescope with the Newtonian of similar aperture, shown on page 24, to appreciate the extreme compactness of this design. The mounting is of the Fork type, which needs no counterweight, and the electric drive is enclosed within the round box at the base of the fork. There is only a limited range of eyepiece movement for all positions of the tube.

such an expert, perhaps through an astronomer friend, or through the local astronomical society. However, should you find yourself without any such aid, the following hints may help.

The telescope's optical components must be well aligned. A catadioptric telescope should have been checked before it left the factory, and should need no subsequent adjustment. If it seems faulty in this respect (in other words, the image is flared on one side, or otherwise unsymmetrical), the instrument should be returned to the agent or manufacturer.

A refractor, too, should already be well aligned because its optical adjustments are the most permanent of any telescope. If the image is showing one-sided flare, or the out-of-focus star disks are elliptical, it will be necessary to adjust the object glass by using the three sets

The optical adjustments of a Newtonian reflector The first stage (a) is to bring the outline of the diagonal mirror squarely in line with the drawtube. When this has been done, by bodily moving the diagonal mounting, the image of the main mirror is brought into the center of the diagonal (b). This is done by tilting the diagonal, and adjustment (a) may be upset and have to be corrected again. Finally, the main mirror is adjusted until the dark silhouette of the diagonal is central, as in (c). The last delicate adjustments of the main mirror should be done using a star and a high-power eyepiece, so that its out-of-focus disks are symmetrical.

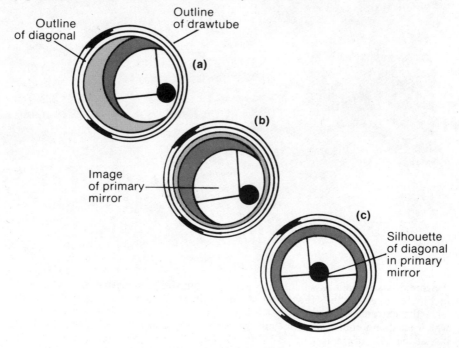

Outline of diagonal

Outline of drawtube

(a)

(b)

Image of primary mirror

(c)

Silhouette of diagonal in primary mirror

of adjusting screws set around the cell. Very carefully screw one set in or out by a turn or two, and observe the result. Proceed by a process of trial and error until the out-of-focus star disks, as described below, are as circular as they will go. If no amount of adjustment can bring them truly circular, there is something wrong with the object glass.

A Newtonian reflector is quite likely to be out of adjustment when it arrives from the manufacturer, or if it has been purchased second-hand. New instruments usually come with instructions on the method of alignment, but a summary will be given here:

1. Remove the eyepiece from the drawtube and rack it well out. Place a cardboard disk over the open end of the drawtube, perforated with a central hole about 5 mm across.
2. Uncap the mirrors and direct the telescope at some bright surface—a white wall, or the daytime sky.
3. Look through the perforation. The outline of the flat will be seen. Ensure that its outline appears concentric with the drawtube—in other words, that the axis of the drawtube passes through the apparent center of the flat. There should be an adjustment device on the flat's mounting for this purpose.
4. Now adjust the tilt of the flat so that the image of the primary mirror appears central within it.
5. Study the image of the primary mirror. The black outline of the flat will be seen, as well as the outline of the flat mounting (sometimes a single strut, more often three or four radiating vanes). Adjust the primary mirror's screws until this outline of the flat appears central in the reflection of the primary.
6. Examine the expanded star disks for symmetry. Any slight errors can be remedied by very delicate adjustment of the screws supporting the primary mirror.

Having brought the optical components to good adjustment, you can begin testing.

There is only one satisfactory test object—a star. For apertures of up to about 150 mm, a 2nd or 3rd magnitude star is ideal—a magnitude fainter, perhaps, for a 300 mm telescope. A very bright star often shows so much atmospheric flare that the test sensitivity is lowered. Don't look at the Moon; almost any telescope will make it look so marvelous that you will be tempted to purchase the instrument on the spot! The planet Jupiter is a good test if the atmosphere is steady and you are an experienced observer. But a star test is useful on almost any night, and even someone who has never looked through a telescope before can make some sort of judgment.

When making a star test, you do not look at the focused star. Turn the focusing knob so that a high-power eyepiece ($\times 200 - \times 300$, depending upon the aperture) moves just a few millimeters inside and outside the position of best focus. The star image will expand into a small disk. What you must now do is compare the appearance of the two disks. The more alike they are, the better the telescope. In a perfect telescope, they will be indistinguishable. A reflector will reveal a little black spot at the center of both intra- and extra-focal disks: this is the outline of the flat or secondary. A refractor has no central spot, but will produce a faintly-colored fringe—the remnants of the chromatic aberration. In a well-corrected object glass the intra-focal disk will have a reddish-purple fringe, while the extra-focal disk will be edged with bluish-green. Use a white test star rather than a red one, to show up these color differences.

It is easier for the amateur to judge a bad telescope than a good one. A certain amount of difference between the two images can be tolerated because the test is an extremely sensitive one. But if the two images are clearly different (one may have a hard, bright edge, while the other has a soft blurred outline, for example), then the optical system is poor, and probably not worth bothering with.

It may seem strange that the focused star image is not used as a test. In theory, a perfect telescope will form an image that consists of a tiny disk (equal in diameter to the resolving limit) surrounded by two or three very faint rings. This disk has a sharp edge, and with a very high magnification resembles a minute planet. However, the air is usually so unsteady that this disk is visible only fleetingly, and is smeared and blurred with false rays. Bad seeing has a minimal effect on the out-of-focus test, which is why it is so convenient a resource for the practical amateur.

The expanded-disk test becomes much less sensitive if they are expanded too much. The writer recently came across commercial literature advising the tester to move the eyepiece so far away from the focus that the disks had a width equal to half the field of view! Such a test would conceal rather than reveal errors. Bringing the star up to about half the apparent diameter of the planet Jupiter is likely to give the best results, at least to begin with, although an experienced eye may prefer to expand it less than this.

Mountings for telescopes Concern over the optical quality of a tele-scope must not lead you into the trap of believing that this is the only important feature of a new instrument. No matter how excellent the image, observation will be nothing but a torment if the instrument wobbles or vibrates every time the observer puts an eye to the eye-piece, or if the image is constantly drifting out of the field of view and cannot easily be recovered. In fact, astronomical work will soon languish altogether!

Whether a perfectly mounted telescope of rather poor optical qual-ity will enjoy any better use is a moot point. However, there is certainly no point in attending *only* to the optical performance of the instrument when making a selection, because this feature can be appreciated truly only if the telescope is steady and well-mounted.

Any mounting must offer the following features:

1. Rigidity of the tube at any angle of elevation. Ideally the image should not shake at all when the eyepiece focusing knob is turned. If it does move slightly, any vibration should damp down the mo-ment the knob is released.
2. Ease of following the motion of an object, caused by the Earth's spin. A star or planet appears to go round the sky once in twenty-four hours, or at a rate of 1° in four minutes. The field of view of a ×250 eyepiece is likely to have a diameter of only about 12′ arc, which means that a celestial object will pass right across the view in less than fifty seconds. Extremely sensitive and responsive tracking is, therefore, essential.

To some extent, these two requirements go together, for it is im-possible to give a telescope a satisfactory fine motion if it is not steady to start with. The difficulty is not so much in holding the instrument firmly, as in combining steadiness with smooth motion. Mechanically, the most rigid mounting for a tube is to secure it at both ends; the least rigid is to try to hold it at one point only. Yet the demand that the tube be freely moveable, so as to be able to aim at all parts of the sky, means that the telescope must be pivoted around one point.*

Altazimuth mountings Most mountings contain two axes, at right an-gles to each other. The simplest type is known as the *altazimuth*

---

* It is possible to have a fixed telescope fed with light by an adjustable mirror. This is known as a *coudé* system, and it can work very well, but there do not seem to be any commercial versions of this design.

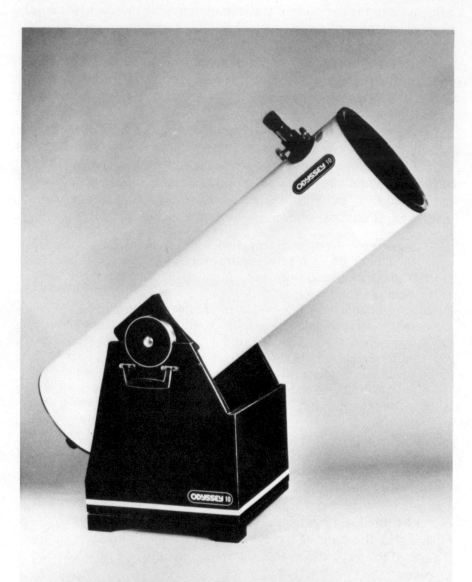

A Dobsonian-mounted Newtonian reflector   By adopting an extremely simple but effective wooden altazimuth mounting, the 330-mm (13.1-inch) reflector illustrated here costs only as much as an orthodox 200-mm equatorial reflector. The short focal ratio of the f/4.5 mirror makes for a very compact system. Manufactured by Coulter Optical, this is an excellent instrument for observing faint objects.

because one axis permits the tube to be moved in azimuth (horizontally, parallel to the horizon), while the other allows vertical motion in altitude. Mechanically, altazimuths have the advantage of simplicity, stability, and compactness. In recent years, indeed, the altazimuth has made something of a come-back in commercial circles, and the box-like *Dobsonian* is now popular for short-focus Newtonian reflectors. These are typically low-power, wide-field instruments used for observing clusters, nebulae, and galaxies, and so do not suffer from the altazimuth's main drawback: that it cannot easily be fitted with an automatic drive.

The author finds the rise of the altazimuth particularly pleasing because it confirms what he has maintained for many years—a mounting need not be finely machined and elaborately finished in order to be effective. The Dobsonian depends for its success on three important factors:

1. The weight of the telescope is "inboard" as much as possible. The less the overhang, the more inherently stable the mounting will be. This is acknowledged by the fact that some of the very latest large telescopes being built for professional use are on altazimuth mountings, with computerized drives to give an "equatorial" motion to the tube.
2. The bearing surfaces are large, being formed from almost friction-free Teflon. Large-diameter bearings encourage stability, and "frictionless" motion leads to minimum backlash. The tube, in a good Dobsonian mount, should respond to finger pressure to move it to a new position, and remain in that position when the pressure is relaxed. Note that this performance cannot be judged by looking at the tube in a showroom, but only by observing a celestial object in its natural surroundings!
3. The "minimum mounting" concept means that larger optics can be purchased for the same total price. With a mirror of 500 mm aperture, one's view of the deep recesses of the sky is transformed when compared with the performance of even a 300-mm reflector.

Equatorial mountings Most commercially-available mountings, however, are of the *equatorial* form. An equatorial mounting directly counteracts the Earth's axial spin, which is what makes celestial objects follow their daily course around the sky. If one axis of the mounting is tilted until it is exactly parallel to the Earth's and the telescope is made to rotate around this axis (the *polar* axis) once a day, in a direction opposite to the Earth's spin, then the telescope tube will remain pointing in the same direction in space. The *declination* axis joins the telescope to the polar axis, and is adjusted only during the initial location of the celestial object. Thereafter, rotation of the polar axis keeps the object within the telescopic field—or should do, if the mounting is accurately set up.

Types of equatorial There are several ways of arranging these axes, and each arrangement results in a mounting of different character.

The principle of the equatorial mounting The Earth, in spinning around its polar axis, causes the stars to appear to circle the sky once a day. This effect can be neutralized if the telescope rotates once a day in the opposite direction, around an axis parallel to that of the Earth.

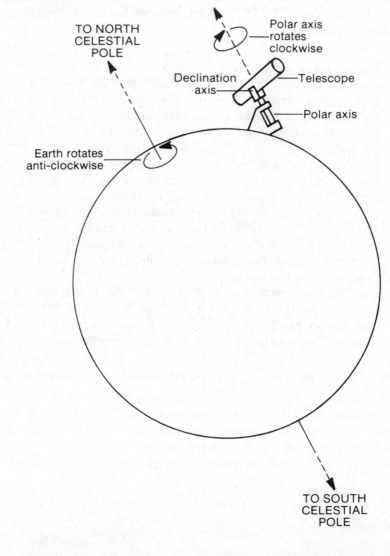

TO NORTH
CELESTIAL
POLE

Polar axis
rotates
clockwise

Declination
axis

Telescope

Polar axis

Earth rotates
anti-clockwise

TO SOUTH
CELESTIAL
POLE

By far the most common is the *German*. The two axes form a T, the vertical member forming the polar axis and the horizontal forming the declination axis. The declination axis carries the telescope tube at one end, and a counterweight at the other. Due to the necessity for a counterweight, the German is not a particularly good design mechanically; but it is fairly compact, and is suitable for both reflectors and refractors. Practically all Newtonians and refractors are sold on German stands.

The *Fork* mounting, as its name implies, carries a large fork at the upper end of the polar axis, inside which the tube can swing. Its advantages over the German are that the telescope is attached to the mounting on both sides of the tube, and no counterweight is needed. On the other hand, it is suitable only for short-tubed telescopes because a long tube could not swing through the fork and point to objects near the celestial poles (see page 46). Not surprisingly, the Fork mounting is favored by the makers of catadioptric instruments.

Many other types of equatorial mounting exist, but they are unlikely to be encountered in the mass market.

Drives and alignment The telescope mounting usually incorporates an electric drive for rotating the polar axis. A synchronous motor, that keeps in exact step with the mains frequency, is used. This frequency is not perfectly stable on an hour-to-hour basis because it can vary by a few percent according to the change of power demand, and the image may slowly drift in the field of view. This difficulty can be overcome by using a hand-controlled oscillator that supplies a steady frequency to the motor. Such oscillators can also be used to make the motor run fast or slow when centering an object in the field of view.

Some sort of slow motion, that can be applied to both axes when making fine adjustments to the direction of the telescope under high magnification, is practically essential. Remember that a movement of just 1' arc may make a noticeable difference to the position of an object in the field of view. Slow motions must, therefore, be smooth and responsive, if the observer is to have confident control over the instrument. Cheaper telescopes often fail on this point, either lacking slow motions altogether or having ones that are much too coarse. Another characteristic of cheaply-made instruments is the poor performance of the motor drive. This drive is communicated to the polar axis via either a worm and wheel or a spur gear, and defective machining or assembly of these turning units will result in non-uniform rotation of the polar axis. This is a particularly serious defect if any sort of long-exposure photography is to be undertaken because the resulting star images will be ellipses, or even lines, if the tracking has not been sufficiently accurate.

The alignment of an equatorial mounting, to ensure that its polar axis is parallel to that of the Earth, is very important if the instrument is to be used to its full potential. Misalignment means that a celestial object will steadily drift through the field of view, even if the motor drive is functioning perfectly. The ideal solution is a permanently mounted instrument that can gradually be brought to good adjustment. The need to realign a portable instrument each time it is used means that a part of each observing session must be devoted to getting ready to observe. The next chapter pays more attention to the problem of aligning an equatorial mounting.

Which telescope to choose? Perhaps the best way of helping the reader to come to a decision will be to summarize the features of some typical telescopes in the refractor, reflector, and catadioptric class.

REFRACTING TELESCOPES

TYPICAL SPECIFICATIONS Object glass aperture 80 mm, focal length 1200 mm (f/15). German equatorial mounting on a tall tripod. Overall weight about 18 kg.

OBSERVATIONAL CAPABILITY An excellent instrument for observing sunspots, and many models include a solar projection screen for safe viewing of the Sun's surface. An enormous amount of detail visible on the Moon—more, probably, than could be drawn in a lifetime. Adequate regular views of Venus and Jupiter, and of Mars when very near the Earth; excellent views of Saturn's rings, but the disk of the planet appears rather dim under the necessarily high magnification. Hundreds of double stars, nebulae, star clusters, and galaxies can be satisfactorily observed, and plenty of variable stars can be followed in their light changes.

ADVANTAGES Only minimum maintenance of the optical parts is required for the telescope to perform to the peak of its capability. Many refractors of this order of aperture, a hundred years old or more, are still in regular use.

DRAWBACKS Expensive and bulky for its aperture. Awkward to use for high-altitude observations, without a zenith prism.

SUMMARY Although refracting telescopes are necessarily limited in aperture because of their long tubes, they are the most efficient design from the point of view of light-transmission—that is, the ratio of the amount of light entering the object glass to that finally concentrated into the image is very high, typically about 76 percent. Light losses occur mainly at the air/glass surfaces, about 4 percent being reflected back out of the telescope at each interface; if these surfaces are coated or *bloomed* (the effect of blooming, which is carried out on most camera lenses to reduce internal reflections, gives the lens a bluish appearance by reflected light) the transmission rises to about 88 percent.

A good refractor gives the "cleanest" star image of any telescope because its optical train is the simplest, and the light scattered around a star, giving it flare or rays and reducing its contrast with the dark sky, is at a minimum. This ignores the effect of secondary spectrum, already mentioned, which normally is so inconspicuous that it will be noticed only around very bright objects.

Nevertheless, it must be recognized that the refractor cannot compete with other designs of telescope where sheer light-grasp is vital. Even a 150-mm aperture refractor would be a massive and expensive instrument.

NEWTONIAN REFLECTING TELESCOPES

TYPICAL SPECIFICATIONS Primary mirror aperture 150 mm, focal length 1200 mm (f/8). German equatorial mounting on a low metal stand. Overall weight about 25 kg. The 200-mm f/6 instrument is also a popular buy.

OBSERVATIONAL CAPABILITY Disk details can be seen on all the planets out to Saturn; even remote Neptune will exhibit a minute disk to distinguish it from a star. Useful for sunspot projection, although the eyepiece may become very hot from the concentrated heat. Many double stars are visible, and more variable stars than one observer could hope to follow come into view. An excellent instrument for observing star clusters, nebulae, and the brighter galaxies in some detail, while numerous fainter objects can be detected well enough for their positions to be recorded. Every year, at least one comet visible with this aperture passes through the sky.

ADVANTAGES The "best buy" in terms of dollars per unit aperture; a representative price list published by a well-known firm quotes a German-mounted, motor-driven, 150-mm Newtonian at $440, and an

80-mm refractor with similar specification at about $500. (It is worth noting that the latter price will also buy a 330-mm, short-focus Dobsonian reflector.) Newtonians are more convenient to use than are refractors, but the eyepiece can assume an awkward position at certain orientations when equatorially mounted, unless the tube can be rotated inside the declination-axis cradle—this feature is worth paying for.

DRAWBACKS The two mirrors must be carefully protected when the telescope is not in use; and, even with the most scrupulous care, the aluminum coating eventually will tarnish and have to be replaced. The optical alignment of the mirrors may also be disturbed in use, and must immediately be corrected, for the image quality is seriously affected by even a small displacement of the primary mirror.

SUMMARY A useful "all-around" telescope, capable of doing justice to most fields of observational work. Its light-transmission, even with brand-new aluminum coatings, is only about 70 percent, falling rapidly to nearer 60 percent in many locations where the air is damp and dirty. On the other hand, this is compensated by the fact that relatively large apertures are available at a reasonable price. However, mirrors scatter a good deal of light, and although they are truly achromatic it may be found that the image contrast is not as good as with a refractor. The arm, or vanes, supporting the diagonal mirror also produce diffraction rays around bright objects. For planetary work, refractors up to about 200 mm in aperture will almost certainly outperform reflectors of the same aperture, assuming, of course, that both instruments are optically excellent. There used to be a rule of thumb that compared the effectiveness of an 80-mm refractor with that of a 150-mm reflector, but many observers would feel that this is a little hard on the reflector!

Although requiring more care and maintenance than other types of telescopes, a good 150-mm Newtonian can give a lifetime's enjoyable and useful service. There is certainly no question of it being simply a "beginner's telescope."

CATADIOPTRIC TELESCOPES

TYPICAL SPECIFICATIONS Aperture 200 mm, effective focal length 2000 mm (f/10). Fork equatorial mount designed to stand either on a table top or on its own tripod. Overall weight of telescope and mounting, but excluding tripod, about 10 kg. This type of instrument is, of course, available in both smaller and larger sizes, but the 200-mm aperture seems to be particularly popular with amateurs.

OBSERVATIONAL CAPABILITY Generally, as for a Newtonian reflector of similar aperture, but with some reservations (see below).

ADVANTAGES Extremely portable and convenient to use. Aperture for aperture it is more expensive than a Newtonian reflector, but still much cheaper than a refractor. Can easily be adapted for photographic use (makers tend to exploit this potential), and because it gives an upright image, it can be used terrestrially for viewing objects only a few meters away. Very steady.

DRAWBACKS Because the optical specification and the mechanical tolerances are very critical, the image quality of different samples of the same model can vary, and readjustment is a matter for the expert. The "sealed tube" design, while protecting the mirrors, tends to trap warm air inside the tube, and this can lead to particularly unsteady images when the telescope is taken out into a cold night. Dewing of the shell or corrector plate can also be a nuisance in damp climates, and a satisfactory dewcap needs to be practically as long as the telescope tube.

SUMMARY As far as sheer definition is concerned, a catadioptric telescope is unlikely to perform as well as the equivalent Newtonian (for example, comparing a 200-mm, f/10 catadioptric with a 200-mm, f/8 Newtonian). One reason is the difficulty of making and mounting optical components to the required specifications—indeed, the marvel of these telescopes is that they perform as well as they do! A second reason is the light-scattering effect of the relatively large secondary mirror, that can have a diameter approaching one-third of the total aperture.* The serious planetary observer, who does not demand extreme compactness, would almost certainly do better with the larger Newtonian that the same money would buy.

It is also probable that a Newtonian (certainly with fresh mirror coatings) will detect fainter stars, nebulae, and galaxies. It is difficult to prevent some skylight or starlight from leaking down through the catadioptric system and brightening the background field of view. In

---

* Most catadioptrics have an extra right-angled reflection behind the primary mirror, to bring the eyepiece to a convenient viewing position. This is an extra cause of scattered light.

practice, however, the reflective coatings of the sealed-tube system should stay in good condition for a very long time.

A catadioptric is a delight to use because its short tube is so stable and can be maneuvered so easily; and, remembering the suggestion that the mounting may be even more important than optical quality, there is a strong argument for buying a catadioptric! After using a swaying Newtonian or refractor, the joy of a rock-solid instrument cannot easily be exaggerated. One's final feeling, however, is that the catadioptric telescope has, to some extent, sacrificed ultimate optical performance for convenience and versatility.

THE DECIDING FACTORS The individual will have to weigh up the merits of the different instruments somewhat along these lines:

Refractor	Excellent definition, minimum maintenance, restricted aperture
Newtonian	Large apertures readily available, good to excellent definition, needs periodic cleaning and adjustment
Catadioptric	Moderate to good definition, reasonable aperture, very convenient and versatile

CHAPTER 2

How the Sky Moves

Let us suppose that you have acquired an equatorially mounted telescope. Before it can usefully be pointed at the sky, its mounting must be set up so that the polar axis is pointing in the correct direction. This involves understanding something about how the Earth spins, and the effect that this has on the motion of the objects in the sky.

The spinning Earth Our globe spins once on its axis, relative to the stars of our galaxy and the other galaxies in the universe, in 23 hours 56 minutes and 4 seconds. This is known as the *sidereal day* or "star day." What this means is that if a telescope were pointed towards any star and left clamped in position, the star would drift out of the field of view in a westerly direction and return to the center of the field after this sidereal-day interval. The stars appear to drift westwards because the Earth is spinning in an easterly direction, making celestial objects rise above the eastern arc of the horizon and set at the western arc.

The axis around which the Earth spins is practically fixed in direction (there is a very slow wobble called *precession,* but one wobble takes about 26,000 years to complete, so that it is of no practical importance in this discussion). If the Earth were bored along this axis and we could sight along it, the same star field would always be visible. Looking northwards we should find, within a degree of the point towards which the axis is directed, the bright Pole Star (*Polaris*). This coincidence is fortuitous, and over the next thousand years precession will carry the direction of the axis past Polaris and away towards much fainter stars. Looking southwards along the axis, no prominent star would be seen; the Southern Hemisphere inhabitants of our globe are denied a bright pole star in their sky.

These points among the stars to which the Earth's axis happens to be directed are known as the *celestial poles*. In one sidereal day, the whole sky appears to rotate around the celestial poles. The illusion is so convincing that people used to think that the Sun, Moon, planets, and stars all rotate around a fixed Earth; and, in day-to-day conversation, we still say that "The Sun *is* rising" rather than "The Sun *appears* to be rising," which would sound very strange to other ears, although an astronomer would appreciate the fine point involved!

The celestial sphere The illusion that the stars revolve around the Earth has led to astronomers speaking of the *celestial sphere*. Imagine that our planet is a very small globe at the center of a huge sphere carrying the stars and the other celestial objects attached to its inner surface, and revolving once in a sidereal day. Two points on this sphere mark the celestial poles. A great circle (which is a line drawn around a sphere, dividing it into two equal parts) can then be traced around the celestial sphere, equidistant from the celestial poles. This is the *celestial equator*. It is analogous to the Earth's equator that divides our planet into two halves, centered on the north and south poles. In fact, the plane that passes through the terrestrial equator also passes through the celestial equator.

To an observer standing on the Earth's equator, the celestial equator always passes through the overhead point (known as the *zenith*), and cuts the horizon at the east and west points. To the north and south, down on the horizon, lie the celestial poles. Once in a sidereal day, every star on the celestial sphere rises and sets in the sky— although, of course, those appearing in the daytime cannot be seen.

Suppose now that our observer sets off on a journey towards the Earth's North Pole. Every nightfall finds the north celestial pole (and the Pole Star) situated higher in the north, while the south celestial pole has disappeared below the southern horizon and lies at a corresponding angle beneath it. The celestial equator, too, no longer passes through the zenith; although it still cuts the horizon at the east and west points, it now passes south of the zenith, and seems to pass lower and lower across the southern sky as the observer's northward journey proceeds. By the time our explorer has reached the North Pole, the north celestial pole lies in the zenith, and the celestial equa-

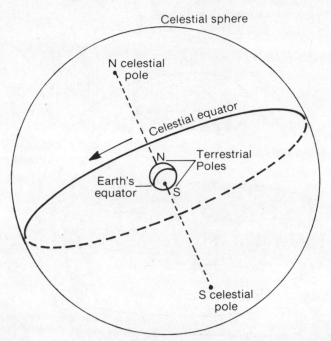

N celestial
pole

Celestial equator

Terrestrial
Poles

N

Earth's
equator

S

S celestial
pole

The celestial sphere This is how the imaginary sphere, that carries the celestial objects on its inner surface, would appear if viewed from the outside. The celestial equator and poles are projections of the terrestrial equator and poles, and the celestial sphere appears to revolve in an east-to-west direction, pivoted on these poles.

tor coincides with the horizon. Exactly half the celestial sphere is permanently above the horizon, so that stars neither rise nor set, but travel horizontally around the sky.

A journey to the Earth's South Pole would produce the opposite effect, with the south celestial pole increasing in altitude above the southern horizon, and the celestial equator sinking northwards in the sky. In fact, wherever an observer is situated on the Earth's surface, the altitude of the celestial pole above the horizon is equal to his latitude, which is also the angle that the polar axis of his telescope must make with the level ground. The azimuth of the upper end of the axis must, clearly, be due north for a telescope sited in the northern hemisphere, and due south for one sited south of the terrestrial equator.

Finding a celestial object The most straightforward way of locating an object on the celestial sphere is to use a good star chart, and to identify the region of sky by using first the naked eye and then the finder attached to the telescope tube. In the writer's opinion this is

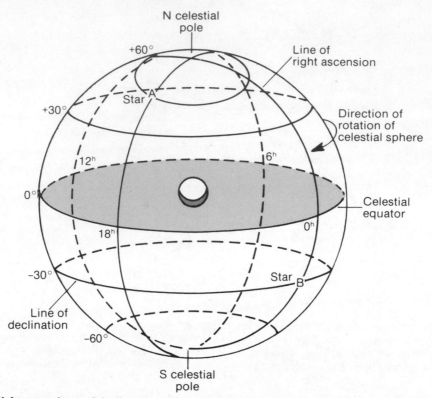

Right ascension and declination   The celestial sphere is divided into a reference grid corresponding to terrestrial longitude and latitude. Right ascension is measured in hours rather than in degrees because it reflects the apparent daily rotation of the celestial sphere.

the best way because each time you look at a star map or at a part of the sky you learn the positions of a few more stars in a constellation.

But if you have no suitable chart, or if the sky is too bright for locating stars to be identified, and you wish to locate an object whose position on the celestial sphere is known, you can literally dial it up and find it in the field of view—assuming that you have an equatorial telescope fitted with what are known as *circles* (i.e., a graduated disk attached to each axis, with a pointer from which the setting can be read), and assuming also that the mounting is set up accurately—which many portable instruments are not!

The position of a star or other object on the celestial sphere is always stated in terms of celestial latitude and longitude—*declination* or Dec. and *right ascension* or R.A., respectively. The declination is the number of degrees that the object is north (positive) or south (negative) of the celestial equator. If the declination axis—the one to which the telescope tube is directly attached—carries a graduated circle, then the telescope can immediately be set to the object's declination. If the instrument were now spun around the polar axis, the object would, at some point, pass through the telescopic field.

To set the telescope in right ascension is not quite so straightforward because the celestial sphere is rotating and objects are being carried steadily across the sky from east to west. For this reason, it is divided up not into 360 degrees, like terrestrial longitude, but into 24 *sidereal hours,* each one 15° away from the next. (Remember that the sidereal day corresponds to 23 hours 56 minutes and 4 seconds of "normal" time, so that one sidereal hour is about ten seconds shorter than a normal hour.)

Therefore the right ascension circle must be set to show sidereal time before the telescope is turned to the right ascension of the object being sought. Usually, there are two pointers on the circle. One is set to show the sidereal time (S.T.) of the moment. Once the motor drive is switched on, the circle moves against the pointer and acts as a little sidereal clock. The instrument is then swung around in right ascension until its own pointer indicates the desired reading on the circle.

It is, perhaps, worth pointing out that the arrangement of circles and pointers varies with different instruments. Some have fixed circles and moving pointers, others have fixed pointers and moving circles. The instructions, if you have purchased a new instrument, should make the method of operation clear. In the case of a second-hand instrument, you may have to work out the way of using the right ascension circle for yourself. The declination circle is always straightforward because the setting is a single operation.

Electronic sidereal clocks can be purchased fairly cheaply. Alternatively, an alarm clock can be adjusted to gain about four minutes per day, so that it will serve as a sidereal timepiece during an evening's work, even though, since sidereal time is reckoned on the twenty-four-hour system, you may have to add twelve hours to the time shown on the dial. To set a sidereal clock, adjust the telescope's right ascension circle setting so that the right ascension of a prominent star is indicated. Turn the telescope onto the star, and the other pointer will automatically indicate the sidereal time of the moment.

It is worth remembering that the sidereal time at any moment is equal to the right ascension of the celestial sphere that is on the observer's meridian. In other words, any object is highest in the sky when the sidereal time is its own right ascension.

The procedure for setting on an object by circles may sound involved. But it becomes straightforward with a little practice, and it is worth mastering, because probably the majority of equatorially mounted telescopes are never used as well as they might be. Their owners tend to set the declination axis, look up at the sky and estimate more or less where the object is, and then "sweep" in right ascension until it comes into the field of view. This may be satisfactory for locating easy objects. However, with an understanding of sidereal time and care in adjusting the circles, an equatorial telescope can be used for surprising feats, such as picking up the planets Jupiter and Mars in broad daylight, or amazing the many friends who believe that not even the brightest stars can be seen in the daytime. Few experiences in astronomy are more satisfying than seeing a celestial object sail into view in the blue sky and exhibiting it to a sceptical visitor!

Why the solar day is longer than the sidereal day   An observer situated on the solid line (midnight on Day A) completes one rotation, relative to the stars, when he returns to the dotted line on Day B. This corresponds to the sidereal rotation of the Earth. However, midnight will not occur until the Earth's rotation has brought him to the solid line again. This interval is the synodic day. The difference of about four minutes, that is greatly exaggerated in the diagram, is caused by the Earth's progress in its orbit around the Sun.

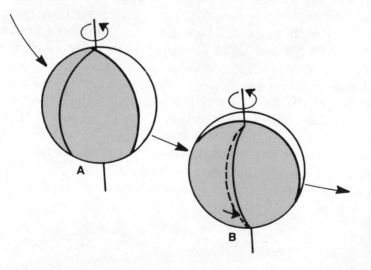

How the Sun moves The fact that the Earth has a rotation period of 23 hours 56 minutes and 4 seconds may come as a surprise to some readers, who will be wondering what has happened to the twenty-four-hour day. The explanation is straightforward enough: we run our lives by the position of the Sun in the sky rather than by the movement of the stars. The Sun appears to take about four minutes longer than a star does to achieve one revolution of the sky because the Earth's *orbital* motion around the Sun effectively slows down its own *axial* motion. This is shown in the diagram.

The Earth's orbital motion makes the Sun appear to encircle the celestial sphere once in a year. With each day that passes it has "moved" $\frac{1}{365}$th of the way around the sky, and after a year it is back at the starting point again. The path that it traces out can be marked on a star map: it is known as the *ecliptic*. The ecliptic is the reflection, as it were, of the Earth's orbital motion, and the plane of the ecliptic corresponds to the plane of the Earth's orbit. It is a great circle dividing the sky into two halves, one half of which is "north" of the orbital plane, the other half of which is "south."

The ecliptic Like two entwined hoops, the great circles of the celestial equator and the ecliptic have to cross at two places. They do so at an angle of 23½°; the angle by which the Earth's axis is inclined from an upright position with respect to its orbital plane. For half the year, therefore, the Sun is traveling along the half of the ecliptic that is south of the celestial equator, while the other six months are spent to the north of the equator. This annual up-and-down journey of the Sun is what brings us summer and winter, with long nights and short days when it is low in the sky (below the celestial equator), and short nights and long days when it is above.

Astronomically, the ecliptic is more significant than the celestial equator. The latter is a reflection of the way in which the Earth happens to be spinning in space—not a particularly significant orientation. But the plane of our orbit coincides almost exactly with the plane of the orbits of the other planets in the solar system, Mercury and Pluto alone excepted. In addition, the Moon's orbit lies very close to this plane. Therefore, when we look at the ecliptic we are looking through the fundamental plane of the planetary system, and neither the Moon nor the other planets stray very far from it. They always lie within an 18° band known as the *Zodiac,* centered on the ecliptic, that passes through the twelve famous Zodiacal constellations.

Each month, on average, the Sun appears to move from one Zodiacal constellation to the next, in an easterly direction (this cannot be observed directly because stars cannot be made out very near the Sun, and the constellations are temporarily invisible). Twice a year, in March and September, it crosses the celestial equator, the first passage being from south to north, when spring comes to the Earth's Northern Hemisphere, and the second being from north to south, when autumn begins. These seasons are reversed for dwellers in the Earth's Southern Hemisphere.

It may be worth mentioning that the crossing points are known as the *solstices,* and that the March solstice is used to mark the zero point of right ascension (0^h R.A.), while the September solstice marks the position of 12^h R.A. These points are in the constellations of Pisces (the Fishes) and Virgo (the Virgin) respectively.

Setting up an equatorial telescope The writer likes to think of a well-set-up equatorial telescope as a planetarium projector in reverse. Switch on a planetarium and starlight streams out, each dot to its appointed place on the starry sphere. Set up an equatorial telescope and starlight should shine back down the tube in accurate response to the setting of its circles.

"Perfect" alignment of an equatorial mounting is an unattainable ideal. What we are seeking is (a) sufficiently precise matching of the telescope's axis with the Earth's axis for a celestial object to remain in the field of view as the Earth turns; and (b) for the circles—if any —to be adjusted sufficiently accurately to allow an object to be dialed up.

Polar axis alignment The first requirement is to have some idea of where the cardinal points (North, South, East, and West) lie, so that the mounting can be set up in approximately the right orientation. If the telescope is being set up by day, a very approximate setting can be obtained by noting the position of the Sun at noon, making due allowance for Daylight Saving Time, if in force. At this time the sun is near that very useful line known as the meridian, that passes through the zenith and cuts the horizon at the north and south points; unless you lie within 23½° of the celestial equator, when the midsummer Sun can pass to the opposite side of the zenith, the Sun is always due south of the zenith as seen from the Northern Hemisphere, or due north as seen from the Southern.

It is convenient to break down the initial, semiaccurate stage of adjustment into two separate tasks. The first is to bring the polar axis into the correct azimuth. Once that is done, its altitude can be adjusted quite easily. Then, for really precise results, fine-tuning must begin.

Orientation in azimuth For obvious reasons, it is helpful if some of the initial work of setting up a permanent mounting can be done by day, and the Sun can, with care, be used to locate the polar axis in azimuth to within a degree or so.

First, turn the polar axis so that the declination axis is horizontal. In the case of a German mount, that has a shaft to act as the axis, this can be accomplished with a spirit level. A Fork mounting may present greater problems, and you may need to construct a simple metal or even wooden "bridge," resting on the two pivots, on which the spirit level can be placed.

Now find the instant at which the Sun will be due south, and, using the precautions described in Chapter 4, twist the whole mounting until it can be brought to the center of the field of view using only the declination adjustment.

In theory, this means that the polar axis must be orientated north–south. However, the telescope tube may not be accurately at right-angles to the declination axis. To allow for this, quickly rotate the polar axis through 180°, bring the declination axis accurately level again, and see if the Sun has returned to the center of the field (you will have to allow a little for the inevitable drift that will have occurred between the two operations).

If it has, then you know not only that the azimuth of the polar axis is good, but that the telescope is accurately fitted to the declination axis. If the Sun does not return to the field, then mark the orientation of the stand and then twist it until the Sun is acquired once more. Make another mark, and twist the stand back until it is orientated midway between the two marks. You know that the machining of the mounting has not been accurate, but you have compensated for the error.

The same procedure can be used in the case of a star, provided you know its right ascension. Wait until the local sidereal time has the same value, and it must be on the meridian.

To find the instant at which the Sun is on the meridian is not quite so easy. You might think that this must be at "noon"—but what is

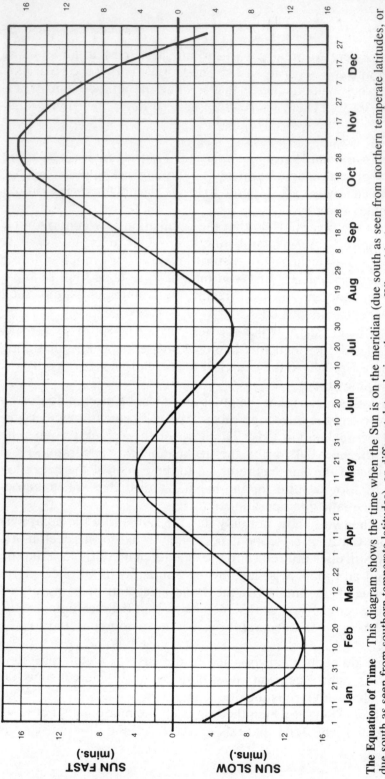
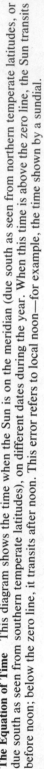

The Equation of Time This diagram shows the time when the Sun is on the meridian (due south as seen from northern temperate latitudes, or due south as seen from southern temperate latitudes), on different dates during the year. When this time is above the zero line, the Sun transits before noon; below the zero line, it transits after noon. This error refers to local noon—for example, the time shown by a sundial.

"noon"? Noon is not 12 o'clock midday, unless your longitude happens to be an exact multiple of 15°. If it is, then your particular Zone Time will reflect the way the Earth is rotating for your station. But if, for example, you live in Pittsburgh, at a longitude of 80°W, the Sun will pass due south twenty minutes after it has passed due south as seen from Philadelphia. Both these cities lie in the Eastern Standard Time zone; Philadelphia's noon does indeed occur almost exactly at 12 o'clock midday, but Pittsburgh's noon will not arrive until 12:20 p.m. The general rule is to add four minutes for every whole degree that you are west of a standard time longitude.

The second precaution involves the motion of the Sun around the celestial sphere. This is far from uniform, due to the eccentricity of the Earth's orbit and its variable orbital velocity, as well as to the inclination of the ecliptic to the celestial equator. At some times in the year the Sun is "fast" and crosses the meridian early, while at other times it is "slow" and crosses late. This error, known as the *Equation of Time,* can amount to over a quarter of an hour. Use the graph in Figure 11 to compensate for this effect.

Adjustment in altitude Once the azimuth setting is to your liking, adjust the altitude of the polar axis until it equals the latitude of your site. An adjustable protractor and a spirit level should permit this adjustment to be made to within a degree or so. This is much the easier of the two adjustments. People who habitually move their telescopes around would be well advised to incorporate a spirit level into the mounting, adjusted so that the bubble is central when the polar axis is at the correct altitude.

Fine adjustment strategies Most people seem to align their equatorials by sighting up the polar axis at the Pole Star (if in the Northern Hemisphere) or at where they think the celestial pole should be (if in the Southern Hemisphere)! The accuracy is unlikely to be very good and will soon betray itself by demanding adjustments to the Dec. axis while observation is proceeding. Even the more careful method described in the above section may well result in an error of a degree or so, which is not good enough for prolonged high-power work or long-exposure photography.

One useful device, an all-in-one sighting device that can be fitted to some catadioptric telescopes, permits you to set the polar axis directly by sighting on the Pole Star. It allows for the small displacement between the Pole Star and the true celestial pole, and if carefully used can give a good and "instant" alignment for portable telescopes.

If the instrument has already been set up in approximate adjustment, there are two checks that can be made by following a star with a high-power eyepiece and cross-hairs (see page 156):

1. Follow the track of a star as it drifts across the meridian (for an hour or so, around the time of meridian passage). If it seems to be drifting in declination, then the *azimuth* of the polar axis needs adjustment. A northward drift means that the upper end of the axis is too far west, and vice versa. The effect is reversed for an instrument in the Southern Hemisphere.

2. Follow the track of a star rising in the eastern sky and at an altitude of about 30°. If it appears to drift northwards in the field of view, then the *altitude* of the polar axis is too great. A star setting in the western sky, at about the same altitude, can also be used, but in this case the significance of any drift is reversed, a southward motion of the star indicating too great an altitude of the axis. Again, the effect is opposite for a telescope in the Southern Hemisphere.

Setting up a portable telescope If the instrument will regularly be set up in the same place, it is well worth making some registration marks on the ground, or on the pillar, so that the mounting can quickly be returned to the same orientation. An accuracy of about ½°, which is good enough for most amateur work—with the exception of long-exposure photography—can, with care, be achieved. For real precision, however, a permanent mounting is practically essential—otherwise you will spend precious clear-sky time getting the mounting aligned before serious work can begin!

Summary Even if you don't own an equatorial telescope—and there is plenty to be said for the altazimuth mounting—reading this chapter will not have been a waste of time. It will help you to read the coordinates of a star map, or to understand why some constellations are visible in summer while others can be seen only in winter. Many people go through their lives without understanding how the Earth's spin, and its orbital motion round the Sun, affect the things we see in the sky. This is a pity. Intelligent man puzzled for thousands of years before the truth emerged; some individuals were banished, while others died, for their beliefs. In our study of astronomy, it is pleasant to think that we can now recognize and appreciate their efforts.

CHAPTER 3

How to Start Observing

You have acquired and set up a telescope: your auxiliary eye. This elegant combination of glass and metal is about to reveal a universe a hundred times brighter and bigger than the one with which you have been familiar. It will certainly be confusing; it may even be a little disappointing. Those color photographs of the best-known celestial objects, to be found in so many books, bear little or no resemblance to the ghostly stains of light revealed in the eyepiece! And is this flickering reddish blob the planet Mars? Where is the polar cap? Where are the dark markings? You may not have expected to detect space-probe-type detail, but surely the telescope ought to be working better than this!

The eye and the telescope The answer is that the telescope may well be working perfectly, but there are two things over which it has no control: the steadiness of the air through which it has to peer, and the sensitivity of the eye it serves. Atmospheric steadiness, or *seeing,* varies from place to place and from hour to hour, and has an enormous effect upon image quality. Individuals' eyes also vary in their sensitivity to fleeting telescopic detail and to dim objects, but regular observing practice makes huge differences to what can be seen. When you put your eye to a telescope's eyepiece, you are using it in a way that is totally different from the style to which is is accustomed, or, indeed, for which it was designed. In normal daily life, the eye is a scanning instrument: it has a very wide field of view and can take in a lot of information at once because the brain is constantly checking on what is happening at different places in this field of view.

It is not used to concentrating its entire power on the small disk of a planet (or even just a part of this small disk), or to laying siege to a tiny area of black sky in the belief that a faint star lurks there! This is an unnatural use of the eye, and proof that the eye and brain have to learn a new task is afforded by listening to a beginner and an experienced observer sharing a telescope. It is not that the latter knows *what* to look for (we hardly ever know what we are going to see when we look at the sky; that is one reason why astronomy is so fascinating), but that he knows *how* to go about looking for it!

In a way, this book is about how to look at the sky. Each different object makes its own demands: it may be easier to see with a low magnification than with a high one; it may be more satisfactory in slightly hazy but very steady air than when the night is transparent and turbulent heat currents make the stars twinkle merrily. Part of one's experience is in judging how to use a night to the best advantage. No reasonably clear night is hopeless for astronomy, if you are prepared to choose from a number of different observing projects. With a new telescope this is no problem, for you will probably want to look at everything.

Sky conditions We must begin with the all-important atmosphere, the veil through which we have to look out into space. Relatively speaking, the effective coating of the air around the globe is as thin as the peel on an apple, but it has an effect upon the radiation entering it that is out of all proportion to its thickness. Most of the energy sent out by the stars and nebulae in our galaxy, and even that of whole galaxies beyond ours, is blocked out. However, in the middle of the narrow "window" through which visible light can pass, it is almost transparent. Astronauts observing with the naked eye from satellites and spacecraft have reported that the faintest stars visible from space are only half a magnitude or so fainter than those visible from the best Earth-based sites.

However, this transparency applies only to the view enjoyed by a terrestrial observer staring up above, towards the zenith. Celestial objects that are low in the sky appear significantly dimmer because their light has had to pass through a much longer atmospheric track. On this count alone, it is always wise to observe objects when they are as high as possible in the sky. But some objects never do rise very high, even when they are on the meridian: the maximum possible altitude is given by the observer's colatitude (90° minus his latitude) added to the object's declination, which means that if the declination below the celestial equator is greater than the colatitude, it never rises above the horizon at all.

In a perfectly clear sky, atmospheric extinction at the horizon amounts to about three magnitudes. The slightest haziness can easily double or treble this amount, so there is not much point in giving a table of atmospheric "extinction," because almost all sites are affected to some extent by haze. Observers in or near towns have also to put up with the dust and fumes that get thrown into the air and are then illuminated by the urban glare. In fact, it is the general street-lighting, rather than the pollution, that dims the stars; this is proved quite dramatically should a power failure occur, when the stars and even the Milky Way may be seen with sensational clarity! What happens is that every suspended particle scatters light, creating a luminous foreground that brightens the field of view and effectively dims the stars. The annoyance and waste of "light pollution" has led to campaigns, on the whole not very effective, although there have been local exceptions, to reduce the amount of street lighting that is thrown up into the air rather than down onto the road. It is a sad fact that the world's urban populations will never again witness the spectacle of a great comet apparently suspended over their rooftops, to match those that awed townsfolk and country dwellers alike as recently as last century; even a brilliant comet like the one of 1882 would be a feeble, unregarded ghost of its true self.

Seeing The transparency of the atmosphere is one thing; its steadiness is another. If the *seeing* is unsteady, a telescopic image flickers and blurs and all fine detail is lost. This may not matter much if we are looking at faint, hazy objects such as nebulae and remote galaxies, but it is of critical importance if a close binary star, or lunar or planetary detail, is being studied. In such cases, seeing is much more important than transparency.

Just a few centimeters' thickness of unsteady air can spoil the telescopic image: try viewing through the open window of a warm room on a cold night! The uprush of warm air rising into the night will make a planet look like a featureless, pulsating blob. Even the heat of a hand, held beneath the object glass of a refracting telescope, can make the image quiver. Atmospheric unsteadiness can ruin the telescopic image regardless of its source or cause—and the causes that are under the observer's direct or indirect control must be suppressed, because there are plenty more, far beyond reach, that cannot.

Local effects and tube currents The first essential is to minimize any temperature difference between the telescope and its appendages and the surrounding air. If you are fortunate enough to have a permanent shelter for the instrument—whether or not it deserves the name of "observatory"—remember that during a sunny day the air inside will become hot, and that if the shelter is enclosed it will remain hot long after the evening coolness descends outside. So open it up and ventilate it well before sunset. It is also a good idea to uncover the optical components, especially the main mirror of a Newtonian, so that the glass has a chance of cooling down with the air—if there is a door in the tube through which the cap is removed, leave it open. Large observatories take particular pains to keep the air inside their domes, and the optical components, as close as possible to the expected night temperature, even to the extent of installing under-floor cooling systems.

If you keep your telescope in the house—which is likely enough if you have purchased a compact portable instrument—don't expect it to perform well the moment it is set up outside. Even the reliable refractor will give poor definition if it contains a tube full of air 10°C above that of the night! The reason for people claiming superior image steadiness for the refractor is partly because the light rays have to pass along the tube only once. Therefore, even if there are minicurrents inside the tube, they have a shorter light-path on which to operate than in a Newtonian, where the rays pass twice, and in a Cassegrain-type system, in which the light makes a total of three journeys along the tube. Furthermore, the definition given by a "warm" object glass is less affected than that of an equally warm mirror, because the errors of the two lens components tend to compensate each other.

On the subject of tube currents, there are both gains and drawbacks in using a telescope with a sealed tube (i.e., a refractor or a catadioptric) compared with an open tube (i.e., a Newtonian). With the latter, the warm air quickly escapes from the mouth of the tube —the trouble is that it escapes so successfully, with such a rush, that until all the warmth has drained from the tube and the mirror the tube currents will be quite severe. This is the reasoning behind skeleton-tube Newtonian telescopes, although few manufacturers offer them, perhaps because of customer resistance (people think a solid tube offers more value for money, and it certainly provides better physical protection for the optics), and probably also because such a tube would be more expensive to make. With a sealed catadioptric or refracting system, there is no violent rush of air, but neither can there be rapid cooling because the warm air is trapped. The most rapid cooling occurs when the tube is relatively narrow, as with a refractor, because there is plenty of tube wall in relation to the amount of air contained within it; the slowest cooler is the stubby tube of a catadioptric.

You will quickly come to recognize the tube-current characteristics of your telescope, and do your best to minimize them by storing it, as far as possible, at "nighttime" temperature. The next source of unsteady air is the warm, daytime ground that gives up its heat as evening approaches and may continue doing so well into the night. Try to observe from a site that does not have large areas of concrete or roof tiles between your telescope and the universe! These materials become particularly hot on a sunny day, and can boil away merrily for some hours. Vegetation is the ideal surrounding because leaves keep relatively cool even in full sunlight, and anyway do not have much capacity to store heat—furthermore, they keep the ground in shade. Large observatories take care to have as much green matter as possible in the vicinity of their domes, even if it is only scrub.

If you intend to observe the Sun, remember that the ideal daytime, nonradiating surface is water: much of the Sun's energy is simply reflected back off the top, while a good deal of the rest penetrates below the surface, which remains relatively cool. One professional solar observatory in the United States is built in the middle of a lake. At night, however, the situation may not be so good because the water surface will not cool down very quickly—but, on the other hand, neither will it radiate heat into the air at a very great rate.

Atmospheric effects It is the unsteady air at heights of thousands of meters that causes the real problems, because nothing practical can be done to cure it, except to change your observing site. The worst effects occur when air masses at different temperature overlap and mix, as they do when a "front" passes across. A warm front is usually accompanied by increasing cloudiness, and so no astronomy can be done anyway, but the succeeding cold front, that may bring very clear skies after the rain has passed, is often accompanied by violent twinkling, a sure sign of unsteady conditions. Now is the time to go hunting for faint nebulae and galaxies—provided they are not too small—for the sky background will be as black as space itself; but the experienced planetary observer won't even bother to uncap his telescope on such a night. Steadiest conditions occur when the upper air temperature is uniform, typically during an anticyclone, a time of high atmospheric pressure. Some regions of the Earth's surface, for example the southwestern United States, enjoy long periods of anticyclonic weather, with both clear nights and steady air. Observers in higher latitudes have to put up with more variable conditions; in western Europe, for example, long and stable anticyclonic conditions are rare, and when they do come they often produce very hazy skies.

Dewing Air always contains a certain measure of water in the invisible vapor state, and warm air can contain more water vapor than when it is cold. There is a limit to the amount that any air can contain, without the water condensing out in droplets; an air mass that contains this maximum water-vapor content is said to be *saturated*. Clearly, if warm saturated air is cooled, water must condense. In clouds, this phenomenon will result in rain. But an equally discomforting event is the deposition of water-vapor droplets, or dew, on the telescope's optical surfaces, for the instrument will immediately go "blind."

Dewing can occur only on a surface that is colder than the air— and therefore, in practice, on a surface that is exposed to the air and loses its heat more rapidly than does the air mass around it. In fact, during an evening's work the air temperature may actually rise if a warm front is approaching, making the tube and mounting run with water. But it must be remembered that dew can be precipitated only if the air is at or near its saturation point, and this will tend to happen more in maritime climates than in desert ones.

The primary mirror of a solid-walled Newtonian will dew up only under the most severe conditions. The object glass of a refractor, or the corrector plate or shell of a catadioptric, is a different matter, and on some nights a dewcap is essential. This is no more than a tube of some insulating material projecting in front of the glass. The idea is that the radiation of heat from the optical surface is reduced—hence helping to keep it above the dew-point—and also that the air within and around the dewcap acts as an insulator from the colder air outside. The typical refractor can benefit from a dewcap equal in length to three or four times its aperture, although the ones sold as standard are often much shorter, and may need to be extended. Long dewcaps for catadioptrics are out of the question, and observing with them on damp nights can be a trial. A brown paper bag converted into a cylinder has been recommended!

It is never wise to wipe a dewed surface. Gentle warming with an electric hair-dryer takes dew away like magic. Before bringing a telescope into a warm house, cap all the optical surfaces so that they do not dew up. Better still, keep the telescope in the garage or garden shed, where the temperature approximates that of the outside air.

Planning a session In some parts of the world, it is possible to consult the diary and arrange for a "star party" on such-and-such a night, with almost perfect confidence that it will take place and be successful. The writer, after many years of observing through British skies, can only envy such people! Nevertheless, you can be a successful amateur astronomer under even the most unfavorable conditions if you are prepared to take a chance on the weather clearing. The secret is to be ready in advance, and to prepare plans for different types of nights, so that you can go outside whenever the sky is reasonably clear and get down to something interesting. Some objects don't require a very long scrutiny, and can be observed perfectly well in a patchy sky—variable stars are a case in point. An enthusiastic variable-star observer will have perhaps a couple of dozen objects on his list above the horizon simultaneously, and because their fields can be located in a few seconds, once they have become familiar, they can be spotted through breaks in the clouds and observed, in the same way as if the sky were totally cloudless. The planets are different because you will want to keep looking at the disk for perhaps half an hour or even longer, and passing clouds are a real frustration.

So, as you plan your observing campaign—and this book is designed to help you to do just that—compile a list of the observing projects that are possible on different nights. It could just be a mental list, but to begin with, it might be even better to write down the actual objects to look at, especially if you are studying a constellation with its collection of double stars, star clusters, nebulae, and galaxies. It is very easy to become distracted when you are out under the stars: there are so many "goodies" on display that it is like being a child in a toy shop, particularly if the night is a really black one and the sky is alive with points of light, looking so near that you could almost touch them. Now, if ever, is the time to stay calm! Remember that there *will* be other nights like this, and stick to your "transparent night" plan, so that after your session you can look through your observing notes with a sense of real satisfaction and achievement.

A "transparent night" plan would probably include some of the following projects:

Steady air	Close, difficult double stars, with one component much fainter than the other
	Planets and their satellites
	Very faint variable stars
Unsteady air	Nebulae and galaxies
	Comets
	Occultations of stars by the Moon
	General naked-eye and binocular work
	Variable stars
	Star clusters
	Sky photography with a short-focus camera
	(These projects could, of course, be undertaken on a night of steady air as well.)

Underneath the "hazy night" heading you can assume that the air will be fairly steady, and the following projects suggest themselves:

Lunar detail
Planetary detail
Bright, close double stars
Variable stars bright enough to be seen well

These lists should not be taken too literally: one person's "hazy night" may be another's "transparent night." Don't forget, either, that the presence or absence of the Moon makes an enormous difference to the appearance of the sky, and that moonlight makes work on very faint objects difficult or impossible; your observing routine will resolve itself into approximately fortnightly spells with the Moon either "present" (approximately the period from First Quarter, through Full, to Last Quarter), or "absent" (from Last Quarter around to First Quarter, through New Moon). "Absent" does not mean that the Moon is nowhere to be seen, but that it sets sufficiently early in the night, or rises sufficiently late, to give some hours of dark sky; "present" means that it is above the horizon all night (as at Full) or at least for most of the night, to the inconvenience of most amateur and practically all professional observers! Work on the planets and bright variable stars, however, is unaffected by moonlight.

In summary, it will pay not to specialize too much on very demanding objects, unless your sky conditions are unusually good. A sensible program of work will include something that can be undertaken under any sky conditions. Having said that, it needs to be sufficiently ordered so that when the sky clears, you know what you ought to be doing.

Comfort A comfortable observer may or may not be a good observer, but an uncomfortable observer is bound to be a bad one! It is worth pointing out, very early, that practical astronomy is physically a very taxing business. You may find yourself turning out to observe at strange hours of the night, exchanging a warm bed for the frosty air, to peer at a faint smudge in the sky that generates enthusiasm in no one in the household apart from yourself! The eyepiece may be at a neck-straining angle, or at such a height that you can neither stand nor sit comfortably. Some observers drive long distances seeking a dark site for their telescope, while others hide from street lamps and illuminated windows in the shadow of trees, hedges, and makeshift screens.

The enthusiast can put up with a lot of inconvenience, but physical comfort when actually making an observation is very important. The observer must be able to concentrate, with absolutely no distractions, on the view in the eyepiece. From this important aspect the catadioptric telescope, with its right-angled eyepiece very close to the polar axis, is a real boon: not only is the eyepiece usually at a handy angle for study, but it moves very little as the telescope is swept across the sky, and can always be reached from a sitting position.

Warmth comes a close second, after physical relaxation, in bodily matters. Remember that astronomical observing is not (or at least should not be) an energetic pursuit. Advertisements for telescopes tend to emphasize the beauty of the product—and sometimes of the beholder, too—without regard for the spartan conditions in which the product may be used. However, the author's experience is that if the hands and feet are warm, the rest of the body can cope reasonably well: a double layer of socks, and thick-soled shoes or boots, will prevent too much body heat from draining out into the ground, while mittens are much more convenient than gloves if any delicate operations have to be performed. Eric Alcock, the English comet and nova hunter, once passed a continuous thirteen-hour winter watch for meteors with his feet in a bag of straw!

A frosty night is one thing; a frosty windy night is quite another, for the wind blows away the fragile cocoon of warm air that develops around the observer's body and acts as an extra, invisible coat. On such a night, the answer may be to seek the lee of a house or tree, or give up the session altogether—which may be hard to do because frosty skies are often sparkling ones.

Indoor observing It is useless trying to observe through an open window, unless the room temperature is the same as that outside. If the window is closed, however, the outrush of air is suppressed, and if the object's altitude is not too great, and it is in a convenient direction, it *is* possible to do some astronomical work from indoors. Ordinary thin window glass is not usually very flat optically, and even binoculars will not define well when looking through it. On the other hand, the thicker plate glass used in sliding patio doors and the like is often of really excellent quality because it is manufactured by pouring it onto the (theoretically) flat surface of a sheet of molten metal. It is certainly worth experimenting, if the telescope is portable, just out of interest if for no other reason. There may be some loss of critical definition, but this does not particularly matter with low-power work. The restricted angle of sky coverage is the most obvious drawback. Eric Alcock, to whom we have just alluded, recently established what is surely a first: he independently discovered the "earth-grazing" comet of 1983, using powerful binoculars, while observing through a plate-glass window in this way!

Dark-adaptation It is always necessary, before starting observational work, to get the eyes dark-adapted. Even the star Sirius, the brightest in the sky, appears about ten thousand million times fainter than does the Sun; a landscape under the Full Moon is 400,000 times fainter than the same view on a sunny day. The eye possesses an automatically adjusted iris that responds to the light level by closing down to about 2 mm in bright conditions and opening up to 8 mm when the light is dim; but the ratio of light passed by these two extreme aperture values is only 1:16, far smaller than the range of brightness in which our eyes are expected to function. To allow our eyes to see both in sunlight and moonlight, and even to detect very faint stars, the sensitivity of the nerve endings in the retina has to change by an enormous amount, as otherwise we should either be totally dazzled in daylight, or else benighted after sunset. So-called "night vision" develops as the nerve endings are sensitized by a substance known as *visual purple* that takes some minutes to become fully active. Do not, therefore, expect to see faint objects immediately upon going out into the dark, but spend a little while looking at the sky and identifying the constellations until you feel that your eyes are working at full power. Physiologists state that visual purple continues to build up the sensitivity of the eyes over several hours, so, if you have to return temporarily to the house, keep your observing eye closed to preserve its adaptation.

Training the eye Which eye *is* your observing eye? Most people, asked to "look through this telescope," will apply their right eye to the eyepiece. But it does not follow that the right eye is necessarily the best one to use. Habit is hard to overcome; so it is a good idea to start your telescopic career by experimenting with both eyes. Spend some time observing the disk of Mars or Jupiter, and decide which eye shows the most detail—you may be surprised at the difference. Remember, also, that when you take up habitual observing the eye that you use will become "educated" to see much better. Even if you start off with the weaker eye, it will rapidly overtake the other one in the challenge of detecting dim stars or fine planetary detail—so how much more sensible to start off with the better one at the outset! Also, try to train the unused eye to remain open but unseeing when observing. This is not difficult in pitch-black conditions, but if the surroundings are rather bright it may help to use a black patch or hood. Shutting one eye and keeping the other open is not helpful to relaxed observation.

Records Every observer will record his or her work in a different way. If you specialize in variable-star observation, you will do much more writing than drawing; if you observe the planets, much time will be spent in sketching their disks, and a book of drawing paper will be useful. It will also be necessary if you draw the fields of nebulae and star clusters. So you may require several notebooks. But it is a good idea to have a single summary book, so that you can note down all the different work you did each night, without going into particular detail. A desk diary, with a page for each day, is ideal because you can make notes of any interesting forthcoming events that might otherwise pass forgotten and unobserved.

The light used for reading and writing at the telescope must be the right compromise between dimness and dazzle. A pocket torch, its bulb coated with red poster paint (or with red cellophane over the glass) can be used; the writer prefers to remove the bright reflector behind the bulb, and so obtain a more even illumination on the note-book or chart. A bicycle rear lamp, as sometimes recommended, is rather too bright for most purposes. If you have an electrically driven telescope, a low-voltage red lamp can be wired up via a transformer and variable resistance, so that the brightness can be adjusted according to the requirements of the moment. Even better is to fit the lamp with a clip, so that it can be attached to a fitting on the telescope tube or mounting, leaving the hands free.

Little luxuries like this, which take minimal effort to provide, can make an enormous difference to the pleasure of work at the tele-scope. So many people become disillusioned when astronomical observing proves to be rather far from the elegance and shirt-sleeved comfort of the telescope showroom! It often *is* cold, and it *does* demand perseverance; all the more reason, then, to do the best we can to make observing that little bit easier and more comfortable. With a plan of campaign, and a handy, comfortable telescope, you will find the available hours too short to do justice to all that the sky contains.

CHAPTER 4

Daytime Astronomy— the Sun

Our Sun seems to be a typical star—if any single star, selected from the vast range of stellar types known to exist, can be called "typical." Solar astronomy, however, is certainly not typical astronomy. We are looking at a star in brilliant close-up, and this nearby view, however exciting, can be a deadly privilege. It is true that the Sun's light and warmth makes life on our planet possible; but it is equally true to say that the Sun is a killer. Expose any living organism to the Sun's raw radiation, and its cells will be broken down and destroyed. Our survival on this planet is dependent upon the apple-peel thickness of atmosphere that absorbs the deadly Sun-rays in sufficient quantity to give life a chance.

Some rays are totally absorbed: these include the extreme short-wave radiation (such as X-rays), of very high energy. But as we come near to the tiny "visible" window where the atmosphere is fairly transparent, so more of the potentially dangerous rays can reach our eyes. Beyond the blue and violet color lurks the invisible *ultraviolet,* the main cause of sunburn: and sunburn is, after all, nothing more than destroyed skin. Beyond the red end of the visible spectrum come the so-called "heat rays," of which the near *infrared* radiation is again transmitted by the atmosphere but invisible to the eye.

An astronomical telescope, when pointed to the Sun, concentrates all this radiation—ultraviolet, visible, and infrared—into its image. Most of the damaging radiation in this package is, in fact, *invisible to the eye.* Therefore, it is no test of a solar filter to say that the Sun

69

looks dim when viewed through it! For example, many observers have found that a piece of unexposed and processed leader from a color transparency film darkens the solar disk sufficiently to permit comfortable observation with the naked eye or a telescope of small aperture. What they do not realize is that this "filter" is almost transparent to heat waves, so that the retina is suffering almost as much damage from this source as if no filter at all was used.

This introduction is not intended to discourage you from observing the Sun! In the writer's view, the Sun is the most amenable of all celestial objects, if viewed safely. What other astronomical body can be observed at a convenient hour, in reasonable warmth, throughout the year, and can be almost guaranteed to have changed its appearance from day to day? And what other body is so important to us here on the Earth? Solar astronomy is a wonderful field of work, and perfectly safe provided that simple and sensible precautions are taken.

The Sun as a star The sun is 150 million km away, and its disk has an apparent diameter of about ½°. Its true diameter, now known to great accuracy, is 1,392, 530 km—109 times that of the Earth. In terms of sheer volume, 1¼ million Earths could be squeezed into its bulk. Familiarity breeds contempt, and the solar astronomer often refers to the size of features seen on the Sun in terms of "Earth diameters," without really appreciating what his expression means. The smallest detail visible on the surface, the so-called *granulation,* is composed of individual cells as large as Texas that form and disperse in the space of about five minutes! The fury of the Sun's forces cannot easily be reconciled with the friendly light-giver that shines in our skies.

The energy source This shining surface or *photosphere* is not the source of the Sun's energy. It is not even particularly hot, by astronomical standards: 6000° C is "only" about three times as hot as a blast furnace, and far cooler than, for example, the core of the planet Jupiter, which is believed to be at a temperature of 50,000° C. The power lies deep towards the Sun's center, where the tremendous pressure has raised the temperature to an estimated 15,000,000°C. Hydrogen atoms break down into their elementary particles, which reform into a more stable atom, that of helium, which does not disintegrate even at this fantastic temperature. The hydrogen–helium conversion is not a perfectly balanced change, however, for there is more energy binding hydrogen than helium. This surplus energy has to be steadily lost from the Sun, or our star would explode: it slowly trickles up to the photosphere and flies off into space in the form of electromagnetic radiation, of which light, heat, radio waves, and X-rays are all different forms.

The corona The Sun does not end at the photosphere. Above it lies a thin shell of cooler hydrogen just a few thousand kilometers thick. This is the *chromosphere,* which, because it is cooler than the photosphere and therefore shines much less brightly, is normally invisible. Then comes the rarefied outer atmosphere, the *corona,* extending for many solar diameters into space. The chromosphere is visible briefly at the beginning and end of a total solar eclipse, when the lunar disk has just covered the brilliant photosphere; and a total eclipse is also the only time when the corona can be seen shining in the sky. Eclipses of the Sun are an event right out on their own in the amateur astronomer's calendar!

Other suns As a star, the Sun occupies a middle place in the hierarchy. The huge majority of stars that we see in the sky belong to an orderly family known as the *main sequence,* that extends from stars having perhaps a million times the Sun's luminosity to other very dim stars that may be only one millionth as luminous. Astronomers have detected numerous stars that seem to be in the same physical state as the Sun. Its very ordinariness is what makes the Sun so important; surely it is not unreasonable to suppose that what we find out about our own star may be true of other similar stars in the galaxy, and, indeed, in the universe?

Of course, we have yet to detect another star with planets revolving around it. There are strong suspicions that one or two nearby stars may be the center of planetary systems. But proof is lacking: no Earth-based telescope can hope to record even a large planet at the distance of the nearest star, which is about 7,000 times as remote as the planet Pluto.

The solar astronomer, therefore, is not just recording facts about the Sun. He is studying a star that happens to be near enough to be covered in great detail. It isn't surprising that solar observation is a popular amateur pursuit; if anything, it is remarkable that even more amateurs don't make the Sun a regular object of study because it is unusual to be able to observe any other celestial object very easily when the Sun is above the horizon.

Observing by projection Insert a low-power eyepiece (about × 50) into the eyepiece tube, hold a piece of white card about 20 cm beyond the eyepiece, and swing the telescope until it is pointing towards the Sun. This does *not* involve squinting up the tube or through the finder, if you value your eyesight; instead, watch the shadow of the telescope tube on the card or on the ground as you hunt—when this shadow is almost circular, you are nearly there. If you are a very systematic person, then the declination circle can be set to whatever the Sun's declination happens to be on the date in question, using the accompanying table, which gives its approximate R.A., Dec., and apparent diameter in minutes and seconds of arc for different dates in the year.

Only a *very* systematic person would bother to use sidereal time to

Projecting the Sun, using a refractor   The shade near the upper end of the tube casts a shadow over the image, improving the contrast.

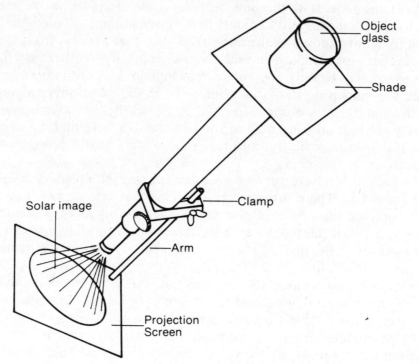

find the Sun! Simply setting the telescope in declination and then swinging it around the polar axis will quickly bring the solar disk into the field of view, when its bright image will be seen on the card. Focus the eyepiece until the edge or *limb* is as sharp as it will go, and there is our local star displayed for inspection, with any spots that happen to be present easily viewable, simultaneously, by a group of people if desired.

		R.A.		Dec.	Dia.	
Jan	1	18ʰ	45ᵐ	− 23°.0	32′	35″
	6	19	07	22°.6	32	35
	11	19	28	21°.9	32	34
	16	19	50	21°.0	32	34
	26	20	32	18°.8	32	32
	31	20	53	− 17°.5	32	31
Feb	5	21	13	16°.1	32	29
	10	21	33	14°.5	32	28
	15	21	53	12°.9	32	26
	20	22	12	11°.1	32	24
Feb	25	22	31	− 9°.3	32	22
Mar	2	22	50	7°.4	32	19
	7	23	09	5°.5	32	17
	12	23	27	3°.5	32	14
	17	23	46	− 1°.6	32	12
Mar	22	00	04	+ 0°.4	32	09
	27	00	22	2°.4	32	06
Apr	1	00	40	4°.3	32	04
	6	00	59	6°.3	32	01
	11	01	17	8°.1	31	58
Apr	16	01	35	+ 9°.9	31	55
	21	01	54c	11°.7	31	53
	26	02	13	13°.3	31	50
May	1	02	32	14°.9	31	48
	6	02	51	16°.4	31	45
May	11	03	10	+ 17°.7	31	43
	16	03	30	19°.0	31	41
	21	03	50	20°.1	31	39
	26	04	10	21°.0	31	37
	31	04	30	21°.8	31	36
Jun	5	04	51	+ 22°.5	31	34
	10	05	11	23°.0	31	33
	15	05	32	23°.3	31	32

Earth closest to the Sun (perihelion, 147 million km) about Jan 4

Sun crosses celestial equator (vernal equinox) about Mar 21

	20	05	53	23°.4	31	31
	25	06	14	23°.4	31	31
Jun	30	06	34	+23°.2	31	31
Jul	5	06	55	22°.8	31	31
	10	07	16	22°.3	31	31
	15	07	36	21°.6	31	31
	20	07	56	20°.8	31	31
Jul	25	08	16	+19°.8	31	32
	30	08	36	18°.6	31	33
Aug	4	08	55	17°.4	31	34
	9	09	14	16°.0	31	36
	14	09	33	14°.5	31	37
Aug	19	09	52	+12°.9	31	39
	24	10	10	11°.3	31	41
	29	10	29	9°.5	31	43
Sep	3	10	47	7°.7	31	45
	8	11	05	5°.9	31	48
Sep	13	11	23	+ 4°.0	31	50
	18	11	41	2°.1	31	53
	23	11	59	+ 0°.1	31	55
	28	12	17	− 1°.8	31	58
Oct	3	12	35	3°.8	32	01
Oct	8	12	53	− 5°.7	32	04
	13	13	11	7°.6	32	06
	18	13	30	9°.4	32	09
	23	13	49	11°.2	32	12
	28	14	08	12°.9	32	14
Nov	2	14	27	−14°.6	32	17
	7	14	47	16°.1	32	19
	12	15	07	17°.5	32	22
	17	15	28	18°.9	32	24
	22	15	49	20°.0	32	26
Nov	27	16	10	−21°.0	32	28
Dec	2	16	31	21°.9	32	30
	7	16	53	22°.6	32	31
	12	17	15	23°.0	32	32
	17	17	37	23°.3	32	33
Dec	22	17	59	−23°.4	32	34
	27	18	21	23°.3	32	34

Sun farthest north of the celestial equator (summer solstice) about Jun 21

Earth farthest from the Sun (aphelion, 152 million km) about Jul 4

Sun crosses celestial equator (autumnal equinox) about Sep 23

Sun farthest south of the celestial equator (winter solstice) about Dec 21

(Due to leap-year adjustments, the dates here are subject to an error of up to one day. The seasons described are those of the Earth's Northern Hemisphere.)

This is one great advantage of the projection method: you can actually point things out to other observers, rather than having to rely on verbal directions. For this reason, most new telescopes incorporate a projection screen among their accessories. This consists of a piece of white aluminum or plastic attached to an arm, so that its distance from the eyepiece can be varied, and with a clamp to attach the arm to the telescope. To keep direct sunlight off the screen, a separate shade is needed. In the case of a refractor or a catadioptric, this can be a piece of cardboard somewhat larger that the diameter of the image, fitted around the upper end of the tube; with a Newtonian, the screen receives only glancing illumination from the Sun, and a narrow shade along the sunward side of the screen is all that is needed.

The size of the solar image This can be almost anything that you wish: the greater the distance between the eyepiece and the screen, the larger the image will be. Image size is also proportional to the magnification given by the eyepiece, but if you use too high a magnification you will not be able to view the entire solar disk at one time. Most people determine the best combination of magnification and screen distance by trial and error, but a formula may be useful here: if the magnification given by the eyepiece is M, and the distance between the eyepiece and the screen is d, then the diameter D of the solar image on the screen, when the Earth is at its mean distance, will be given by

$$\frac{d(M-1)}{107}$$

so that the required value of d is equal to

$$\frac{107D}{(M-1)}$$

D changes by about 1½ percent on either side of the mean value during the course of the year, so that d must be altered to compensate for the changing Sun–Earth distance if a constant-diameter solar image is required.

Applying this formula to a × 50 magnification eyepiece shows that an image measuring 150 mm across will be obtained by positioning the screen 330 mm beyond the eyepiece. This is a convenient arrangement, and 150-mm diameter images of the Sun are used by amateurs all over the world. If you set the screen to this size, remember that our own planet, on the same scale, would have a diameter of slightly less than 1.5 mm, and you may care to draw a black spot of this size somewhere on the screen to act as a sobering reminder of our status!

Sizes and positions If you are ever measuring the size of a sunspot group, to which we shall return later, it is useful to remember that on the 150-mm scale a distance of 1 mm corresponds to 9300 km of photosphere near the center of the disk, but considerably more, due to foreshortening, towards the limb. Halfway from the center to the limb, the scale has increased to 10,700 km per mm; two-thirds of the way it has become 12,500 km, and three-quarters of the way it is about 14,000 km per mm. Measures beyond this distance become less and less reliable because the foreshortenir increases dramatically towards the solar limb. (Remember that foreshortening only affects distances measured along a radius of the Sun; distances measured parallel to the limb are unaffected.)

Projecting the image on the screen has another benefit besides allowing a group of people to observe simultaneously: with the Sun, as it were, laid out for physical measurement, all sorts of interesting observing projects can be undertaken. Not only the dimensions of the sunspot groups and other solar features, but also their positions, can be recorded; and the daily (sometimes almost hourly) changes are fascinating. If you are the sort of person who likes to observe methodically, and enjoys making accurate measurements, then this branch of astronomy may be just the thing. Solar astronomy sets a whole range of challenges, from obtaining the most continuous possible record of sunspot frequency over weeks, months, and years, to precise observation of the changing details of a single spot. Some advanced amateurs have built true solar observatories, where the Sun's disk shines out in a darkened room and can be viewed, in ease and comfort, in much better detail than that possible with a simple screen and shade out of doors. You can also, when you reach this stage of sophistication, turn your hand to examining our star spectroscopically, and charting the distribution of the clouds of hydrogen and calcium across its surface. If the Sun exerts its powerful influence on you, and you have the facilities, then you may well end up by observing indoors, inside your own solar laboratory, rather than outside in the pleasant fresh air!

Drawing the projected image Commercial projection screens are rarely very substantial. Furthermore, as they are attached to the telescope tube, any interference with the screen tends to make the image move. There is, therefore, no question of making a direct copy of the solar image by drawing round the sunspot images on a piece of paper laid over the screen, unless the arrangement is unusually rigid. The usual technique is to prepare a master outline, consisting of a circle of the correct size divided up into a network of small squares—ten or more each way, depending upon the accuracy being sought—drawn in very fine pencil or ink so that they obscure as little surface detail as possible. A similar circle, drawn with heavier lines, is also prepared, and the actual drawing is made on a piece of semitransparent paper that is laid over this second circle and grid. The thin paper is marked only with a circle divided up into four quadrants, and the location of the individual spots in relation to the grid on the projection screen is then copied on to the drawing.

In the writer's experience, it is often very difficult to detect small spots on a gridded projection screen because the lines distract the eye. It is also common to find small blemishes being mistaken for solar markings. A useful ruse is to have a piece of thin white card, which is held just in front of the screen (so that the solar image is still in focus) and given a rapid to-and-fro motion. If you keep your eyes fixed on the image, the motion blurs out any defects on the card and allows tiny features to be detected with certainty. Then, keeping your attention fixed on the feature, remove the card, and try to identify the spot on the grid. It may have vanished, and repeated forays with the card will be necessary before its position can be established with certainty. But it is well worth taking trouble with minute spots, because, being so hard to see, they may well be missed by others.

Orientating the image A solar drawing is of little or no use—except for merely establishing the number and size of the sunspot groups— unless it is correctly orientated. In other words, the North–South and East–West axes must be shown. If this is not done, it is merely a collection of sunspot outlines, with no logic to their arrangement. The Sun appears to rotate upon its axis in 27.28 days, and one of the main interests of solar work is in watching for any new groups appearing at the eastern limb. It is also useful to keep a record of sunspot activity in each hemisphere. To do this, the "cardinal points" must be marked, and this is the purpose of the crossed lines on the drawing.

The easiest way of orienting the drawing is to stop the driving motor if you have one (and it is extremely difficult to make a solar drawing if you have not, although it can be done), and to twist the master projection blank on the screen until a sunspot appears to drift accurately along one of the grid lines. This drift is precisely in an *East-to-West* direction. Now arrange the cross on the thin copy paper so that it coincides with the two central grid lines on the circle underneath. One line can be marked E–W, the other N–S. If you are using an inverting telescope somewhere in the United States and are observing around noon, then the North point of the image will appear towards the top, and the East will be to the right, but you can confirm these directions by moving the telescope slightly and watching the way in which the image is shifted. For example, if you move the direction of the telescope slightly towards the North, then the Sun will appear to move towards the South direction on the screen.

If your equatorial mounting is aligned with sufficient accuracy, the East–West direction can be established by moving the telescope slightly in right ascension, and letting a suitable spot skim along a line. Remember that the daily drift of the spots on the Sun, caused by its rotation, will be in an East-to-West direction.

Some telescopes, particularly refractors, arrange for the projected image to be turned at right angles to the tube, using a zenith prism or mirror near the eyepiece. Direct sunlight now hits the projection screen at a glancing angle, as with a Newtonian. The solar image will be reversed compared with the direct arrangement, and its orientation must be established by observation, as described above.

Observing by direct vision Projection is certainly the easiest way of obtaining sunspot positions, and is a delightful method of exhibiting the Sun to a group of people. Its drawback is the loss of contrast and clarity compared with the direct view. Even though the screen is shaded, there is still abundant sky-light flooding the image and turning its blacks to grays. The projection screen can be almost totally enclosed with a light-proof box to improve the contrast, but the use of a screen still imposes a certain loss of detail. If you are interested in observing the finest possible texture of a sunspot, then direct viewing is the best way of going about it.

To observe the Sun in safety, a proper filter *must* be used. It must not only dim the brilliant disk down to a viewable intensity; even more important is the removal of the dangerous side radiations that cannot be detected by the eye. We have already seen that unexposed and developed color film, often used as a temporary defense of the eye, is practically transparent to heat rays, even though it "looks" safe.

Before the modern introduction of very thin and durable optical-quality plastic, solar filters were small pieces of deeply dyed glass that fitted over the eyepiece. If you should ever come across one of these accessories, hurl it far into a convenient pond in case some innocent amateur should be tempted to use it. Not only is dyed glass far from safe: the region of the eyepiece is the focus of all the Sun's heat, and the most concentrated point of all is just beyond the eye lens, at the so-called *exit pupil,* where such a filter has to be placed. If such a "sun-cap" absorbs sufficient solar energy to make observation safe, it is likely to crack with the heat; if it does not get hot enough to crack, then it is passing this energy through to the eye!

Most telescope manufacturers now offer a much safer and more sensible alternative. This is a plastic material known as *mylar,* coated on both sides with a metallic film that absorbs some radiations and reflects others away out of the telescope. This is placed over the corrector plate or object glass, or over the open mouth of a Newtonian. It looks somewhat like a sample of the metallized space blankets used by adventurers, and the first reaction is usually one of disbelief that anything at all could be seen through this wrinkled remnant. But its virtue lies in its thinness: the radiation that does manage to pass through (about 0.01% of the incident light) is unaffected optically, even though the plastic sheet is not perfectly flat. Probably the best-known solar filter material is Solar-Skreen, which can be obtained in sizes to fit all apertures. It turns the Sun's disk a blue color. The well-known firm of Questar produces their own safe filter for use with their Maksutov-Cassegrain instruments, and the Sun appears orange. The Questar transmits more infrared radiation than does Solar-Skreen, but both have been declared safe for visual use, provided there are no surface defects. These will show up as small light areas when the filter is held up to a bright light, or to the daytime sky. Even a minute blemish can raise the total amount of transmitted solar radiation by a dangerously high factor, and a filter showing any defect at all should be rejected as unsafe.

Aperture reduction Solar observation is often carried out in rather unsteady conditions. On a hot day, when the surrounding ground has

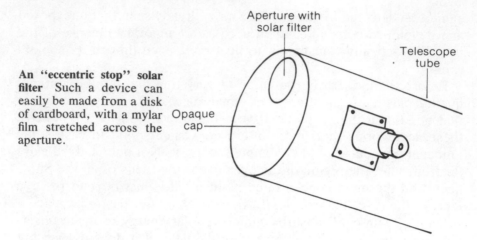

An "eccentric stop" solar filter Such a device can easily be made from a disk of cardboard, with a mylar film stretched across the aperture.

heated up, the air will be "boiling" violently, and the telescope will not perform to its resolution limit, or anything near. Under normal conditions, it is rare for details beyond the resolving power of apertures between about 100 mm and 150 mm to be made out, and often an aperture of only 75 mm is sufficient to do justice to the view. Therefore, some suppliers offer "eccentric-stop" filters for use with Newtonians and catadioptric instruments. These take the form of a cap fitting over the tube, with a circular aperture cut out towards one side and a solar filter fitted into this opening. The aperture is of such a diameter that it fits in between the edge of the true telescope aperture and the outline of the diagonal or secondary mirror; in a 250-mm aperture telescope, for example, there would be room for an eccentric stop of about 100 mm diameter.

At first sight, an eccentric stop looks certain to give "eccentric" definition. In fact, and probably surprisingly, it gives a purer image than does a standard reflector or catadioptric because the central obstruction, that scatters light away from the object being defined and lowers the image contrast, has been abolished. The focal ratio of the telescope is increased because the focal length remains unchanged but the aperture has been reduced by perhaps 2½ times. This is a further virtue, since long-focus instruments are renowned for sharpness of image and quality of contrast. The large light loss is of no account when viewing the Sun, and, as we have seen, daytime definition is often inferior to that of only moderate apertures, so that no detail is lost; in fact, it may well be gained. If you own a telescope of even 200 mm aperture, experiment with an eccentric stop of, say, 80 mm in aperture, and you may be surprised at the result.

Other ways of reducing the intensity of the solar image, such as using uncoated mirrors in special eyepiece attachments, have been put out of court by these full-aperture solar filters: a step to be welcomed.

The features of the Sun The Sun's disk, seen under steady conditions with a magnification of about ×75—so that the whole star can be seen in the field of view at one time—is a superb sight. The whole surface is faintly mottled, but so elusive is this effect that the individual details seem to vanish when you try to concentrate on them. This is not the true granulation, which is visible with magnifications of about ×250 or more, and with apertures of perhaps 150 mm, but an effect caused by the eye grasping half-perceived details. Towards the solar limb, the intensity of the photosphere drops very markedly. This is the action of the thin atmosphere or chromosphere, which is cooler than the visible surface and so absorbs energy being radiated outwards; the effective thickness of this atmosphere is much greater at the Sun's limb than near the center of the disk because we see light from the limb that has passed at a shallow angle through the chromosphere; the effect is analogous to the absorption of starlight by our own atmosphere when looking at an object near the horizon. Towards the limb, also, bright filamentary patches known as *faculae* become easy to see. These are regions of hot hydrogen suspended only a few hundred kilometres above the photosphere. Their light, therefore, has to pass through a slightly smaller depth of chromosphere, and they are less affected by limb-darkening, even though they are really no more luminous than the photosphere itself.

Faculae indicate activity—magnetic activity—in the lower regions, and this is what produces sunspots. Turbulence inside the Sun produces very powerful, local magnetic fields. The hot solar material responds to these fields and is forced into whirling cyclonic patterns, visible with special instruments. At the center of the magnetic field, the energy is so strong that it blocks the outflow of radiation from the solar depths, and creates a local cooling on the photosphere.

This cool region is a sunspot. The center of the spot appears practically black: the region known as the *umbra*. This appearance is misleading, being an effect of contrast with the blindingly bright photosphere, for the temperature is about 5000°C, which is still hotter than that of the surface of many stars! An annulus around the umbra, the *penumbra,* is much lighter in tint, and often shows a great deal of fine structure. Sunspot-plotting is the main diet of the solar observer.

Observing sunspots If you view the Sun on a "typical" day, you may notice up to five or six individual spots, or groups of spots close together. Some umbrae will be much larger than the Earth; these will have conspicuous penumbrae, and probably there will be other smaller spots in the vicinity. Expect also to see some groups consisting of a pair of significant spots (these are loosely termed *bipolar*), while others are conspicuously single or *unipolar*. There may also be some very small dark pinpoints on the disk. These are known as *pores*. A tiny group of two or three of these can appear in a few hours, and they may mark the beginning of a new sunspot group.

Sunspot counts It is a simple but very rewarding task to count the number of sunspot groups that are visible on the disk each day. Ignore the individual component spots, and concentrate upon the groups themselves. This is known as the *active area* (AA) count. It is easy enough to count AAs when there are only a few well-scattered groups on the disk, but when the Sun is active, and long chains of spots appear, it can become difficult to decide where one group ends and another begins. Often this must be a matter of judgment because the certain test (which is to study their magnetic fields) can be made only with special equipment. Formally, AAs are deemed to be separate if they are more than 10° of solar longitude apart (see page 86), but a single large group may be more than 10° long. In general, spot groups develop in solar longitude (roughly East–West) rather than in solar latitude (roughly North–South), and the major spots of a bipolar group usually have a sprinkling of much smaller spots between them, to indicate their connection; but it can become very difficult to disentangle a crowd of spots near the limb, where they seem to be almost touching each other but may really be many Earth-diameters apart!

If you keep up an active-area count over several weeks or months, the activity will seem to be going up and down in a random way. Over a period of years, however, the Sun is seen to undergo a steady change, betraying its eleven-year cycle of activity. This is manifested most obviously by the number of sunspot groups that are visible, although the cycle also affects other solar behavior, such as the intensity of the invisible "wind" of atomic particles being pumped out into space, as well as the shape of the corona. It is often referred to, however, as the *sunspot cycle*. The last period of maximum activity occurred in 1980–81, and minimum is expected to be reached by

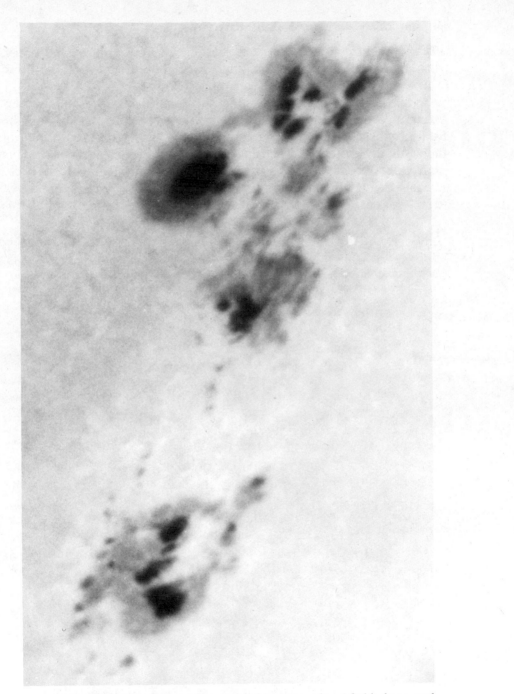

A sunspot group The dark umbra and lighter penumbra of this large and complex group can be clearly seen, together with faint mottling on the surrounding photosphere. (*N. W. Scott, 300-mm Cassegrain*)

1987. Typically, the rise to maximum takes only about 3½ years, whereas the fall to the next minimum takes about 7½ years, but these values do vary considerably from one cycle to the next.

Simple "eyeballing" of the solar surface, then, can produce interesting data for the observing book. After each calendar month, the total number of active areas observed (including reobservations of the same group on different days) is divided by the number of days on which observations were made. The result is the Mean Daily Frequency (MDF) for that particular month. The great advantage of MDF work is that it doesn't take up much time: a maximum of five

The changing presentation of the Sun's axis and equator   These two diagrams represent the solar disk viewed on the first day of each month (January–June in [a], and July–December in [b]). A tracing can be made for the date of observation (interpolating between dates if necessary), and laid over the drawing or on the projection screen for immediate identification. Ensure that the cardinal points are correctly set; it may be necessary to reverse the tracing to achieve this.

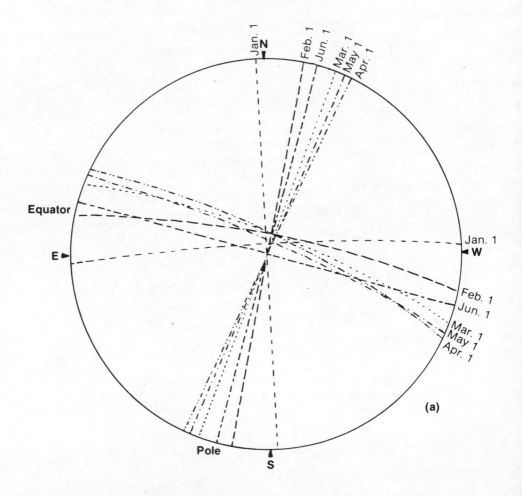

minutes will permit a thorough observation of the Sun's disk. This means that a useful observation can be made in a brief break in the clouds, or in a few moments snatched in the course of a busy work schedule. The advantage of solar work—that it is done in daylight—is also a potential drawback if you spend the day at the office! But it is perfectly possible to make an active-area count before breakfast, or before setting off for work, or even at lunchtime if you care to take a portable telescope along with you. There is no reason at all why interesting solar observation should be the prerogative of the retired or redundant amateur, although it is fair to say that more comprehensive observations may take a prohibitive amount of weekday time, particularly in winter, when the days are short, and therefore will have to be confined to weekends and holidays.

Solar latitude and longitude. If you want to refine this work, the next step is to decide in which solar hemisphere the active areas are located. To do this, you need to know the position of the equator

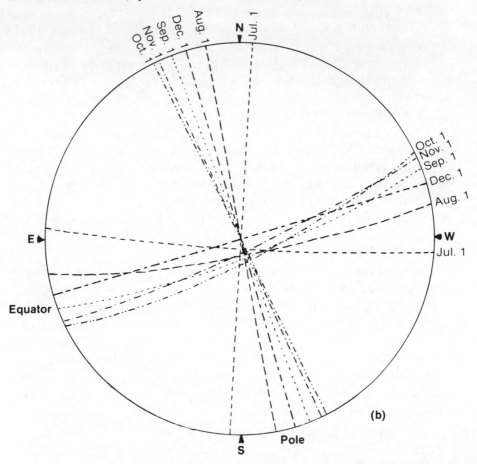

because it never coincides with the East–West diameter of the Sun. Although the plane of the Earth's orbit passes through the center of our star, the Sun's axis is not exactly at right angles to this plane, but is inclined at an angle of 7°.2 to the "upright" position. The North Pole of the Sun is tilted at its maximum extent towards the Earth on September 10, while the South Pole is at maximum presentation on March 7. Between these times, on June 7 and December 9, the Earth passes through the plane of the solar equator, which then appears as a straight line crosssing the center of the disk. (These dates are subject to a variation of a day or so from year to year.)

A tilt of 7° may not sound much, but it is sufficient to displace the equator quite noticeably, around the equinoxes. However, the Earth's own axial tilt of 23°.4 has a much more significant effect upon solar coordinates because it makes the Sun's axis appear to swing in a large arc from side to side as the year progresses. At the beginning of the year (January 6) the axis appears to lie truly North–South. By April 8, however, it has swung some 26° in a clockwise direction. It returns to the North–South orientation by July 7, reaching its maximum counterclockwise swing of 26° on October 10. (Again, these dates are subject to a slight annual correction.)

The result of these combined noddings and swingings is summarized in the diagrams on pages 84 and 85. To decide definitely in which hemisphere any particular active area lies, you must make an accurate positional drawing using the method already described, and then superimpose the position of the Sun's axis in relation to the North–South line. The equator, in its correct inclination, can then be drawn in, freehand. The following table gives values of P (the inclination of the axis from the North–South line) and B_o (the solar or *heliocentric* latitude of the center of the disk, which is the same value as the amount by which one of the poles is inclined towards the Earth), at five-day intervals throughout the year.

Sunspot positions Sunspots usually avoid the 5° or so of solar latitude on either side of the equator, so you will not have to be ultra-accurate in plotting its position for the purpose of dividing the disk into two hemispheres, unless a rare spot is seen in an unusually low latitude. Nor are spots often seen at latitudes greater than about 40°, so, again, the odd exception is of considerable interest. A tendency of spots to appear most frequently at higher latitudes at the beginning of a cycle, and for the zone of activity to work equatorwards as the cycle progresses, is known as *Spörer's Law*.

THE PRESENTATION OF THE SUN'S DISK
THROUGHOUT THE YEAR

		P	B₀			P	B₀
Jan	1	+ 2°.3	− 3°.0	Jun	25	− 5°.7	+ 2°.2
	6	− 0°.1	3°.6		30	3°.4	2°.7
	11	2°.5	4°.1	Jul	5	− 1°.1	3°.3
	16	4°.9	4°.6		10	+ 1°.1	3°.8
	21	7°.2	5°.1		15	3°.4	4°.3
Jan	26	− 9°.4	− 5°.5	Jul	20	+ 5°.6	+ 4°.8
	31	11°.6	5°.9		25	7°.7	5°.2
Feb	5	13°.6	6°.3		30	9°.8	5°.6
	10	15°.5	6°.6	Aug	4	11°.8	6°.0
	15	17°.2	6°.8		9	13°.7	6°.3
Feb	20	− 18°.9	− 7°.0	Aug	14	+ 15°.5	+ 6°.6
	25	20°.3	7°.1		19	17°.2	6°.8
Mar	2	21°.7	7°.2		24	18°.7	7°.0
	7	22°.8	7°.2		29	20°.1	7°.1
	12	23°.8	7°.2	Sep	3	21°.4	7°.2
Mar	17	− 24°.7	− 7°.1	Sep	8	+ 22°.6	+ 7°.2
	22	25°.4	7°.0		13	23°.6	7°.2
	27	25°.9	6°.8		18	24°.5	7°.2
Apr	1	26°.2	6°.6		23	25°.2	7°.0
	6	26°.3	6°.3		28	25°.7	6°.9
Apr	11	− 26°.3	− 5°.9	Oct	3	+ 26°.1	+ 6°.6
	16	26°.0	5°.5		8	26°.3	6°.4
	21	25°.6	5°.1		13	26°.3	6°.0
	26	25°.0	4°.7		18	26°.1	5°.7
May	1	24°.3	4°.2		23	25°.8	5°.2
May	6	− 23°.3	− 3°.7	Oct	28	+ 25°.2	+ 4°.8
	11	22°.2	3°.1	Nov	2	24°.5	4°.3
	16	20°.9	2°.6		7	23°.5	3°.8
	21	19°.4	2°.0		12	22°.4	3°.2
	26	17°.8	1°.4		17	21°.0	2°.6
May	31	− 16°.0	− 0°.8	Nov	22	+ 19°.5	+ 2°.0
Jun	5	14°.1	− 0°.2		27	17°.8	1°.4
	10	12°.1	+ 0°.4	Dec	2	15°.9	0°.8
	15	10°.1	1°.0		7	13°.9	+ 0°.2
	20	7°.9	1°.6		12	11°.8	− 0°.5
				Dec	17	+ 9°.6	− 1°.1
					22	7°.2	1°.7
					27	4°.8	2°.4

(*P:* a positive value means that the Sun's North Pole is displaced to the West. *B₀:* a positive value means that the Sun's North Pole is presented to the Earth, and that the center of the disk lies in the Sun's Northern Hemisphere.)

It is obvious that any observations involving making a drawing are going to take far longer than a single scan of active areas. However, there is no objection to making a simple "weekday" scan and supplementing this by a more detailed "weekend" observation in which the active areas are accurately plotted. Because many sunspot groups last for a week or more, and do not change their heliocentric latitude and longitude very much in the meantime, you may be able to use some of these as reference points from which to judge the hemisphere in which other, more transient groups and pores lie.

Accurate positional work If you have the time, it is clearly best to make a detailed drawing on every possible occasion. From such a drawing you can not only derive an active-area count and a hemisphere analysis: you can also measure how large the sunspots are, and find out their position on the Sun's surface in terms of heliocentric latitude and longitude. If you do this, then any slight *proper motion* or drift on the hemisphere relative to other sunspots can be detected—if the observations are sufficiently accurate!

If you are making a positional drawing, remember that the alignment of the master outline, with its North–South and East–West lines, is of crucial importance. If this is wrong, then all the resulting positions will be wrong. So check the setting, by using the solar-drift method, both before and after the drawing is made. If the second check reveals an error, correct the cardinal points on the drawing by half this error, for it means that the orientation of the solar image has somehow changed during the observation.

Another point to bear in mind is that on a positional drawing the position and general outline of a spot group is much more important than the amount of fine detail shown. Attempts to include detail often result in the spot being drawn too large, or "wandering" from its correct position. You must also decide which part of the group to use as your reference point for determining position. With a single spot, its center is the obvious choice. If the spot is bipolar, remember that the leading or western component usually outlives the following or eastern component. However, a large and active group can change in form so much in the course of a few days that the original reference point may lose its identity, and you may decide to go back and re-measure the drawings in a different way.

There are several ways of reducing a drawing to heliocentric latitude and longitude. One involves laying the drawing (made, remember, on translucent paper) over one of a set of master blanks that represent the Sun's presentation at different times of the year. The coordinates can then be read off directly. However, the advent of home computers and programmable calculators has made the straightforward algebraic method very quick and reliable. You need not record anything other than the North–South and East–West lines on the drawing, and all that you need are a ruler and a protractor. The formulae given here may seem frightening, but the work is minimal. Even without a memory facility, a spot group can be reduced in a couple of minutes. The method is as follows.

From an almanac obtain the following values for the date and time of observation (or P and B_o can be obtained to sufficient accuracy from the table given in this chapter):

1. The position angle of the North Point of the Sun's axis, $P;$
2. The heliocentric latitude of the center of the disk, $B_o;$
3. The heliocentric longitude of the Sun's meridian—the straight line joining its North and South Poles, L_o.

The value of L_o is changing all the time, to keep pace with the Sun's mean rotation period—as seen from the Earth—of 27.2753 days, in which time it spins through 360° of heliocentric longitude. This rate means that it turns through 1° of longitude in 1 hour 49 minutes, and each passing day carries its features 13°.2 further West.*

You must then measure the position of each spot on the drawing, as follows:

1. Its distance from the center of the disk, r. Divide r by R, the radius of the disk drawing. Call this value, $\dfrac{r}{R}$ sin ρ (rho). You will also need to use cos ρ, but the value of ρ itself is not used.
2. Its position angle, measured clockwise, from the northern point of the Sun's disk (i.e., *not* the northern point of its axis) to the line joining the spot and the center of the disk. Call this angle θ. Then, by allowing for the day's value of P, derive θ, which is the position angle of the spot measured from the Sun's true axis.

---

* The Sun does not rotate as a solid body would do, but spins more slowly with increasing latitude. The rotation period given here is the average for a latitude of 15° North or South of the solar equator. The equator itself spins about ¼° faster per day; at a latitude of 30° the rate is about 0°.7 slower.

To find the spot's latitude, B, use the formula

$$\sin B = \frac{\cos \rho \sin B_o}{\sin \rho \cos B_o \cos \theta} \qquad (1)$$

And to determine its longitude, L, use

$$\sin (L_o - L) = \frac{\sin \theta \sin \rho}{\cos B} \qquad (2)$$

Notice that formula (2) gives the *difference* between the longitude of the Sun's central meridian and that of the spot group, and that the solar longitude at the center of the disk *decreases* as time passes. If the value of $(L_o - L)$ is *positive*, then the group is that number of degrees *East* of the Sun's meridian, and if it is *negative*, then the group must be to the *West*.

Although a considerable number of people look casually at the Sun, and perhaps even count the number of active areas visible, few seem to make detailed drawings with accurate orientation, and still fewer go to the extent of calculating the positions of the sunspot groups. No doubt they are discouraged by the amount of work involved. Certainly it is no small labor to produce an accurately orientated drawing, and at first sight the calculation appears very involved. In compensation, however, let us emphasize that there are very few chances in amateur astronomy to make accurate measurements. Much observational work is *qualitative* (comparing two or more things in terms of each other, as in variable-star work), or *descriptive* (describing what one sees in terms of form, color, etc., as in the case of planetary work). The Sun, because of its brightness and size, and the very nature of its surface markings, offers the amateur something more: something to which he can put a ruler.

It works in reverse, too, because in measuring the Sun the amateur is also measuring his or her own performance as an observer. Although the positions of sunspot groups do show a real change of latitude and longitude with time, it is a slow change, and if they appear to jump around on the photosphere from day to day, you can be fairly sure that the reason lies at the Earth-facing end of the telescope! By investigating the cause of these errors, and improving your technique, you will gain the double satisfaction of increasing the value of the data and knowing that your own standing as an observer has been enhanced.

Sunspot sizes The area of a sunspot can be estimated by placing a piece of tracing paper, marked off in small squares, over the accurately drawn outlines of a spot group, and counting the number of squares that are either entirely covered or more than half-covered by the umbra and penumbra, ignoring the rest. Professionally, sunspot areas are calculated in terms of millionths of the visible hemisphere. On a 150-mm diameter image, an area of one square millimeter encloses 28½ millionths, which will enable a reasonable calculation to be made if the group is fairly near the center of the disk. The overall length of a group can be measured by timing how long it takes to drift past a line on the projection screen or in the eyepiece, the telescope being stationary; with every second that passes, the Sun appears to move 17,500 km through space, or almost 1½ times the Earth's diameter.

These estimates must be corrected for foreshortening if the group is away from the center of the disk because the apparent shape of the spots will have been to some extent squashed. To compensate for this effect, multiply the derived value of length or area by sec ρ, where $\sin \rho$, as before, corresponds to $\frac{r}{R}$.

Other solar phenomena If the atmospheric conditions happen to be particularly steady, and you have a direct-vision filter available, then use a high magnification and study any large sunspot group that happens to be visible. The contrast of tones and textures, the graceful curves, the surrounding splashes of umbral material over the photosphere, all seen with the razor-sharp definition that a good telescope and excellent seeing will give, conspire to produce a sight as memorable as any in the universe. Every now and then, in moments of perfect seeing, the photosphere will crystallize into a mass of tiny points that seem to disperse as soon as they are perceived, only to return again with startling suddenness a few seconds later. It is a pity that amateurs spend so much time looking at our star with a low magnification, so as to include the whole disk at one view, that they fail to appreciate the glories of the individual spots.

If a period of excellent seeing comes along, and you have the time to spare, spend an hour scrutinizing a well developed bipolar group. Noticeable changes can happen in this time. Unfortunately, long sunny periods and excellent seeing do not normally go well together because the moment the Sun begins to shine strongly its radiation builds up heat around the telescope and sends rippling air currents across the view. No doubt this is the reason for so much low-power work; so seize the chance of a high-power view when you can!

If you have the time and inclination, draw in the faculae, or estimate their positions, when you make a routine observation. These "clouds" first develop as relatively small patches, usually less than a day before the appearance of the pores that mark a new sunspot group. After the group has died away, however, the faculae may remain in position for several solar rotations before fading away. Therefore, you can keep track of a lost active area by recording its associated faculae.

The difficulty of observing faculae anywhere but near the solar limb, where derived longitudes become rather uncertain, makes it difficult to derive precise positions for these features. But it is certainly interesting to try to make the connection, and to see if a group of faculae seen at the eastern limb can be recovered, after its week or more of invisibility during the disk passage, as it approaches the western limb.

Faculae are sometimes confused with *flares*. A flare, however, is a very violent, short-lived phenomenon: a brilliant outburst of luminous hydrogen over a sunspot group that lasts for only a few minutes. A flare produces a surge of radiation of many different wavelengths, and can cause auroral displays, compass twitchings, and radio fade-outs on the Earth, but only the most violent ones are seen without using sophisticated filtering equipment to suppress the Sun's dazzle without dimming the flare itself. During the last solar maximum, only two so-called "white-light" flares were recorded by amateurs, on April 24, 1981, and June 4, 1982. There is not much point in looking for white-light flares as an observing program, but the possibility of noticing an intensely bright patch of light appearing in an active sunspot group should always be borne in mind.

Eclipses of the Sun It is hard to point to any obvious "coincidences" in the universe. So many apparent coincidences may have an underlying reason which we have not yet fathomed, so that to call them coincidences may only reveal our ignorance (for example, a human being stands roughly midway in size between the diameter of an atom and the diameter of the observable universe). But one remarkable feature of the skies does seem to be due to chance and nothing more: the Sun and the Moon, though differing vastly in real size, appear of almost identical diameter.

Total eclipse of the Sun, 11 June 1983   Two photographs taken on Ektachrome 200 with a 90-mm aperture f/11 Maksutov telescope. (a) Taken with an exposure of 1/125 second at the onset of totality. (b) Taken with an exposure of 1¼ seconds, to record the outer corona. (*Michael Foulkes*)

In fact, the average diameter of the Moon appears to be slightly smaller than that of the Sun (about 31½′ arc compared with 32′ arc), so that if it chances to pass directly in front of the Sun—which is what causes a solar eclipse—a hair-like ring of photosphere remains visible. If this happens, the eclipse is said to be *annular,* or "ring-like," and none of the hidden glories of the Sun—the reddish prominences leaping from the limb, and the pale corona—come into view. Fortunately, both the orbits of the Moon and the Earth are slightly eccentric, so that the Sun in June appears smaller than average, and the Moon, every month, passes from its remotest or *apogee* to its closest or *perigee*. This delicate interplay of changing distances can result in eclipses occuring when the Moon appears to be larger than the Sun, and so can hide the photosphere completely.

Between 1985 and 1999 inclusive there will be twenty-one solar eclipses, of which eleven will be total. However, adding all the time of totality together amounts to only thirty-five minutes of darkness before the end of the century, for the average duration of a total eclipse is much less than five minutes.

Astronomers have calculated, well in advance, just which parts of our globe will lie in the black patch of shadow that the Moon casts on the Earth at the time of a total eclipse. This patch is usually less than 200 km wide, and it hurtles over land and sea at a rate of about 3000 km per hour, engulfing all within its path in the seconds, or minutes, of the most awesome spectacle in nature.

If you want to observe a total eclipse of the Sun, you will have to make a journey, for the United States will see no more this century. The most favorable forthcoming eclipse is that due to occur on July 11, 1991, the path of which crosses Central America and Brazil, and it will last for the exceptionally long time of six minutes, fifty-four seconds to an observer located in the middle of the track. However, a number of travel firms have now caught on to the idea of "eclipse cruises," and organize trips either to a landfall within the path of totality, or else to an ocean rendezvous with the shadow. Astronomically, the sea offers flexibility in choosing a good-weather site, but it cannot provide terra firma for telescopes!

Fine eclipse photographs have now become a commonplace, and if you intend traveling to observe an eclipse, and have a portable telescope, you will find no lack of information in past issues of astronomical magazines such as *Sky & Telescope*. Using 400 ASA film with an f/11 instrument, exposures of about ⅟₂₅th of a second will show the inner corona and prominences well, while longer exposures will record a larger corona but overexpose and hide the inner details. Using 35 mm film, a focal length of 1000 mm will comfortably fit the outline of the Moon, with the inner corona, into the frame.

Unlike a total eclipse, which can be seen only from a very restricted track, a partial eclipse is visible from a large portion of the Earth where the Sun is above the horizon. It is interesting to watch the serrated edge of the Moon steadily advancing across the Sun's disk, occulting any sunspots in its path. In fact, this kind of observation is rarely made on the occasion of a total eclipse because everyone is too excited preparing for the total phase to pay more than superficial attention to the preliminary stages, and after totality no one cares anyway! Nevertheless, no partial eclipse can compare with the tension and drama of a total one, and the writer hopes that some readers of this book will be among the future groups of adventurers who chase and thrill to them.

CHAPTER 5

The Dead World of the Moon

The Moon is the Earth's closest neighbor, and the sunlight reflected from its surface takes only 1½ seconds to reach our eyes. In another sense, however, that surface is more remote in time than anything else visible in the sky. Even if we could detect a star situated on the farthest edge of the Galaxy from the Sun, its distance would be no more than 80,000 light-years or so—we should be seeing it as it was when the light was emitted 80,000 years ago. The distances to other galaxies are to be measured in millions of light-years, and at a distance of say 100 million light-years even a large and luminous galaxy appears rather faint in a moderate telescope. At 3,000 million light-years, a galaxy is too faint to be seen in any telescope, although it can be detected photographically; but, if it could be seen, the observer would be looking back 3,000 million years in time. This, effectively, is the time-gulf being bridged when the Moon, just 1½ light-seconds away, is being observed.

In cosmic terms, the Moon was stillborn. From its birth to its virtual death, or at least sleep, may have been only a few hundred million years. The fact that the *Apollo* astronauts found pieces of rock that had not undergone any remelting process during the past 4,000 million years or so suggests that no global upheavals have happened in the meantime. On the Earth, geologists have to search very hard to find rocks that are as much as 3,000 million years old, and these lie in a few quiescent areas of its crust. Mountains and continents still move, and volcanoes pour forth lava. What is the exception on the Earth is the rule on the Moon.

Its ravaged face tells of primeval violence. As the planets condensed from accumulations of solid particles inside a huge dark cloud surrounding the young Sun—the solar nebula—smaller aggregations plunged into them, building the protoplanets up to their present size. Very probably, the dense metallic matter condensed first, producing the nickel-iron cores that are evident in the four inner planets (Mercury, Venus, Earth, and Mars), and which may well exist in the outer giants Jupiter, Saturn, Uranus, and Neptune as well. Over these cores floated the lighter elements, particularly the silicon and carbon that went to form the mantle and the surface crust. To begin with, the planets were probably like great blobs of viscous liquid flying through a dark cloud of protoplanetary matter. One of these blobs had a smaller one accompanying it: the Earth and the Moon to be.

The Earth and the Moon The different course of these two bodies' histories is a reflection of their size. The Earth has eighty-one times the mass of the Moon, but only nine times its surface area. In space, a body loses heat by radiating it away from its surface; therefore, the greater its surface area, the more rapidly can heat be lost. If two bodies begin their careers at the same temperature, the amount of heat energy contained in each one is proportional to its mass. Therefore the Earth, when formed, had far more heat energy than the Moon, and proportionally far less surface available from which this energy could be lost. In other words, the Moon cooled down much more quickly than did the Earth. It probably formed a markable solid crust within a few million years of its recognizable origin, and so the impacts of the last large solid condensations within the solar nebula began to leave permanent (i.e., still recognizable) craters at a time when those being formed on the Earth were still melting back into the hot surface.

Another important difference between the two worlds is the presence or absence of an atmosphere and liquid water. The Moon's very small gravitational pull can be scorned by any gas molecules, which will leak into space. Therefore, in the vacuum to which its surface is exposed, any surface water would immediately vaporize and vanish. With no "weather" to blur and erode it, the surface has been preserved perfectly through an appreciable fraction of the history of the universe.

Movements and phases of the Moon Everybody knows that the Moon keeps the same face permanently turned towards the Earth. Astronomers call this "captured rotation." When formed, it was probably spinning in just a few hours. But the rhythmic pull of the Earth on the once hot and liquid sphere has deformed it slightly. The Earth-facing crust is thinner than the crust covering the averted hemisphere, and lava has crept through ruptures in this more fragile crust, forming the relatively smooth *maria* ("seas"), which cover roughly half the Earth-turned face. The opposite hemisphere has no large maria.

In fact, we can see rather more than half the surface because the Moon's axial spin is constant, so that it rotates exactly once with respect to the Earth in the lunar month of 29½ days. However, its orbital velocity is not constant, being greater at perigee than at apogee, and so the Moon does not appear to move around the sky at a

The Moon, 12 hours before First Quarter Note the distinctive craters Aristoteles (upper) and Eudoxus, toward the northern cusp. Just below the equator lies the badly ruined Hipparchus (best seen at this phase), with the much more distinctive Albategnius immediately to the South. (*James Muirden, 90-mm refractor*)

The Moon, 2 days after First Quarter Plato is the dark, elliptical crater near the northern cusp, while the magnificent Copernicus is very distinctive, a little North of the equator. Further South, Bullialdus catches the morning sunlight from its position in the Mare Nubium (*James Muirden, 90-mm refractor*)

constant rate. At perigee it is a little ahead of its mean position, and we are able to see a little way beyond the eastern limb; near apogee it falls behind, and the western limb is slightly advanced. Also, since its axis is tilted slightly from the upright position with respect to its orbital plane, we enjoy alternate views beyond the lunar poles. Known as *librations,* these small swings can have an appreciable effect upon the presentation of objects near the lunar limb. In all, about 59 percent of the surface is detectable from the Earth at one time or another.

The change in size of the Moon's disk from perigee to apogee is quite significant because its distance from the Earth at these times changes from about 367,000 km to 405,000 km. At perigee its diameter is some 10 percent greater than at apogee, and a perigee Full Moon sends us almost 25 percent more reflected sunlight! To appre-

The Moon, 2½ days after Full   The mountainous border of the Mare Crisium makes it look like a gigantic crater lying near the eastern limb. Elsewhere, the distribution of ray-craters (note especially Copernicus and Tycho) and some unusually dark patches can be well seen. Brilliant Aristarchus is also prominent. (*James Muirden, 90-mm refractor*)

ciate this effect, however, requires a direct comparison, and this can be done most successfully by photography. Some of the publications listed in Appendix I give future dates of perigee and apogee.

During its 29½-day journey round the Earth, the moon waxes from New to Full and then wanes from Full back to New again.* The line dividing lunar day from night is known as the *terminator,* and during the first half of the lunar month we are seeing the sunrise terminator moving steadily across the disk from the eastern towards the western limb. When it reaches the western limb, the disk is fully illuminated, and we are seeing the Moon, as it were, with the Sun behind us. The sunrise terminator now passes on to the averted hemisphere, and the

* This interval is known as the *synodic period*. The *sidereal period* is the time taken for the Moon to make one circuit of the ecliptic, and is about 27⅓ days long.

sunset terminator makes its appearance at the eastern limb, to pass in its turn across the face of our satellite.

The lunar shadows There is virtually no color on the Moon, and its surface is a series of gray tints. When the sunlight is shining down from a high altitude, we see only a confusing patchwork of detail; there is no way of distinguishing high-altitude features from low ones because the observer is effectively hanging above the surface and the contours cannot be seen, except at the extreme limb. But when the Sun is low in the lunar sky, and shadows are therefore long, the heights stand out in superb relief. Because the terminator marks the region of the rising or setting Sun, it is here that the shadows are longest and our view of the surface most spectacular.

The phases of the Moon As the Moon revolves around the Earth, we experience different views of the sunlit hemisphere and see the phases as shown.

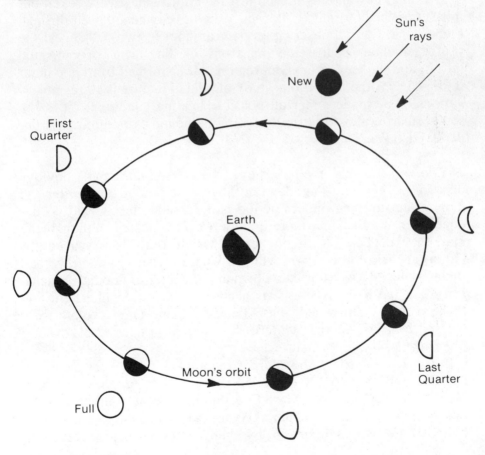

It follows from this that different regions of the lunar surface are seen under the most favorable illumination at specific stages of the lunation. Features near the eastern limb are under low-light conditions when the Moon is visible as a young crescent in the evening sky (sunrise illumination), and again under sunset illumination just after Full. Near the center of the disk, shadows are longest at First Quarter (local sunrise) and at Last Quarter (local sunset), while towards the western limb the view is most detailed just before Full (local sunrise) and again just before New Moon, when our satellite is a crescent rising before dawn. Because the crescent phase is difficult to observe well, the Moon always being rather low in the sky, the eastern limb is usually examined under local sunset illumination and the western limb under local sunrise illumination, both of which occur near Full Moon phase.

The lunar day and night each last for a fortnight (two weeks), but it is surprising how quickly the shadows move, when viewed through a telescope. Careful observation over an hour or two will reveal the tips of mountain peaks emerging from the long night, or vanishing into it; and the view of a given region is transformed from one night to the next. Although serious study of a lunar feature requires observations to be made under all possible lighting conditions, it is the spectacular terminator strip that attracts the most attention, and understandably so.

The lunar maria With the naked eye, the maria are the most obvious markings. These are huge lava outflows that poured forth after the main crater-forming impact period had passed—this is easy to deduce from the fact that large craters are very scarce, although the maria are peppered with tiny ones, visible only from spacecraft. Often, at their borders, you will see where earlier craters have had their "seaward" walls broken down and the interior flooded: Fracastorius, in the Mare Nectaris, is a classic instance, while here and there very low "drowned rings" betray ancient, almost totally submerged craters. Another feature, low wrinkle ridges, look exactly like frozen ripples in the surface.

Some seas, such as the Mare Crisium—that lies very near the eastern limb—are totally encircled by the bright cratered uplands. Most, however, from the Mare Foecunditatis in the southeast to the huge Oceanus Procellarum in the West—have flowed into each other, bounded not by mountains but by changes of color and surface tex-

The Mare Nectaris area of the Moon This photograph was taken about two days after Full. Note the chain of four huge craters extending down the terminator; the uppermost is Langrenus, 130 km across (*Martin Mobberley, 355-mm Cassegrain*)

THE LUNAR MARIA

Name	*Meaning*
Mare Aestatis	Summer Sea
Sinus Aestuum	Seething Bay
Mare Anguis	Serpent Sea
Mare Australe	Southern Sea
Mare Autumni	Autumn Sea
Mare Crisium	Sea of Crises
Palus Epidemiarum	Marsh of Epidemics
Mare Frigoris	Sea of Cold
Sinus Gay-Lussac	Gay-Lussac Bay
Mare Hiemis	Winter Sea
Mare Humboldtianum	Humboldt's Sea
Mare Humorum	Sea of Moisture
Mare Imbrium	Sea of Rains
Mare Incognito	Unknown Sea
Sinus Iridum	Bay of Rainbows
Mare Marginis	Border Sea
Sinus Medii	Central Bay
Lacus Mortis	Lake of Death
Palus Nebularum	Marsh of Mists
Mare Nectaris	Sea of Nectar
Mare Novum	New Sea
Mare Nubium	Sea of Clouds
Mare Orientale	Eastern Sea
Mare Parvum	Little Sea
Oceanus Procellarum	Ocean of Storms
Palus Putredinis	Marsh of Decay
Sinus Roris	Bay of Dew
Mare Serenitatis	Sea of Serenity
Mare Symthii	Smyth's Sea
Palus Somnii	Marsh of Sleep
Lacus Somniorum	Lake of Dreams
Mare Spumans	Foaming Sea
Mare Tranquillitatis	Sea of Tranquillity
Mare Undarum	Sea of Waves
Mare Vaporum	Sea of Vapors
Mare Veris	Spring Sea

(Not all of these names are "official," but they may be found on some lunar maps.)

ture. (It is worth pointing out, by the way, that these curious names, which are translated on the opposite page, date back to the 17th century, when the dark patches were believed by some astronomers to be true seas.) One of these areas is particularly famous: the Mare Tranquillitatis, where the first men to land on the Moon descended in July 1969, establishing "Tranquillity Base." The chart shows in detail just where this landing occurred, so that you can exhibit the spot to interested friends! Some of the smaller plains have been given other fanciful labels, for example *Sinus* (bay) or *Palus* (marsh).

The craters Point a telescope, even a small one, towards our satellite, and the craters jump into prominence. Most of the Southern Hemisphere, and the northern polar regions too, are imprinted with impact marks dating back to the earliest days of the solar system. The walls of primeval rings, some of them huge, can just be detected beneath the later peppering—and it might be an interesting and worthwhile task to try charting the positions of these old, overlaid craters, either from photographs or at the telescope—while other large craters such as Clavius, near the South Pole, are sufficiently recent to be well-preserved and yet to contain smaller but still substantial craters that have impacted on their floors and walls. Probably all the craters above a few kilometers in diameter were formed with a central mountain mass, caused by the surface rebounding back from the shock of impact, but in the older features these mountains may have been reduced by later surface flows: Ptolemaeus, near the center of the disk, is a conspicuous example. The largest crater on the Earth-turned hemisphere, Bailly, has a diameter of about 300 km, but is very poorly placed for observation, near the Southwestern limb.

Two craters find their way into every lunar observer's "hit parade." One is Copernicus, a 90-km formation surrounded by mare surface, in the Western Hemisphere; the other is Tycho, of similar size, in the southwestern highlands. Both are the center of extensive *ray systems,* immense streaks of glassy droplets that were, presumably, sprayed out as vaporous silica when these catastrophic impacts occurred. To judge from their perfect preservation, they must have been formed fairly late in the Moon's history.

Other features Valleys and mountains abound. Unlike the Earth's mountain ranges, which have been produced by gentle crustal folding (that is still going on), the lunar mountains seem to be the relic of

Ptolemaeus, Alphonsus, and Arzachel A conspicuous chain of great craters, lying near the center of the Moon's disk. Ptolemaeus (upper) measures about 140 km across (*Martin Mobberley, 355-mm Cassegrain*)

huge collisions—notice, for example, how the large arc of mountains to the North and East of the lovely Mare Imbrium gives the mare itself a resemblance to an enormous partly drowned crater, which, perhaps, is what it is. Some of the peaks in this Apennine range rise to five kilometers or more above the floor, and they form a beautiful sight at First Quarter, when their summits catch the sunlight while the land below is still in darkness. Occasionally, they can be detected with the naked eye as a bright projection into the dark hemisphere. One of the most spectacular ranges borders the Sinus Iridum, not far away. This is the surviving wall of a large crater that now seems to form a cliff overlooking the mare surface below. All these features, as well as great valleys and clefts, await the telescopic explorer.

Observing the Moon Despite being the easiest object in the sky to look at, the Moon is very far from being the easiest to observe properly. The mass of detail is bewildering. Where to begin? The task seems hopeless, and most people, having experienced the first thrill of seeing just what a powerful telescope can reveal, keep the Moon as one would a special antique, to be shown to friends, admired, and then put back into its case.

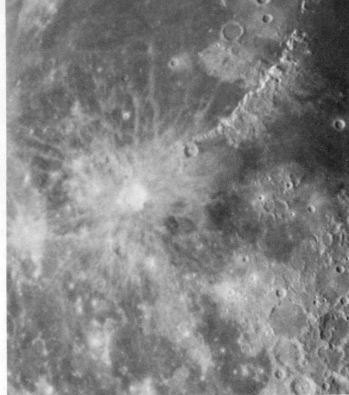

The Mare Imbrium area of the Moon This fine photograph shows the region from the Moon's North Pole to some way South of the equator. Note the curved Apennine range forming the southeastern border of the Mare Imbrium and the bright, ray-crater Copernicus left of center. (*N. W. Scott, 300-mm Cassegrain*)

Because the Moon *is* the closest approach to an antique in the Earth's vicinity, this attitude is not surprising. A few observers have claimed that occasional colored glows are seen, proving that the Moon is not entirely dead, and the evidence, though scanty, has some support. To the great majority of telescope-owners, however, the Moon is a museum of ancient history, and lunar observation is concerned with apparent change rather than real change: the effects of light and shade that can make a region appear rutted and rock-strewn on one night and a featureless silvery carpet on another.

Resolving lunar detail Any aperture and any magnification will give a good view of the Moon, but they will show different things. Remember that the resolution limit of a telescope applies to features on our satellite as well as to double stars, although the relationship is rather more complicated. This is because lunar detail does not consist of tiny star-like dots, but rather of areas of brightness and shadow. The greater the contrast between adjacent areas, the easier they are to resolve; so a narrow, shadow-filled crevasse (that appears as a dark spider-web crossing the light surface) may be made out quite easily, whereas a pair of craterlets, although separated by more than the telescope's resolving limit, may be nothing more than a tiny smudge, or even utterly invisible, if there is no shadow inside them.

Schiller A curious elongated formation about 190 km long, but only half as wide; it has the appearance of two adjoining craters whose connecting wall has collapsed. (*Martin Mobberley, 355-mm Cassegrain*)

There is another reason, too, for clefts being easier to detect than craterlets. The eye responds to lines particularly well. Take a lunar valley that can just be made out as the aforementioned spider-web line, and erase most of it, leaving just a few short sections. It will probably disappear from view altogether because insufficient nerve cells in the retina are now being stimulated. Restore the missing sections, and the valley will come back into view as a complete line. The table, based on one appearing in *The Moon,* by H. P. Wilkins and P. A. Moore (1955), gives some idea what may be detectable on the lunar surface using telescopes of different aperture, assuming that the objects are near the terminator and therefore cast a shadow:

LIMITING LUNAR DETAIL WITH DIFFERENT APERTURES

Aperture	Smallest craterlet	Narrowest cleft
25 mm	15 km	800 m
50	7	400
75	5	270
100	3.5	200
150	2.5	140
200	1.8	100
250	1.5	80
300	1.2	70
400	0.9	50
500	0.75	40

The narrowness of the "narrowest cleft" may surprise some people. Remember that this does assume that the object is filled with black shadow, and seen against a bright surround. Many clefts occur on rather dark mare surface, and will be much harder to detect; others are half-lost in mountains and mounds that cast their own shadows over them. So do not take this table too factually, but be guided by what you *do* see rather than by what you think you *ought* to see.

Magnifications for lunar work Naturally, to detect objects near the limit of visibility, a high magnification must be used. The Moon is a good object for high-power work because the shadow-contrast is so high; there are plenty of sharp black-and-white edges to focus on, which is helpful, because at high magnification the focusing is critical

Plato The famous dark-floored crater, 90 km across, lies on the northern shore of the Mare Imbrium. (*Martin Mobberley, 355-mm Cassegrain*)

—altering the position of the eyepiece by less than a millimeter can spoil the definition. The planet Jupiter, for example, is much more difficult to focus accurately because its markings are much vaguer than those on the Moon, and its limb tends to fade off into the sky. Viewing the Moon, using a 100-mm telescope, magnifications of ×250 can be used effectively if the seeing is sufficiently steady, while powers of up to ×350 or ×400 may bring the finest details into better view if the aperture is from 200 mm to 300 mm.

The golden rule in deciding which magnification to use is never to use a power higher than necessary. The fact that a crater appears twice as large does not automatically mean that twice as much detail can be seen inside it! Before settling down to use a particular eyepiece, study a small region with different powers and decide which one, on that particular night, does the job best. If the atmosphere is reasonably steady, with occasional really perfect moments when the high-power image freezes into crisp and glorious detail, it may be worth staying with this power. If such moments are rare, and the really powerful eyepiece gives a generally slightly blurred and de-

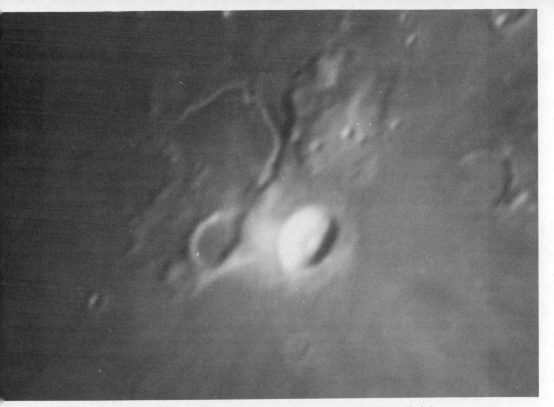

Aristarchus and Schroter's Valley Aristarchus, 45 km across, is the brightest crater on the Moon's surface, and shows faint radial bands on its inner walls, which can be seen in the photograph. The long U-shaped valley can be made out with a small telescope. (*Martin Mobberley, 355-mm Cassegrain*)

graded image, you will do better with a lower magnification and an apparently sharper image (it is not really sharper—it just appears so because the scale of the view has been reduced).

The visibility of different phases You will not need to be reminded how the current lunation is progressing; if you are a lunar observer you will be eagerly keeping track of the terminator, while if your prime interest is faint galaxies you will be anxiously counting the nights before the bright-sky period around Full!

However, it is not simply the phase that affects the lunar observer. Remember that the Moon moves around the sky once in a month, more or less along the ecliptic. Therefore, at certain points on its path it is either particularly high in the sky (when it occupies the position taken up by the Sun in midsummer), or particularly low (when it is in the midwinter Sun's position). In the Northern Hemisphere, these points correspond to the Moon being in the constellations Taurus/

Gemini and Ophiuchus/Sagittarius respectively–for Southern Hemisphere observers, the positions are reversed. Concerning the visibility of the Full Moon, which is opposite the Sun in the sky, we can draw the obvious inference that winter Full Moons are going to be much higher in the sky than are summer ones.

The following table, drawn up for observers in the Northern Hemisphere, indicates the most favorable seasons at which to view different phases, as well as their appearance at other times.

VISIBILITY OF THE LUNAR PHASES

Season	First Quarter	Full	Last Quarter
Vernal equinox (March)	*Very favorable*	Fairly favorable	Unfavorable
Summer solstice (June)	Fairly favorable	Unfavorable	Fairly favorable
Autumnal equinox (September)	Unfavorable	Fairly favorable	*Very favorable*
Winter solstice (December)	Fairly favorable	*Very favorable*	Fairly favorable

The best time at which to observe a thin crescent in the evening sky is early in the year (February–March), when the Moon's declination is considerably more northern than that of the Sun, allowing it to be seen at the greatest possible altitude above the sunset place. The Old Moon is best seen in the October–November dawn, when it rises at the earliest possible time before the Sun. In general, then, the summer months are the least favorable for lunar observation. Again, observers in the Southern Hemisphere will need to reverse the months and seasons indicated here.

A lunar program   To provide a comprehensive guide to just the more obvious lunar features would involve writing a small book. Many observing guides include an outline map of the Earth-facing hemisphere, together with a gazeteer describing the more important seas, craters, and mountains. Others take the observer through a lunation and discuss the objects that will be well placed for observation at

The Theophilus chain This well-preserved 80-km crater has two noticeably eroded companions, Cyrillus and Catharina, of about the same size. These also appear at the left-hand edge
of the photograph on page 103. (*N. W. Scott, 300-mm Cassegrain*)

different phases. Both approaches have their merits. But they also have the weakness that they must, necessarily, be so general that the observer is not being encouraged to get down to the job of making a real lunar *observation*.

Sightseeing is all very well, and the writer is certainly not suggesting that you should resist the temptation to spend many nights with your telescope simply enjoying the miracle of this unimaginably ancient, unbelievably fresh-looking wreck of a planet. But it is also hoped that you will follow the honorable path of many generations of lunar observers, or "selenographers" as they used to call themselves, and put pencil to paper to make some sketches. It is only in the course of making a sketch, and struggling (mentally) to decide what it is that you *are* seeing, that true "observing" begins to supersede mere "looking."

Crater-drawing To give a stimulus to this work, we can take advantage of the fact that space-borne cameras have photographed the surface of the Moon in much finer detail that can ever be made out from the Earth. The *Orbiter* probes, that circled our satellite in 1966 and 1967 as a preliminary to establishing the *Apollo* landing sites, covered the entire lunar surface, resolving details as small as 65 meters across in the "close-up" frames. What this means is that craterlets and surface clefts that are beyond the visibility of amateur instruments can be mapped with precision, and a master chart of an interesting area can be drawn, against which to compare telescopic observations. Therefore we include an outline chart, on a scale of 1 km per millimeter, of several craters, traced from an appropriate *Orbiter* photograph, and corrected for the inclination of the center as viewed from the Earth.

It is suggested that initially you simply trace the outline of the crater into the observing book, perhaps showing just the crater rim. Then, at the telescope, fill in as many details as you can see. As a general guide, it is sensible to insert the black shadow edges first. Once this is done, try to locate the more obvious features (hills, small craters, clefts) in their correct position. Finally, when you are satisfied that the general proportions of the drawing are correct, tackle the finer markings. If the seeing is at all unsteady—which will, almost invariably, be the case—you will now be experiencing sudden sharp awareness of markings that, almost before they have been grasped, disappear as if they had never existed. Some seconds will pass, and then they will reappear, perhaps just long enough for the position and

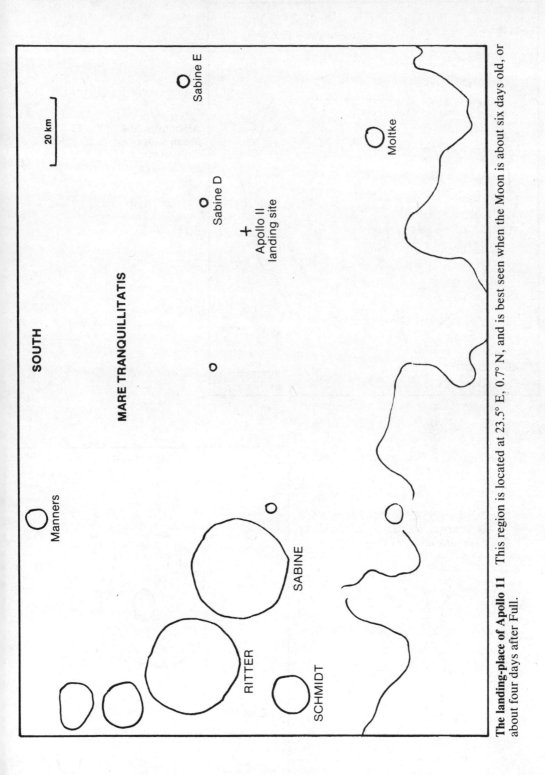

The landing-place of Apollo 11 This region is located at 23.5° E, 0.7° N, and is best seen when the Moon is about six days old, or about four days after Full.

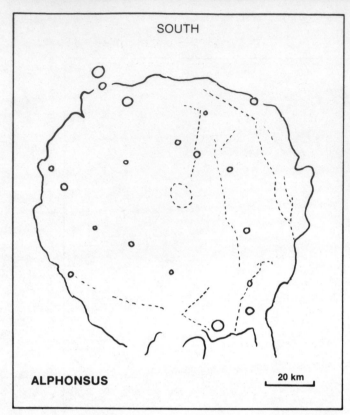

SOUTH

ALPHONSUS

20 km

Alphonsus (04° W, 13° S). Moon's age at sunrise, 7½ days; at noon, 15 days; at sunset, 22 days.

Cassini (05° E, 40° N). Moon's age at sunrise, 6½ days; at noon, 14 days; at sunset, 21 days.

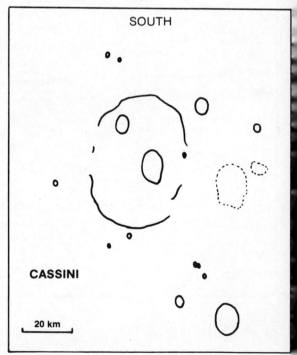

SOUTH

CASSINI

20 km

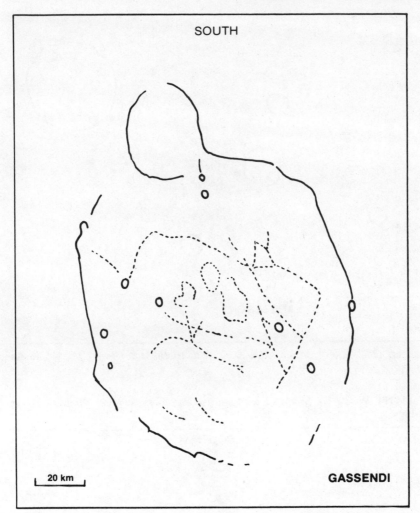

SOUTH

20 km

GASSENDI

Gassendi (41° W, 17° S). Moon's age at sunrise, 11 days; at noon, 18 days; at sunset, 25½ days.

form of one to be memorized and committed to paper. Then they will vanish again, and another wait ensues. Part of the technique of lunar and planetary observation is to learn to "store" details in the memory, so that you can concentrate fully on what is in front of your eye during the moments of excellent seeing, and extract this information on to the sketch, or into the notes, as soon as it disappears.

When you are satisfied that the drawing is as complete as it can be, compare it with the master outline. You will not, of course, have

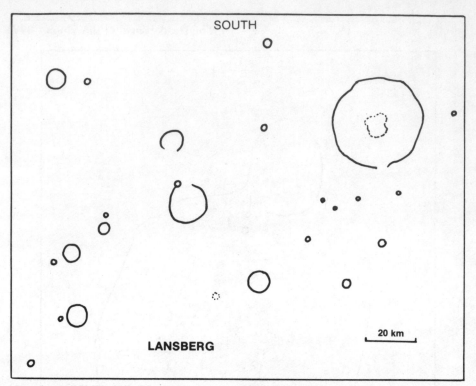

Lansberg (27° W, 00°). Moon's age at sunrise, 10 days; at noon, 17 days; at sunset, 24 days.

Pitatus (14° W, 30° S). Moon's age at sunrise, 8½ days; at noon, 16 days; at sunset, 23 days.

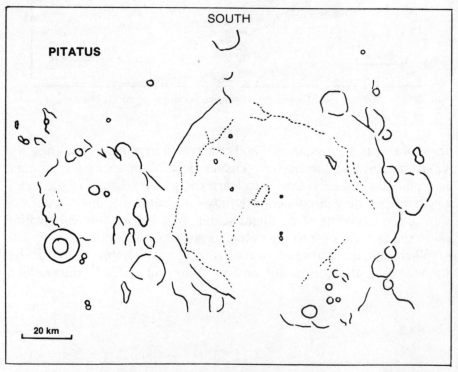

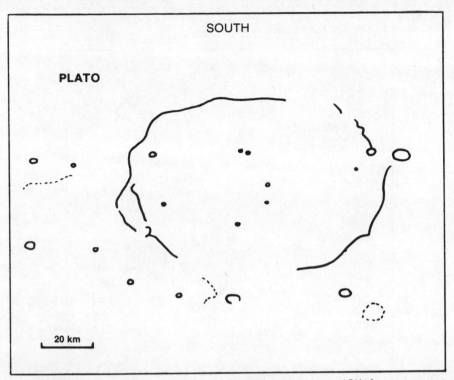

Plato (09° W, 52° N). Moon's age at sunrise, 8 days; at noon, 15½ days; at sunset, 22½ days.

Posidonius (29° E, 32° N). Moon's age at sunrise, 5 days; at noon, 12 days; at sunset, 19½ days.

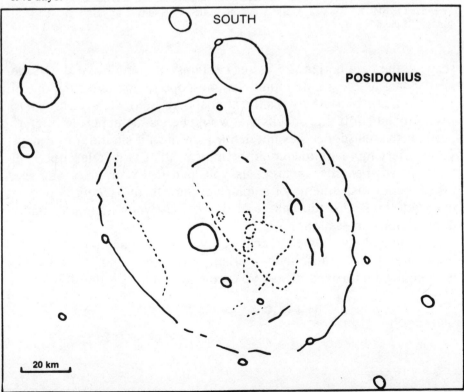

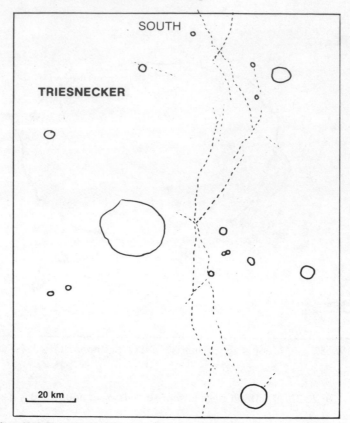

Triesnecker (04° E, 05° N). Moon's age at sunrise, 6½ days; at noon, 14 days; at sunset, 21 days.

seen *all* the details that the crater contains: at some point down the descending feature scale either the telescope or the eye runs out of recording capability. Perhaps the atmosphere is not steady enough to reveal minute dots and specks that would be visible in perfect seeing; perhaps the telescope does not define as well as it should; or perhaps your observing technique could be improved. If your telescope is at fault, it will become apparent as you gain in experience, and use other, better instruments of similar aperture; if the seeing is poorer than you think, this too will be made obvious when a night of really good seeing comes along!

It is quite likely, however, that you yourself are as much to blame as telescope or atmosphere—perhaps even more so. Could your technique be improved? Have you really extracted all that the image

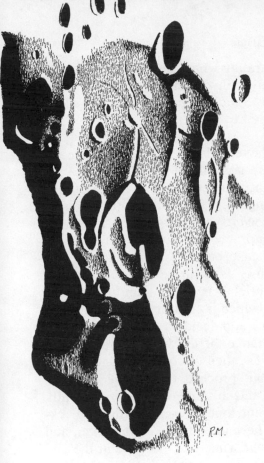

Metius, Fabricius, and the floor of Janssen Seen under evening illumination on 24 October 1983. (*Rob Moseley, 150-mm Newtonian, x60-x120*)

R.M.

The Sirsalis Cleft Seen under morning illumination on 17 December 1983. (*Rob Moseley, 150-mm Newtonian, x120-x240*)

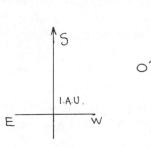

contains? Have you recorded accurately what you *have* seen? Is the position of that craterlet right, or have you displaced it slightly? Refer to the master and make a judgment. Should you have seen that cleft? Refer to the telescope and have a look!

Interpreting the view You should never take the master outline, or indeed any photograph or chart, as "gospel." If you think that your drawing represents the features that you see better than the outline does, it would be criminal to change it. The task before you is to represent accurately and honestly *what you have observed.* As the observer and artist, you have the supreme right to say what it is that you have observed; but you must also back up this right with the conviction that your representation is as accurate as you can make it. If the master suggests that craterlet B is larger than craterlet A, and so should be seen more easily, which is contrary to your impression, then refer back to the telescope and decide if craterlet B has, in fact, been misrepresented. If so, change the drawing. But if, in your judgment, craterlet A *is* easier to see, then of course you will leave your interpretation as it stands, but proceed to ask the question "Why?" Does B hold less shadow, and so stand out less prominently for this reason? Or is it of a different tone, merging with the background much more than A does? There must be an explanation, and you should not rest until you find out what it is, because at the end of this chain of observation, question, reobservation, and perhaps an answer, lies a discovery. It may not be much of a discovery, compared with the discovery of a comet or a nova, if you value such things by the general attention they earn; but if it is a piece of new knowledge it is a discovery nevertheless, and you may now know more about this small feature of the Moon's surface than does anyone else in the world.

Quite possibly you will have to observe the crater on many occasions before a satisfactory, consistent picture emerges. The sunset view will be very different from the sunrise one, and the noon view different again, for now the surface tone, rather than the relief, is seen. (Observed under a high Sun, the lunar landscape may be too glaring for comfort. A neutral-density "Moon filter" may be appreciated, although some observers find a green tint more restful than gray.) Try to document the way the visibility of features changes from night to night. When does that cleft lose its shadow? Does it disappear from view at lunar noon, or can its outline still be traced?

At what point in the lunation does that bright patch become prominent? And so on.

Educating the eye At the beginning of this exercise, when you select a crater and first view it through your telescope, you may be appalled at how small and featureless it appears, compared with the outline in this book. Remember this impression—and, if possible, write it down! The task seems hopeless, to begin with. It may still seem so, an hour later, after struggling to get the few details that you *can* see right on the sketch. But if you carry on, tackling other objects as the terminator moves across the Moon, and return to your original crater one lunation later, almost certainly you will be astounded at how

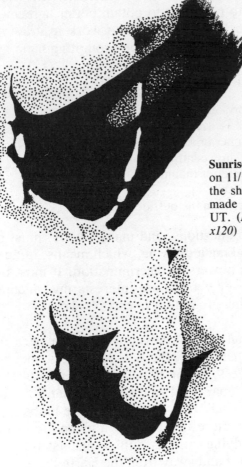

Sunrise over Prinz Two drawings made on 11/12 May 1984 showing the retreat of the shadows in two hours. The first was made at 22.15 UT, the second at 24.10 UT. (*Rob Moseley, 150-mm Newtonian, x120*)

much more can be seen. You may be tempted to throw the first sketch away, in embarrassment. Don't: it is a record of what you saw on that first night: it is the beginning of a voyage of discovery to discover your own potential as an observer, and it is a promise of further improvement to come.

You may decide to continue with this work, drawing as many lunar craters as possible. If you do, try to obtain an outline from a photographic atlas, as a starting point. This is not "cheating"; it is a sensible and helpful practice to get as accurate a basis as possible before filling in the details. However, outlines must always be treated with reserve because there is, in truth, no such things as a crater's outline; its apparent shape may change a good deal with the rising and the setting Sun. Nevertheless, it is helpful to have something with which to work, even if you decide to modify it to obtain better agreement with the telescopic appearance. When you set to work to draw the details, a series of questions like this should be prompting hand and eye:

- Is the crater circular? If not, what shape is it?
- Is the wall continuous or broken?
- Is the shadow edge smooth or irregular?
- Which looks steeper, the inner or the outer slope?
- Is the interior mountain (if any) central? If not, where is it placed?
- How many peaks has the interior mountain?
- Is the rest of the floor smooth?
- Is the floor darker or lighter than the outside country?

. . . and so on. Each of these questions, and many others, must be answered in the process of making a drawing, which means, in turn, that the eye must improve its powers of discrimination. It must become educated to see. However your future interests may incline, the Moon is a wonderful training ground for telescopic work.

Some other programs Individual crater-study is an obvious field of work, but there are several others. One, the search for very ancient, almost totally obliterated craters, might prove fruitful; it could, at least, lead to a totally original lunar map! Another is a polar libration study. Because both lunar poles are heavily cratered, the limbs here are very irregular. It might be interesting to draw a given arc of the limb from night to night, studying how its outline against the sky changes as the pole nods toward and away from the Earth during the lunar month. The East and West limbs are much smoother, but could also be tackled.

In the past quarter-century there has been much interest in the possibility of fleeting, local glows occurring on the Moon. This began with a famous observation by the Soviet astronomer Nikolai Kozyrev, who observed a reddish glow inside the crater Alphonsus in 1958. Although the glow seems to have lasted for only a few minutes, its effects have lingered. During the 1960s, amateur observers began searching systematically for fleeting glows and brightenings. A few—notably inside the craters Aristarchus and Gassendi—have been recorded by several observers working independently, but most have not.

Hunting for transient lunar phenomena (TLPs) is not work for the beginner, because these effects—assuming that they are truly lunar and not atmospheric, as many sceptics claim—are so faint that they would not be detected by anyone who was not thoroughly familiar with the appearance of the surface features in their "normal" aspect. Nevertheless, they should serve as a reminder that the lunar surface may not be as dead as is widely supposed, and that the observer should always be prepared for the unexpected.

Lunar occultations The Moon travels around the celestial sphere at an average rate of about 1° every two hours, and in the course of this journey it will pass in front of, or *occult,* any star that happens to lie in its path. The positions of most of the stars, referred to by coordinates of right ascension and declination, are known with very great accuracy: something of the order of 0.01″ arc, which is equivalent to the thickness of a hair seen from a distance of 250 meters. Therefore, if the instant when the Moon's limb either obscures or reveals a star is noted, we have a precise "fix" on the position of that part of the limb, at that particular moment, on the celestial sphere. The star is being used simply as a reference point.

The original purpose of occultation work was to measure the position of the Moon in its orbit and to correct the lunar predictions published in almanacs. After all, the Moon is moving around the Earth at a velocity of about 2.5 km per second, so that (in theory) an occultation timed to an accuracy of one tenth of a second will define the position of its limb to within 250 meters. However, modern lunar theory is so precise that there is no real need to apply corrections on a year-to-year basis, so that this aspect of occultation work is largely redundant. Instead, there is the interesting possibility of seeing if the Moon keeps to schedule over periods of decades or centuries. A quantity called *G,* the *gravitational constant,* appears in all equations that describe how astronomical bodies revolve around each other. Is *G* truly constant, or is it very slowly changing with time? If, for example, it is gradually becoming smaller, then orbiting bodies will slow down, even though their distance apart remains unchanged. The Moon's orbit is the most accurately known in the universe, and is the best test of the constancy or otherwise of *G:* occultation observations help monitor it. A changing *G* would have profound implications for the past and future of the universe. Up to this time, however, there is no direct evidence to support a variation of the gravitational constant.

This is one reason for observing occultations. Another is that the Moon's blade-like limb, cutting across a star image, "sees" details that are beyond the ability of the telescope to resolve. In angular terms, this limb advances across the stars at a rate of 0.55″ arc in one second of time.* Every star in the sky has a true angular diameter much smaller than this. Therefore, if a star being occulted by the Moon consists of two components only 0.1″ arc apart, they will be too close to be resolved in any normal telescope, but the lunar limb could take an appreciable time to cover them up—as much as one-fifth of a second, if the line joining the two components is parallel to the Moon's direction of motion. This may sound a very brief interval of time, but it is a very noticeable "fade" compared with the instantaneous snap of an ordinary occultation. Several extremely close, hitherto unsuspected binary stars have been discovered in this way, and others are sure to reveal themselves in due course.

Yet another reason is a strange reversal of the original one. Although most stars have had their position in the sky measured with great accuracy, some have not. Occultation timings can improve them, by using the lunar limb as the standard marker!

Above all, occultations are fun to watch. Some of the publications listed in Appendix I include predictions for stars down to about the 7th magnitude, but if you are particularly enthusiastic you may be able to apply for individually prepared predictions, for your own station, down to a fainter limit. Your local society may be able to

---

* This is the rate at which the western point of the Moon's limb appears to move. If an occultation occurs near one of the poles, the motion is almost tangential to the limb, so that the star is covered up at a slower rate.

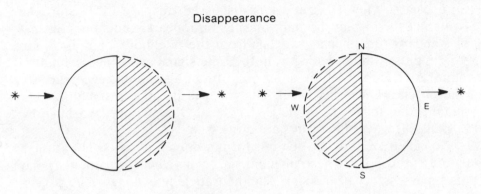

Disappearance

Reappearance

How the Moon occults a star   Before Full (left), disappearance takes place at the dark limb, and reappearance occurs at the bright limb. After Full, the sequence is reversed.

help here. Remember that each observer on the Earth's surface views an occultation event from a slightly different position, so that no two timings are likely to be the same; the predictions issued in almanacs are for a few standard latitudes and longitudes, but the conversion to your own station will not be difficult.

The usual policy is to observe only dark-limb events, which means disappearances into occultation at the western limb when the Moon is between New and Full, and reappearances at the eastern limb during the period from Full back to New. With most telescopes, stars fainter than the 3rd magnitude are very hard to make out at the bright limb, and so useful timings cannot be made.

Probably 90 percent of all occultation observations are of disappearances. There are two reasons for this: first, they are observed during the first half of the lunation, when the Moon is in the evening rather than the morning sky; second, it is much easier to time a star vanishing than one flashing into view. It follows from this that the profile of the western limb is known better than that of the eastern limb, and that good observations of reappearances are particularly valuable.

To observe an occultation you need a prediction, a telescope, and a stopwatch. A high magnification is not normally necessary, unless the star is very faint, when a powerful eyepiece could help bring it into clearer view (this effect is discussed on page 215). Start the watch at the instant the star vanishes or appears, and stop it against a standard time signal, whether from the telephone or radio. Subtracting the watch reading from the time signal gives the occultation time. An accuracy of the order of a fifth of a second is necessary for the observation to be useful, but you will be able to obtain further details, and forms, from the U. S. Naval Observatory, Washington, DC 20390, where the results are processed.

So-called *grazing* occultations form a branch of work that attracts some enthusiasts. These phenomena occur when the Moon's polar limb passes very near a star. The theoretical line along which a passing contact will occur is plotted on a map, and a team of observers are distributed across this line, so that some may see the star disappear entirely for a minute or two, while others may see it flicker behind invisible peaks on the Moon's limb. Grazing occultation expeditions are, or should be, enjoyable social occasions, as well as producing results of particular astronomical value because they document the limb profile and also give a very accurate measure of the Moon's declination.

Eclipses of the Moon About three times in every two years, on average, the Full Moon passes through the Earth's shadow and is eclipsed. This eclipse will be total, or partial, according to how central the passage happens to be. At the distance of the Moon, the dark central *umbra* of the Earth's shadow is about 6000 km across, considerably larger than the Moon's diameter of 3476 km. In fact, the duration of total immersion in the shadow can be nearly two hours, although most total eclipses are only half as long as this; with another hour being added on for completely entering the umbra and a further one for completely leaving it, the total umbral stage can occupy some four hours. The much lighter *penumbra* surrounding the central umbra, being the region where sunlight is only partly cut off by the Earth, is so faintly shaded that the Moon's dimming is barely noticeable.

If the Earth had no atmosphere, the umbra would be utterly dark and the Moon would disappear altogether at totality. Indeed, this is the impression you may have during the period shortly after the so-called *first contact,* when only a small part of the lunar disk lies in the umbra: it looks exactly as if a bite has been taken out of the Full Moon. But this is simply an effect of contrast with the brilliant uneclipsed portion. As the shadow extends over more of the surface, the eclipsed area begins to shine like old copper: sunlight is being refracted by the Earth's atmosphere, turning the shadow reddish because air transmits red rays more strongly than any other color of the visible spectrum. In a sense, we are now looking at the Full Moon through a deep red filter, once *second contact* has occurred and the Moon is totally immersed in shadow. (Third and fourth contacts correspond to the ending of totality and final emergence from the umbra, respectively.)

This red filter is, however, of an uneven and unpredictable nature. If the Earth's atmosphere is unusually murky, the eclipse will be darker than usual. Whether or not ordinary clouds can affect the shadow's brightness is doubtful because most of the sunlight passing through these lower reaches of the atmosphere must be completely absorbed. It is the much thinner air in the stratosphere, at altitudes of between 15 km and 40 km, that does most of the transmitting, and clouds do not reach as high as this. Other material can, though: in 1963 there was a violent eruption on the island of Bali in Indonesia, and a huge amount of very fine dust found its way up into the stratosphere, producing the exceptionally dark total eclipse seen in December of that year. Even more spectacular was the eclipse of July 6, 1982, that occurred three months after the tremendous El Chichon explosion in Mexico. During this eclipse the Northern Hemisphere of the eclipsed Moon appeared much darker than did the Southern, apparently due to the predominantly northern distribution of volcanic dust in the stratosphere. This was a striking demonstration of how efficient minute solid particles of matter can be in intercepting light.

To observe an eclipse, it is best to employ a low magnification, so that the whole lunar disk can be seen at one time. Little is to be gained from the use of high powers because the edge of the shadow is so ill-defined. Try making timings, to the nearest minute, of the moment when the shadow edge crosses certain craters, and make written notes of the appearance of the shadow at different stages of the eclipse. Which craters are particularly noticeable in the shadow? Which are hard to see? Do any areas appear particularly dark or bright? Are any spectral colors seen at the edge of the shadow, or does the umbra exhibit any anomalous brightenings?

With the Moon being so dim, very faint stars can be seen in its

FORTHCOMING LUNAR ECLIPSES

Date	UT	Type	Magni-tude	Dura-tion (mins)	Longi-tude* (°)
1985 May 4	19.57	Total	1.20	70	60 E
Oct 28	17.43	Total	1.10	42	90 E
1986 Apr 24	12.44	Total	1.20	68	168 E
Oct 17	19.19	Total	1.30	74	67 E
1987 Oct 7	03.59	Partial	0.01		63 W
1988 Aug 27	11.04	Partial	0.30		166 W
1989 Feb 20	15.36	Total	1.28	76	129 E
Aug 17	03.09	Total	1.60	98	45 W
1990 Feb 9	19.12	Total	1.08	46	76 E
Aug 6	14.13	Partial	0.68		149 E
1991 Dec 21	10.33	Partial	0.09		159 E
1992 Jun 15	04.58	Partial	0.69		74 E
Dec 9	23.45	Total	1.28	74	3 E
1993 Jun 4	13.01	Total	1.57	98	165 E
Nov 29	06.27	Total	1.09	50	99 W
1994 May 25	03.32	Partial	0.28		53 W
1995 Apr 15	12.16	Partial	0.12		176 E
1996 Apr 4	00.04	Total	1.37	84	1 W
Sep 27	03.04	Total	1.24	72	46 W
1997 Mar 24	04.36	Partial	0.93		69 W
Sep 16	18.52	Total	1.22	66	77 E
1999 Jul 28	11.28	Partial	0.42		172 W

* This gives the longitude on the Earth's surface from where an observer will see the Moon on the meridian at the central phase of the eclipse.

vicinity. The well prepared occultation observer can have a good time: during the total eclipse of July 1982, one observer—using, admittedly, a 450-mm reflector—was able to time forty-eight disappearances and twenty-five reappearances! The accompanying table lists those lunar eclipses that are due to be seen before the year 2000.

CHAPTER 6

Observing the Planets

When a planet shines prominently in the sky, telescopes are drawn irresistibly towards it. Whatever your main interest happens to be, the sight of bright Jupiter or Mars, or of Venus sparkling in the twilight sky, is a distraction. White or orange disk, or silver crescent, here are bodies whirling—like the Earth—around the Sun: tiny globes bound to their course by the invisible, unbreakable grip of gravity. So near, compared with even the nearest star, that they seem almost touchable; so far that a journey at the bewildering speed of a spacecraft can take months or years.

Unlike our regular Moon, they are not almost constant members of the nightly host. They may be lost from naked-eye view in the vicinity of the Sun for weeks at a time; few moments in the observer's diary are more memorable than the first sighting of a planet in the dawn sky, after conjunction with the Sun. Again, they swing closer to the Earth at some times than at others, so that their disks appear to change in size; and the innermost planets, Mercury and Venus, pass through a cycle of phases. The planetary sky, therefore, is one of continual change, and the configuration of the planets on any night will not be repeated even approximately in a long lifetime.

Distinguishing the planets Remember that the Earth is one of the nine orbiting major planets, and that its own whirling around the Sun would create an apparent whirling of each planet, even if that planet were fixed in position on its own orbit. When, in addition, the planets all have their separate motion, the picture becomes so complicated

131

that astronomers spent thousands of years trying to untangle the mystery!

Fortunately, celestial mechanics need not concern the amateur. The planets' movements around the celestial sphere have now been calculated for centuries ahead, and although you will need a star map to plot the positions of the dim outer worlds Uranus, Neptune, and Pluto before they can be identified, the original five (Mercury, Venus, Mars, Jupiter, and Saturn) can be spotted by anyone with a nodding acquaintance with the sky, because they are all easy naked-eye objects. In fact Venus is always much brighter than the brightest star, Sirius; Jupiter usually is, and Mars sometimes is. Mercury, too, can almost attain the brightness of Sirius, although it is well camouflaged in the twilight sky near the Sun, from which it never escapes. Only Saturn, the most glorious sight of all, hides its beauty in the guise of a "star" of the first magnitude.

The Earth's station, third planet outwards from the Sun, divides these neighbor worlds into two very unequal groups. Planetary geologists see similarities between the four inner planets (Mercury, Venus, Earth, and Mars) and the four outer planets (Jupiter, Saturn, Uranus, and Neptune)—Pluto being a world apart—but the observer finds greater contrast between Mercury and Venus on the one hand, and the rest of the planets on the other: the *inferior* and *superior* planets respectively.

The inferior planets Because their orbits lie between that of the Earth and the Sun, the inferior planets can never appear far from the Sun in our sky. That is the first thing to distinguish them from the other members of the solar system. The second important difference is that they both pass through a complete cycle of phases. When they are nearest to the Earth, at *inferior conjunction,* the night side is turned towards us and they are virtually invisible. On the other hand, when the illuminated hemisphere faces us, they are on the far side of their orbit, at *superior conjunction,* and appear shrunken by distance. In both positions, of course, they are either in line or almost in line with the Sun, drowned in the glare (although Venus is so bright that it can be seen even on the day of inferior conjunction, if circumstances are favorable).

In practice, the best views of the inferior planets are to be had in the phase range from fairly thick crescent to gibbous, when they are at a reasonable angular distance or *elongation* from the Sun. As the

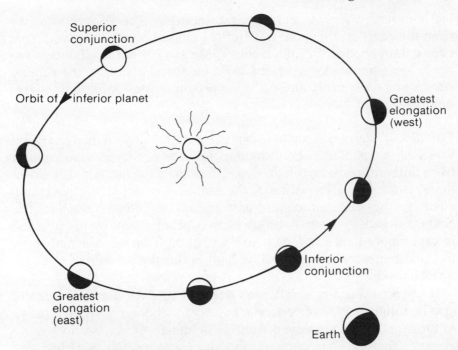

How an inferior planet appears to move   The motion of an inferior planet is shown here relative to the Earth, which is assumed to be stationary in its orbit. From being a thin or invisible crescent at inferior conjunction, the planet shrinks to a distant full disk at superior conjunction. Half-phase occurs at the time of greatest elongation from the Sun.

diagram shows, they are at half-phase (looking like the Moon at First or Third Quarter) when they are at *greatest elongation,* which in the case of Mercury amounts to between 18° and 28° from the Sun (this large range occurring because its orbit is very eccentric, and some elongations occur with the planet near perigee, and others when it is near apogee). In the case of Venus, which has an almost circular orbit, the range is only between 46° and 48°. Mercury never appears to be farther from the Sun than does a two-day-old crescent Moon, whereas Venus can swing out to twice the distance, setting or rising in a dark sky outside the period of twilight.

MERCURY

First, find your planet! Mercury is the shyest of all the planetary bodies. Although it is not particularly faint, its periods of visibility are very fleeting, and it is a great moment in any amateur astronomer's career when Mercury is first spotted.* The further you live from the equator, the more difficult the chase becomes. In fact, from sites within about 35° of the equator, Mercury is not at all difficult to see in the twilight sky when it is near elongation because the ecliptic stands at a reasonable angle to the horizon at the sunrise and sunset points.

Visibility of Mercury To understand this more fully, remember what was said in Chapter 2 about the appearances of the celestial equator from different terrestrial latitudes. The higher the latitude, the lower it sinks; the lower the latitude, the higher it rises. At the terrestrial equator, the celestial counterpart passes directly overhead. The ecliptic, which is the path followed in a general way by the planets, is never more than 23½° away in declination from the celestial equator, and therefore it must also be high in the sky as seen from equatorial latitudes.

However, Mercury and Venus demand that another requirement also be fulfilled. We are concerned always with their position relative to the Sun around the time of sunrise or sunset. If that portion of the ecliptic separating Mercury and the Sun forms a small angle with the horizon, then Mercury will appear much lower down in the twilight than if it forms a large angle. Because of the way in which the celestial equator and the ecliptic intertwine, evening elongations of Mercury form the greatest angle with the horizon when they occur in spring, and the smallest in the autumn; morning elongations are highest in the autumn and lowest in the spring.

From equatorial sites, the difference is negligible. If you are located at a latitude of about 50° or more, however, Mercury may be virtually undetectable at an unfavorable elongation, and the "evening-spring, morning-autumn" rule must be applied if you hope to sight it.

Mercury does not share its meager favors equally between our two hemispheres. Through the chance arrangement of the planetary orbits, it happens to be near perihelion when the best viewing time occurs for observers in the Northern Hemisphere, and near aphelion at the most favorable time for southern observers. This means that observers in the Southern Hemisphere can have favorable views of

---

* By a strange coincidence, the writer was working on this chapter on the morning of October 6, 1983. Looking out at the dawn sky he noticed, just above the dawn arc, a prominent "star," which turned out to be Mercury!

the planet when it is an extra 10° away from the Sun. This is a huge difference. Nevertheless, practically all the important early telescopic work on Mercury was done from the Northern Hemisphere because that was where the observers happened to be!

For most people, the main excitement is the chase. First, await a favorable elongation from the Sun, and begin searching for it perhaps a fortnight before greatest evening elongation, or about a week before greatest morning elongation. (These suggested intervals are different because it seems to move out much more rapidly from the Sun in the morning than in the evening sky, although it closes in more slowly afterwards.) The best way of hunting for it is to use binoculars, starting to sweep over the Sun's place about half an hour after sunset, or perhaps forty minutes before sunrise. It will probably appear pinkish because its light is reddened by the passage through the low atmosphere, but its surface is really of the same color as the Moon's, and when observed high in the sky it appears white. Once located with binoculars, it should be easy to pick up with the unaided eye.

Telescopic appearance If you turn a telescope onto Mercury when it is a naked-eye object low down in the West or East, you will not see much except proof of our unsteady atmosphere! The seeing becomes dramatically worse within about 20° of the horizon, and a naked-eye Mercury is never higher than this; under all but exceptional conditions you will see only a flaring blob of light, probably multicolored due to the color dispersion of our nonachromatic envelope, in which something resembling a small disk may occasionally be distinguished. Under such conditions, even to see the phase is something of an achievement. It is gibbous during the first part of an evening elongation, becoming a crescent as it moves back in towards the Sun, its apparent diameter during this time increasing from about 5″ to 9″ arc. At half-phase (greatest elongation), with a diameter of 7½″ arc, it appears only as large as a lunar crater 15 km across. To see details in an object as small as this, steady seeing is as important as a reasonable aperture and high magnification.

Mercury was photographed in great detail by *Mariner 10* in 1974 and 1975, and there is practically no scope for "useful" telescopic work: it is a crater-strewn wilderness, rotating on its axis once in 58.65 days, so that it completes 1½ rotations with respect to the Sun in its eighty-eight-day "year," or exactly three rotations in two years. Pre-space-age telescopic work from the Earth's surface was unable to establish either of these facts.

Daylight observation of Mercury If you have a telescope of about 200 mm in aperture, and an equatorial mounting with graduated and accurately set circles, then you should be able to find Mercury in broad daylight, and enjoy more favorable view than could ever be obtained in twilight, at least as far as steadiness is concerned. This was the method adopted by the few great planetary observers of the past who studied the planet seriously—Schiaparelli, observing from Italy just a century ago, used a 220-mm refractor. But the sky must be very blue; that is essential. The slightest haze will obscure so faint a spark as daytime Mercury, and even blazing Venus can be hard to see if the sky has a whitish cast.

It is suggested that you begin by locating Mercury in the ordinary way at a morning elongation, and follow it in the telescope until well after sunrise. This will enable you to see how the planet does appear in the daytime sky: easy enough to make out when you know that it is at the center of the field, but dauntingly difficult to find by sweeping acres of blue. If you move the telescope too fast, it can easily sail through the view without being seen. Experiment with an orange photographic filter over the eyepiece; because Mercury is white and the sky is blue, the filter will dim the sky proportionally more than the planet, and may aid the visibility of the latter.

Under steady conditions, the daylight Mercury is a completely different prospect from that flickering twilight blur. It resembles a minute Moon. At greatest elongation, indeed, a magnification of ×240 makes it appear as large as a half-Moon seen with the unaided eye. Try this magnification, if you can, and judge for yourself. Probably, it will appear smaller than the Moon, but this is a physiological effect, perhaps something to do with the Moon Illusion, that makes our satellite look larger when rising or setting than when high in the sky.

Finding Mercury by day, rather than tracking it into the daytime sky, is a real challenge. First of all, your equatorial mounting must be set up so that the axis is within a degree, or better, of the celestial pole. With a permanent set-up, this will already have been done. But if you use a portable telescope, and normally set up and sight on the Pole Star, or use other star-setting techniques, the approach must be different because the Sun is the only star in the sky! There is, fortunately, an adequate way of achieving alignment under these circumstances: point the telescope towards the Sun, and work backwards. First of all, set the R.A. and Dec. circles on the mounting so that they correspond to the Sun's position on the date and at the time in question, as though you were seeking a nightime star. Then, ensuring that the base of the mounting is level, with the polar axis at the correct altitude as measured from the base, twist the mounting in azimuth until the Sun is in the field of view, or as near as it will go. If the error is only one solar diameter or so (½°) it can be ignored. If it is much larger than this, you will have to find out what is wrong.

Once the polar axis is in good adjustment, set the telescope on Mercury by noting the difference in right ascension and declination between the Sun and the planet. Bring the Sun squarely into the field of view (taking appropriate precautions), note the reading on the two circles, and then change them by the appropriate amount. By doing this, you tend to eliminate errors in setting up the mounting because you know that the Sun is in its proper position and can safely be used as a signpost or reference mark!

It is a great help if the telescope and the observer are in the shade when trying to make any sort of daylight observation. If you are inside an observatory, close the roof sufficiently to put as much as possible of the telescope in shadow. If you must use the Sun to acquire the planet, have a screen of some sort to block off the brilliant source once it has served its purpose. Solar dazzle on the object glass of a refractor, or on the corrector plate of a catadioptric, can easily make the planet invisible. Add some sort of hood (a dewcap, or a black paper tube) to shield the front optical surface from scattered light.

Paradoxically, the best time to look for Mercury in daylight is when it is on the far side of its orbit, in the gibbous phase; not only is its disk smaller and the surface brightness more concentrated, but it reflects more light towards the Earth than when in the crescent phase. When searching, use a magnification of about ×50 to ×75, very slowly scanning the target area of sky in overlapping sweeps about 5° long, so that you can be sure that the planet's position has been covered. If it does not appear, await a bluer sky and try again. In general, it is hardly worth looking for the planet when it is less than 10° from the Sun because sky glare becomes severe and will drown the pale disk.

Transits of Mercury When it passes through inferior conjunction, Mercury can sometimes appear projected against the Sun. These transits are not particularly common: the last one occurred in 1973, and the remaining 20th-century events will be on November 6, 1993, and November 15, 1999. The first will be well seen from Australia and New Zealand, and the second will be partly visible from these locations and from the United States. None will be visible from Africa or Europe because the Sun will be below the horizon when the phenomenon occurs.

A good way of observing a transit is to project the Sun's image onto a card, as in routine solar work, so that the minute black dot of Mercury can be seen steadily traversing the disk. It is always a revelation to compare the size of this planetary body—admittedly a small one—with any sunspots that may be visible, and to remember that its true relative size is even smaller than appears because the solar surface is about half as distant again as the globe of Mercury.

VENUS

The movements of Mercury are imitated by its companion Venus, but on a far grander scale. Instead of hugging the twilight, Venus can swing so far from the Sun that it rises or sets in a dark sky. Even at its faintest, Venus is eight times brighter than Mercury can ever be. When brightest, which occurs in the crescent phase when the planet is some five weeks from inferior conjunction (either waxing or waning), it has been recorded as casting a shadow!

Turn even a small telescope onto the planet at such a time, and you will see a miniature searchlight. Even if the atmosphere is steady, the glare will defeat any attempt at critical observation. The cloud layers of Venus reflect sunlight like a mirror, about fifteen times as efficiently as does rocky Mercury, or the Moon. Such brilliance shows up all the avoidable and inevitable defects of telescope and eye that scatter light away from the true image: these include dust on the optical surfaces, obstructions in the light path, lack of perfect achromatism of the optical elements, and even diffraction effects at the edge of the object glass, mirror, or corrector element. They are there all the time, but only with an object of the intensity of Venus, observed against a dark background, are they thrown into prominence. It is not surprising that the planet is often quoted as a good telescopic test-object, particularly for a refractor. If the disk of Venus comes up to a clean edge with nothing but a hint of greenish-purple about it, then the objective is certainly a good one. However, the lack of a clean edge could simply be due to bad seeing, and you will need some experience to disentangle the effects of atmosphere and instrument. A star test, as described on page 34, is the best of all because it is practically independent of atmospheric conditions.

The brightness of Venus The markings in this cloudy envelope being nothing more than faint shadings, you have to bring the intensity of the disk to a level where these shadings have the best chance of being made out. Intensity, in this sense, is a relative thing; it depends upon how sensitive your eye is, as well as upon the brightness of Venus itself. If you observe in deep twilight, or in a dark sky, the retina will be operating in its "dim light" phase, with visual purple flooding the nerve endings. Compared with the sky background, the planet's brilliance is overwhelming. Of course, it can be dimmed with a filter, and you can experiment with different neutral-density filters placed over the eyepiece. On the other hand, if you observe in daylight, or in bright twilight, the retina is not in an ultrasensitive state. The brilliance of Venus is the same as in the former case, but it does not appear so bright because the sky background is now much stronger. Since a daylight Venus is certain to be higher in the sky than a deep-twilight one, the former time offers a better chance of securing a steady view.

In fact, you may find the disk of Venus a little too faint in the bright daylight sky, preferring the transition period when the Sun is on the horizon. If you have an equatorial telescope, though, you should certainly experiment with finding Venus in daylight. Use the same method as described for Mercury. At its brightest phase, Venus can readily be made out in full daylight with the naked eye, particularly if its position has first been pinpointed with binoculars. Try following the narrowing crescent into inferior conjunction, at the end of an evening elongation; on a really transparent day the hairline crescent may be picked up within a few degrees of the Sun, and it will then appear enormous, about one minute of arc across. Of course, you must be particularly careful not to let the Sun creep into the field of view, and an effective shading system to shield the optical elements from direct sunlight will also increase the chances of finding the planet.

The movements of Venus Although Venus takes only 225 days to orbit the Sun, it appears to take much longer than this because the Earth's orbital motion in the same direction effectively slows it down. The average interval between repetitions of the same phase is 584 days. This slowing-down effect is greatest when the planet is near superior conjunction, small and full-phase, on the far side of the Sun. It occupies about four months in passing from an angular distance of 15° west of the Sun (i.e., at the end of a morning elongation) to the same distance east of the Sun (at the beginning of an evening elongation). But it takes less than three weeks to cover the same arc of sky when passing through inferior conjunction. At this time, it is closer to the Earth than any other major planet can ever come—only about 41 million kilometers away, compared with Mars' closest possible approach of 57 million kilometers.

Therefore, for about 440 days out of the 584, Venus is more than 15° from the Sun, and can readily be picked up in the daytime with almost any equatorial telescope. Few people seem to realize this, and so confine their attention to the period when it is obvious to the unaided eye; instead, set yourself the challenge of finding the planet on every possible day, and see how well you fare, paying particular attention to the neglected gibbous phase.

Phase estimates Having located Venus, and "adjusted" its brightness to suit your eye, what is to be seen? The most obvious feature is the phase itself. Almanacs list this phase in advance, such as "0.435," "0.492," and so on. This means that the theoretical width of the illuminated portion, measured along the planet's equator, is respectively 0.435 or 0.492 of its diameter. Therefore 0.435 represents the very thick crescent phase, while 0.492 resembles almost a perfect half. On another occasion the value might be 0.966, meaning that the disk appears almost completely illuminated, only a sliver 0.034 of a diameter wide being dark. This is the gibbous phase—in fact, almost full—near superior conjunction. These values are often represented as percentages, in other words 43.5 percent, 49.2 percent, and 96.6 percent respectively.

So, before looking to see what markings are visible, you will need to draw the correct phase outline. You *could* draw the expected shape just using the almanac value—but don't assume that this will represent the phase as seen through the telescope! The almanac assumes that Venus is a matt white globe, like a table-tennis ball. In fact, it is a roasting, cloud-covered, gloomy inferno dripping with fuming acids. Near the terminator, clouds reflect sunlight in a different manner than a matt white surface; there is less scattering, and so

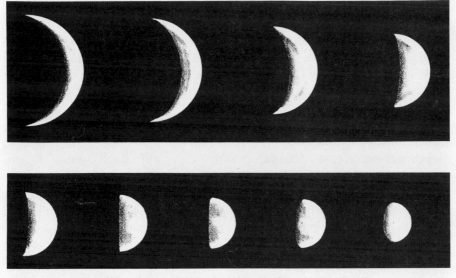

Venus   A series of drawings made during the autumn of 1983 showing, to scale, the planet's apparent reduction in size as it swung away from the Earth. (*James Muirden, 90-mm refractor, x140*)

less light is reflected back at an angle towards the Earth. This results in a noticeable falling-off of brightness before the true day–night boundary arrives. Therefore, because we cannot normally see this true boundary, the observed phase of Venus tends to be smaller than the predicted value. One interesting investigation is to see what this difference is. Does the amount vary from one elongation to another, or from one decade to another, or even from one day to another? If the nature of the clouds changes, this could have a corresponding effect on the phase appearance.

Allied to this effect is the shape of the terminator itself. In theory, it should be a smooth ellipse from cusp to cusp. In practice, it sometimes appears slightly deformed. Presumably this effect is caused by particularly dark or bright terminator features revealing themselves as dents or projections. These are likely to be changeable on a day-to-day time scale—after all, the currents in the atmosphere have been found by space probes to move at a rate of one rotation every four days, in remarkable contrast to the 243-day rotation of the solid globe beneath. There is, therefore, no reason why changes on the scale of a few hours should not occur, and perhaps every reason to expect that they will. One of the many benefits of daylight observation is that you can keep Venus in view for literally hours at a time, and try to follow such changes.

Such terminator deformations are usually reported when the planet is near the half-phase, or a crescent. At such times, a magnification of ×100 or ×150 gives a reasonable view of the disk, and the number of reports may have something to do with the fact that this is the most popular stage of the elongation with observers! Far fewer observations are made of the gibbous phase, when its disk is much smaller—by the time the 75 percent phase is reached, Venus appears only 14½" arc across, compared with 24" arc at half-phase and 40" arc when it is a 25 percent crescent. This is a very large change, but, on the other hand, observations of Mars often have to be made when its disk is less than 14" arc across. So do your best to cover the neglected gibbous phase, and extend the range of normal amateur endeavor.

Disk drawings It is worth pointing out here that the "international" convention for drawings of Venus, Mars, and Jupiter is to make the equatorial diameter 50 mm across. In the United States, the scale has formerly been two inches. Therefore, if you are going to send your drawings to an astronomical organization for inclusion with the work of other amateurs, be sure that you make them of the appropriate size. Otherwise, of course, you can choose your own size. In the writer's opinion, there is something to be said for choosing a varying disk size to correspond, approximately, to the apparent diameter of the planet at different positions in its orbit. This would seem particularly appropriate in the case of Venus and Mars, which regularly swing near the Earth and then shrink back into the depths of space. But this is something for individual taste, and certainly should not be done if complications would be caused at a later stage, when drawings are being compared with those by other observers on a different scale.

Therefore, the first task in making an observation of Venus is to get the terminator right. It is worth taking some trouble over this because if the terminator is wrong, everything else will be wrong as well. Perhaps the best starting-point is a simple semicircle drawn in the observing book using a metal or plastic template (you can use compasses if you like but the dent in the middle of the semicircle is a giveaway—on the real Venus you don't know where the middle is, so why should you on the drawing?). This semicircle is the planet's limb. Then apply pencil and rubber until you are satisfied that your terminator matches the telescopic view. Once this is right, you can start looking for shadings on the disk. Almost always, the region near the terminator seems to shade away, and the extreme limb often looks particularly bright, as do the cusps when the planet is a crescent. This may be a contrast effect, if the planet appears particularly bright against the sky, but it may be real: photographs taken from space do suggest that the polar regions of the planet, which lie at the cusps, often have bright cloud caps; but never forget that your task as an observer is to record what you see, and not what you think ought to be there. If you ever succumb to preconceptions, your observation is not just worthless, but thoroughly misleading. No matter what you are observing, do all in your power to confirm that you really have seen all that you record. Assume nothing. If you are doubtful about a feature, by all means include it, but underline the doubt in your notes.

The cloud markings of Venus usually come into the "doubtful" category. You may be sure that a hazy gray streak or patch exists, but where does it begin and end? Is it darker or lighter than another marking? Often it is impossible to decide. Sometimes the "markings" appear only when the air is unsteady, and disappear when the disk settles down into a sharp outline. When this happens, you know that our atmosphere rather than that of Venus is the culprit. A bright, glaring image can easily produce false effects by dazzling the eye. The writer must confess to doubts about the reality of some of the Venusian markings recorded on amateur drawings!

Colored filters A considerable literature has grown up around the subject of Venus seen through filters, and it is interesting to experiment. Remember that the human eye is most sensitive to yellow light—the color that the Sun sends out in greatest abundance. Remember also that the atmosphere of Venus, like that of the Earth, tends to reflect blue light back into space and let red light pass through to the surface. In theory, the cloud features should be visible more clearly with a blue filter than with a red one because the blue filter passes the light that is reflected by the cloud tops, where the most contrasty detail is to be found, and eliminates the red light, that brings back messages from the more amorphous lower layers, or even from the surface of the planet itself—except that the atmosphere of Venus is so thick, about ninety times the density of the Earth's, that practically no sunlight reaches the surface anyway.

The reality is rather different, or at least not so straightforward. Bear in mind that Venus is being viewed through the blue foreground of our own atmosphere—which is blue for the very reason that it scatters and reflects blue light. The most obvious effect of a blue filter is to decrease the contrast between Venus and the sky, particularly at the shaded terminator. You will, therefore, be likely to find that the phase appears smaller when the planet is viewed through a blue filter than when it is observed in "white" light. The amount of difference may depend upon how dark the blue sky is, and can be expected to be greater when Venus is being observed in bright daylight than in twilight.

A red filter increases the contrast between the planet and the sky by suppressing the blue background. Therefore, this may make the terminator appear more definite, and perhaps enhance the phase value. But, again, local circumstances such as time of day, transparency of sky, and so on, may have a significant effect upon the result. The only way of coming to a conclusion is to experiment. The two filters that have become almost standard for this work are the Kodak "Wratten" filters No. 25 (red) and No. 47 (blue). These can be purchased as small gelatin squares, from photographic dealers. For routine work, the yellow Wratten 8 is often used, to improve the contrast of the planet against the sky.

Other phenomena Two other phenomena of Venus are worth mentioning briefly. When in the crescent phase, its night side has sometimes been recorded, shining like very faint Earthshine on the Moon. This has been termed the *Ashen Light*. Unfortunately, it is very easy to imagine that the dark side is visible, and confirmed sightings are rare.

Mention must also be made of transits of Venus across the Sun. These occur in pairs, separated by eight years, after which none are to be seen for over a century. The next one will occur in the year 2004, and presumably no one now living on this planet has any recollection of the last, which occurred in 1882!

Observing the superior planets In most respects, the planets situated beyond the Earth—Mars, Jupiter, and the other superior planets— are easier to observe than are their inferior counterparts. They appear to travel round the ecliptic independently of the Sun, and their sunlit hemispheres are turned more or less towards the Earth whenever they are visible, so that they do not pass through a cycle of phases.

A superior planet embraces the scurrying Earth inside its huge orbit, from which it follows that it moves much more slowly round the Sun. Even Mars, the innermost of the superior planets, takes 687 days to orbit the Sun once, while the next planet, Jupiter, takes almost twelve years; Pluto takes 2½ centuries! Therefore, the way in which these planets appear to move around the sky with respect to the Sun is very largely an effect of our own moving platform.

The motion of a superior planet The diagram shows the sequence of events for a "typical" superior planet. Let us start with the Sun, Earth, and planet in a line, with the Earth between the other two bodies. This is known as *opposition* because the planet is diametrically opposite to the Sun in the sky: as one rises, the other sets. Opposition is the phase around which the planetary observer's schedule hinges, because this is the time when a superior planet is closest to the Earth. Its sunlit hemisphere is turned accurately towards the observer, and the best telescopic views are to be obtained at this time. On the diagram, the positions of the Earth and the planet are represented by A and A'.

How a superior planet appears to move The arcs A–D and A'–D' represent the distances traveled by the Earth and the superior planet in similar intervals of time. The more remote the planet, the smaller is the arc A'–D' and the shorter the time between the Earth-based observer seeing the planet at opposition (position A) and conjunction (position D).

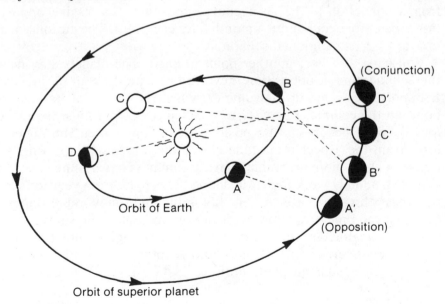

Orbit of Earth

Orbit of superior planet

Now let the Earth continue on its way around the Sun, and three months later it will have reached B; but the planet, moving much more slowly in a far larger orbit, has only reached B', so that it is no longer opposite the Sun in the sky, but appears to have moved around the sky, so that it now sets sometime around the middle of the night, and is already well above the horizon at sunset. Six months after opposition, with the Earth at C—having achieved half a circuit of the Sun —the planet is at C', not yet on the far side of the Sun, but setting in the west soon after the sky has become dark. Finally, at D and D', the two planets and the Sun are in a straight line again, but this time the Sun is in the middle. This position is known as *conjunction,* and the superior planet is invisible, hidden in the Sun's glare.

After this, the pattern proceeds in reverse. The planet reappears from conjunction as a "morning star," rising in the dawn. As the weeks pass, so it rises at a greater and greater interval before the Sun, until finally another opposition occurs, when the Earth has caught up with the planet again.

If a superior planet were stationary in its orbit, oppositions would be annual affairs, occurring on the same date each year. The closer and faster-moving the planet, the greater the interval between oppositions. Outermost Pluto takes 248 years to circle the Sun, and oppositions are only 1½ days later each year, whereas in the case of Jupiter the Earth has to spend about a fortnight passing from C to D, and another extra fortnight over the next six months when passing from conjunction to the succeeding opposition. Successive oppositions therefore occur about a month later each year. The curious case of Mars will be mentioned separately.

You must also bear another point in mind. The planets stay near the ecliptic, and at opposition they are opposite the Sun in the sky. If opposition occurs at midsummer, when the Sun is at its highest point on the ecliptic, it is easy to see that the planet must be at the *lowest* point, occupying the position of the midwinter Sun. Therefore, from the aspect of altitude above the horizon, winter oppositions are much more favorable than are summer ones. As the superior planets gradually work their way along the ecliptic from year to year, they pass through "seasons" of favorable and unfavorable oppositions. At the time of writing, both Jupiter and Saturn lie south of the celestial equator, and they will not pass north again until 1987 and 1995 respectively, while Uranus and Neptune will remain in the southern celestial hemisphere until the 21st century!

MARS

Mars's orbit is relatively close to that of the Earth, and it circles the Sun in a much shorter time than do the other superior planets, so that its motion around the sky is rather different from that of its more distant companions. If you look at the diagram, this should become clear. At A and A', the Earth and Mars are at their closest, and Mars is at opposition. However, by the time the Earth has accomplished its three-month journey to B, Mars has followed closely in its footsteps, and is at B'. Even after six months, with the Earth at C, Mars has reached C' and is, as yet, nowhere near conjunction. In fact, the Earth takes about thirteen months to shake Mars off its tail and bring it to conjunction, and another similar interval to catch up with it again at the following opposition. Successive oppositions therefore occur at intervals of, on average, 780 days, or two years and two months.*

This means that Mars can be seen more continuously than can any other planet. For example, it was at opposition on March 31, 1982, and a year later it could still be seen in the western sky, low down, after sunset! However, by this time its disk looked no larger than that of the distant planet Uranus. The season for really useful work with small or even moderate apertures (up to 300 mm or so) lasts for only a few weeks around the time of opposition because the distance between the two worlds changes so quickly.

You must also remember that successive oppositions of Mars are not going to give an equally close-up view. Its orbit is markedly eccentric, the solar distance varying from about 207 million kilometers at perihelion to some 250 million at aphelion. The Earth's solar distance, however, is almost constant at 150 million kilometers. It therefore follows that the Earth–Mars distance at opposition ranges from 57 million kilometers, if Mars happens to be at perihelion at the time, to 100 million at aphelion. At an aphelic conjunction it will be 400 million kilometers away, exhibiting a disk only 3.6" arc across: smaller than that of remote Uranus.

The recent "low point" for observers of Mars was the opposition of Feburary 1980, when its disk never exceeded 14" arc in diameter. Fortunately, we are now seeing opposition distances steadily decrease as the favorable apparitions of 1986 (23.1" arc) and 1988 (23.7" arc) approach. However, the disk of Mars will appear larger than 20"

---

* The interval between successive oppositions or conjunctions is known as the *synodic period*.

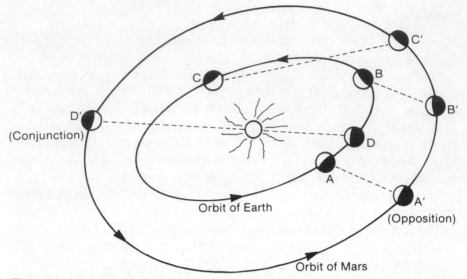

The motion of Mars Being the closest to the Earth, Mars moves faster than any other superior planet, and keeps up with the Earth for a longer time after opposition. Over a year elapses before the Earth reaches D and Mars is brought to conjunction.

arc across, which is only half the apparent diameter of Jupiter, on only about a hundred nights during the current decade! To get an idea of the largest size it can ever appear, draw a circle 100 mm across and view it through a small telescope, or binoculars, from a distance of 40 meters. This is the perihelic opposition view, as seen through an astronomical telescope magnifying 20 times as much as the instrument you are using (for example, × 160 if you are using × 8 binoculars). Remove the paper to a distance of 75 meters, and you have the aphelic opposition view. It is certainly not very large; and it appears even smaller than this when viewed through a real telescope on a night of typical seeing conditions, when the details flicker and blur so tantalizingly at the limit of vision. However, under good conditions, and with a practised eye, even a 60-mm refractor can reveal the polar cap and its hazy hood before it vanishes in the Martian spring. A dark shading or two should also be seen, including the famous Syrtis Major. Mars rotates on its axis in 24 hours 37 minutes, so that the disk presents the same aspect just over half an hour later on successive nights.

The surface of Mars Mars is certainly a most difficult planet to observe. But the fascination that attracted the 19th-century observers, some of whom saw it as the abode of intelligent life, is still strong. It is, after all, the only planet in the solar system to offer us even a reasonable view of its true solid surface. It has Earth-like winds, and might, once, have had Earth-like running water, although now it is a frozen waste. It is thrilling to catch a glimpse of the glistening polar cap, and the warm-looking red-ochre disk belies the terrible chill. When Mars shines in the sky on a brilliant winter's night, it glows with a richness of color unmatched by any star.

The wonderful results of the *Mariner* and *Viking* spacecraft completely eclipse what surface detail can be made out by Earthbound telescopes. Part of its surface is heavily cratered with ancient impact markings, but elsewhere the old features have been destroyed by severe vulcanism, and mountains and valleys on a scale unmatched on our own world are to be found there. Consider the highest of all, Olympus Mons, 25,000 meters high (2½ times the height of Everest), or the Tharsis ridge, 10,000 meters high and extending for 4000 km; or, again, the vast complex of valleys known as Valles Marineris, many of which are half as deep as Everest is high, extend for hundreds of kilometers, and would totally swallow our own much-vaunted Grand Canyon in Arizona! It is testimony to the difficulty of making out detailed features on the planet that no astronomer had suspected the existence of this violent topology because the shadows cast even by these tremendous peaks and crevasses are undetectable from the distance of the Earth.

If Mars showed a phase like the Moon, so that its terminator could be well seen, there might have been a chance. In fact, the planet can show a slight phase (like the Moon when about three days from Full) some three months before and after opposition. But even then our view of the terminator region is very poor. We must content ourselves with observing the general forms of the lighter and darker regions of the planet, as shown on the accompanying map.

The contrast of the Martian features is quite high, and at an average opposition, with a magnification as low as ×100, some dark areas should be visible towards the center of the disk. You will notice how the edge of the disk appears particularly bright and sharp, unlike the limbs of Jupiter and Saturn, which fade off towards the sky. This is because the surface of Mars is a poor reflector compared with the frosty haze that builds up near the limb (the terminator region, at which the Sun is either rising or setting), where the temperature is some 80°C below zero. Sometimes, white "clouds" are seen in this region. They can appear as bright as a polar cap, but are never seen

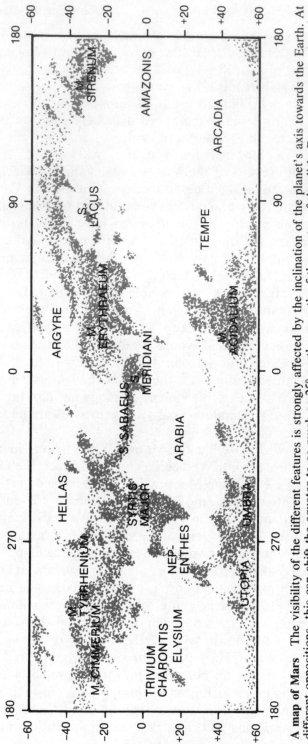

A map of Mars The visibility of the different features is strongly affected by the inclination of the planet's axis towards the Earth. At different oppositions, this can shift the equator to as much as 25° north or south of the center of the disk. Dust storms can also alter outlines and visibility.

fully on the disk, and presumably are a sunrise or sunset phenomenon. Remember that the thin atmosphere (about $\frac{1}{100}$th of our own in density) is almost entirely made up of carbon dioxide, which can freeze solid in the very low Martian night temperatures. The polar caps themselves, however, are made up principally of frozen water, over which an extra coat of carbon dioxide may freeze in the bitter winter.

Martian seasonal effects What you will observe on Mars, then, is not so much the permanent surface detail as the progress of the seasons. Two factors affect the Martian 687-day year. The first is its axial tilt of 25°, which is only very slightly greater than that of the Earth. Therefore the Northern and Southern Hemispheres enjoy alternate summer and winter seasons, and, depending upon circumstances, one or other hemisphere will be tilted towards the Earth at opposition time, giving us a preferential view of one of the poles.

The second important factor is the eccentricity of the orbit. Perihelion happens to occur when the South Pole is turned towards the Sun, so that the southern summer is much hotter than the northern one, which occurs at aphelion. The Earth experiences a similar arrangement, but the eccentricity of our orbit is so slight that the difference in heating effect between the two hemispheres is inappreciable. However, at every Martian perihelion, so far as we have been able to observe—and some occur when the planet is near conjunction, unobservable—tremendous dust storms are whipped up. At the perihelic opposition of 1956, even the shrinking polar cap disappeared, and for a few nights the planet was a featureless ochre disk. When *Mariner 9* went into orbit in November 1971, another dust storm was temporarily obscuring the surface. These storms can be monitored with apertures of 100 mm or even less, if Mars happens to be well placed at the time; they seem to begin in three different elevated regions in the Southern Hemisphere (Hellespontus, Noachis, and the Solis Lacus), spreading and multiplying: the impression is one of primary and secondary whirlwinds on a tremendous scale.

The coming views of Mars The following list gives a summary description of the circumstances of the remaining oppositions of the 20th century.

1986: Opposition will occur on July 10, with the planet in the far southern constellation of Sagittarius. Although its disk will be 23.1″ arc across, it will be very low in the sky for observers in most parts of

Europe and the northern United States. The disk will have an apparent diameter of 10″ arc or more for the period from the first half of April until the end of October. The South Pole will be slightly inclined towards the Earth, but the opposition view will be practically equatorial.

1988: This will be the closest opposition before the end of the century. It will occur on September 28, with the planet near the celestial equator, in Pisces. The maximum disk diameter will be 23.7″ arc, and it will be more than 10″ arc across from the second half of May until the end of the year. The South Pole of the planet will be well presented.

1990: Opposition will occur on November 27. The planet will be in the northern constellation of Taurus, and its maximum disk diameter will be 17.9″ arc. Its diameter will exceed 10″ arc between the middle of August and the end of January 1991. At opposition, the South Pole will be slightly inclined towards the Earth.

1993: Opposition will occur on January 7. Mars will be in Gemini, and high in the sky for nothern observers, but the maximum disk diameter will be only 14.9″ arc, and it will be larger than 10″ arc only between the middle of October 1992 and the end of February 1993. The equator will be presented at opposition time.

1995: This very distant opposition will occur on Feburary 12, in the constellation Leo, when Mars (magnitude − 1) will form an attractive pair with the bright star Regulus. Greatest disk diameter will be only 13.8″ arc, and the period during which it will be larger than 10″ arc will last from the middle of December 1994 until the end of March 1995. The North Pole will be fairly well presented.

1997: Mars will come to opposition on March 17, in Virgo, a little north of the celestial equator. The maximum diameter of its disk will be 14.2″ arc, and it will appear more than 10″ arc across between the middle of January and the first part of May. This opposition will see the North Pole of Mars turned at almost its maximum possible amount towards the Earth.

1999: At opposition on April 24, the disk will appear 16.1″ arc across, and the planet will again lie in Virgo, somewhat south of the celestial equator. It will appear to be more than 10″ arc across between the beginning of March and the first part of July, and the Northern Hemisphere will again be well presented.

It may also be of interest to know the dates of perihelion and the times of the Martian seasons, so that these can be related to any observations that are made. This information is included in the following table:

	Perihelion	Northern midsummer	Southern midsummer
1984	Nov 9		Dec 6
1985		Nov 28	
1986	Sep 26		Oct 23
1987		Oct 15	
1988	Aug 13		Sep 11
1989		Sep 2	
1990	Jul 1		Jul 30
1991		Jul 21	
1992	May 18		June 15
1993		Jun 6	
1994	Apr 4		May 3
1995		Apr 24	
1996	Feb 21		Mar 20
1997		Mar 12	
1998	Jan 8		Feb 5
1999		Jan 27	
2000	Nov 25		Dec 24

The rotation of Mars This has the effect of steadily bringing a given region into view, and then away again, on successive nights. With every night that passes, its rotation has fallen behind by 37½ minutes, or 9½° of Martian longitude. Therefore, if you reobserve the planet at the same time of night after a week has gone by, the feature that was formerly on the planet's central meridian will now be invisible, near its hazy eastern limb, and you will have to wait for some hours before it swings into good presentation again. After a lapse of three weeks it will be on the far side of the planet at this time, and therefore totally unobservable during the hours of darkness; the original presentation will not recur until a month and a half after the first observation. This means that a spell of bad weather around the time of opposition could ruin observation of a certain span of longitude. The serious observer of Mars cannot afford to miss any chances of a clear sky and steady seeing, no matter what the hour of the night!

The satellites Its two minute moons, Phobos and Deimos, are about 23 km and 13 km across, though very irregular in shape, as proved by the *Viking* close-up photographs. Neither is particularly faint (at an average opposition, Phobos appears of magnitude 11.6, and Deimos is about a magnitude dimmer), and, were they stars of this brightness, they could be detected with a telescope of 150-mm aperture. However, the glare of the planet makes them difficult to see, and few observers have ever recorded either moon.

THE ASTEROIDS

It is believed that all the planets in the solar system grew from much smaller lumps of material that coalesced together—the smaller lumps coming, in turn, from the fine dusty material that collapsed in nebular form to produce the solar system. Although the vast majority of these condensations have, indeed, gone to form the planets, or have become planetary satellites, some found themselves in orbits safe from planetary collision. The huge zone between the orbits of Mars and Jupiter is where most of these little bodies have found refuge, and some amateurs have made a personal hobby of detecting and following these strange worlds.

Well over two thousand asteroids, or minor planets, have been discovered since the first and largest (Ceres) was found by the Italian observer Piazzi on January 1, 1801. The brightest, Vesta, can sometimes reach a magnitude of 6 and be visible with the naked eye at opposition; Ceres itself never appears brighter than the 7th magnitude. Because Vesta is only half the diameter of Ceres (about 540 km as against 1000 km), it must be much more reflective. Indeed, it is now known that the asteroids fall into two main classes, determined by their reflectivity or *albedo:* the so-called C-type, which are dark, and the S-type, that reflect up to a quarter of the light falling on them.

Asteroid classes The C-type asteroids are, presumably, as close as we can come to bodies made of "original" solar-system material. Their surfaces appear to consist mainly of carbon and its compounds, and apparently never became particularly hot in their formation; if they had undergone any sort of melting process they would have been transformed into the type of rock found on the surfaces of the terrestrial planets. The S-type, on the other hand, contain a high proportion of silicates, that are indeed found in rocks that have been melted at some point in their history. Such small bodies could never, by their own internal heating, have become sufficiently hot to process rocks in this way, and the inevitable assumption is that the S-type asteroids are the fragments left from ancient collisions of much larger bodies that had already condensed and heated up sufficiently to form a crust.

The reflectivity or albedo of these two classes are about 2–5 percent in the first case, and 15 percent in the second. Vesta, at perhaps 25 percent, is surely highly metallic. It certainly seems to be unusual among the asteroids, and it is one of the easiest to see because of its brightness.

Identifying asteroids Only Ceres, and one or two others, have been seen as perceptible disks, even with the greatest telescopes. Their sizes and reflectivities have had to be calculated by indirect methods. For the amateur (and indeed the professional) the task of identifying them rests on their position in the sky rather than their appearance in the telescope, for to all intents and purposes they look like ordinary stars.

You could, in theory, identify an asteroid by consulting an ephemeris, noting its position on the celestial sphere on the night in question, pointing the telescope to that position by the methods already described, and looking at the center of the field of view. However, because your pointing is almost certain to be a few minute of arc in error—if not more—you will be confronted with a field of stars, any one of which could be the planetary candidate. Knowing the brightness of the quarry does help, and the yellowish tint of the typical asteroid can be a clue if it is bright enough for color to be detected. But you are very unlikely to be able to make a rapid positive identification in this way.

There are two practical ways of spotting asteroids: by comparing the telescopic view with a star chart and so finding the intruding "star," or by watching to see which "star" is moving slowly across the sky.

Using a star chart Few asteroids are ever brighter than the 8th magnitude, even at a favorable opposition. If you are interested in identifying even a reasonable "bag" of asteroids, therefore, you will require a very comprehensive chart of the ecliptic region, showing stars down to at least the 10th magnitude. Some useful atlases showing stars down to this order of faintness are listed in Appendix. I.

Even then, however, the task is not as straightforward as might appear at first sight. Using a publication such as the *Observer's Handbook,* which contains asteroid ephemerides, the position of the object is plotted on the chart, which is then compared with the telescopic view. Possibly there will be so few stars in the vicinity of the asteroid that it can be identified immediately. Often, though, there will be some doubt about whether this or that "star," both of which are near the right position and appear to be of the right brightness, is indeed the candidate. This problem often arises where the ecliptic crosses the Milky Way, in the northern constellation of Gemini or the southern one of Sagittarius. To make an absolutely positive confirmation of its planetary nature, you will almost certainly have to observe its motion in front of the stars.

Identification by orbital drift Around the time of opposition, an asteroid will be moving across the celestial sphere at a rate of about ¼° (half the Moon's diameter) per day. Therefore, if your candidate is near a convenient reference star, and you reobserve it after an interval of about two hours, it will have moved by 1' arc relative to its companion. Such a drift should be obvious, if a high magnification is used.

Clearly, this method can be used to identify an asteroid if you have no comprehensive star chart at all. In fact, if the body is very dim, orbital drift may be the only feasible method. You then locate the field of view in which the body is known to lie, using bright stars or the telescope's circles, and draw the positions of all the suspect objects. Reobservation on the following night—or perhaps on the same night, after an interval of some hours—will reveal a difference. In theory, you will find that one of the "stars" has moved from its original place to a new place, so betraying its planetary nature. In practice, you may find that a star has vanished and presumably left the field of view altogether, hence requiring a fresh search in the adjacent region, or that a "new star" has appeared, but that you cannot be positive that it wasn't just overlooked on the previous occasion! Such problems are part of the fun of asteroid-hunting and make the final identification all the more satisfying.

In practice, it is extremely difficult to draw a crowded star field so accurately that the motion of one of its members can readily be identified. At the second view, all the stars may appear to have moved! It may, therefore, be worth trying the "transit" method. Arrange a single hair across the North–South diameter of the field of view of a low-power eyepiece.* The orientation can be secured by turning the telescope from side to side in declination, and twisting the eyepiece in the drawtube until a star remains on the hair during its traverse of the field. Then, having drawn all the field stars, label them A, B, C, etc. Turn off the telescope drive, and, using a stopwatch, time the

* See page 13.

moment at which the stars are carried past the hair by the Earth's orbital motion. Repeat the observation at a later time, and the asteroid's orbital motion will have altered the time difference between itself and the other stars. The approximate rate of change, for an asteroid near opposition, is 2½ seconds of time per hour, so that an interval of a few hours will reveal a clear difference.

Retrograde motion We have not yet referred to a strange phenomenon of planetary motion that can be observed in all the solar-system bodies, but is particularly noticeable in the case of the superior planets and the asteroids. We are seeing these bodies from a rapidly moving platform, the Earth. However, the direction in which the Earth is moving, relative to the planet being observed, undergoes a great change during each apparition. For most of the time, the planet appears to be moving along the ecliptic in the "correct" direction, in other words from West to East in front of the stars. This is the direction in which its orbital motion would carry it if we, as observers, were situated at some stationary point near the center of the solar system. However, as opposition approaches and the Earth draws closer to the planet, our own much faster orbital velocity has to be taken into account. We are, in fact, overtaking the superior planet on an inside track, and this overtaking effect is at a maximum at opposition. The planet, at this time, appears to be moving backwards against the stars, in an East-to-West direction. This phenomenon is known as *retrograde motion*.

The duration of retrograde motion increases with the planet's distance from the Sun. At a perihelic opposition, Mars retrogrades for about a month on either side of opposition date, whereas Neptune retrogrades for almost half the year. The point where the planet's motion changes direction is known as the *stationary point*. Asteroids reach their stationary point about forty days before and after the date of opposition. At this time their motion in right ascension drops to zero, but they may still show a shift in declination because retrograde motion takes the form of a loop in the sky rather than a simple back track. It would be pointless using the transit method when a body is at or very near its stationary point, and the best time to search for an asteroid is, clearly, at the fast-moving and bright period around the time of opposition. An inspection of the paths shown in Appendix III will reveal the curious forms that these retrograde loops can take.

Of the "Big Four" asteroids, Ceres, Pallas, Juno, and Vesta—the first to be discovered—all but Juno can readily be detected with binoculars. The track of Pallas can take it far from the ecliptic at times because its orbital plane is inclined to that of the Earth at the unusually large angle of 35°.

A few amateurs have become particularly interested in logging asteroids, searching assiduously for the fainter ones to improve their "score." Because only a few dozen ever become brighter than the 11th magnitude, this is a task for moderate apertures and much patience.* A typical 200-mm catadioptric or Newtonian could, with care, reach several hundred; but for most amateurs a sighting of the Big Four will be enough to satisfy their curiosity, and *Sky & Telescope* gives sufficient information to help you find them.

JUPITER

Turn even a pair of binoculars on to the planet Jupiter, and it shows a small but perceptible disk. With a telescope of as little as 60 mm in aperture, and a magnification of ×30, cloud belts can be seen; if you have an aperture of 75 mm or more, there is always something new to see on the disk. Jupiter, without doubt, is the amateur's planet. It not only *is* large, but it *looks* large (its apparent diameter is over three times that of an average opposition view of Mars, and twice that of Saturn), and its rich cloud markings are always changing. It spins on its axis in less than ten hours, so that different regions of the globe will be seen at the same time on successive nights, while a long continuous watch will reveal the whole surface. It does not hide itself away in the Mars manner, shrunken and dim, for long periods, but is bright and easily observable for some nine months in the year. Finally, it possesses four satellites which are bright enough to be seen in the smallest telescope. What more could the observer wish for?

Jupiter's make-up Everything about the king of the solar system is on a grand scale. One thousand Earths could fit into its bulk, and one cloud feature alone—the Great Red Spot—is larger than our planet. Its thick white and yellow clouds of frozen ammonia crystals float over a darker background of methane ice. Below this, hidden from

---

* Frederick Pilcher, of Illinois, has observed over one thousand asteroids, down to magnitude 14½, in the past thirteen years. Most of the observations were made with a 356-mm (14-inch) Celestron.

view, must lie the hydrogen body of the planet, forced by the tremendous pressure into liquid form. The chemical composition of Jupiter is very similar to that of the Sun, and, despite the surface chill of $-40°C$ at the methane level and about $-150°C$ in the bright ammonia clouds, its core is estimated to have a temperature of perhaps 50,000°C. It is worth noting that the cloud surface of Jupiter is warmed more by the slow convection of inner heat than it is by the distant Sun, and, if the Sun were suddenly to fade out, its surface would be the warmest accessible place in the solar system.

The cloud features near the equator take about 9 hours and 50 minutes to spin round once. Higher than Jovian latitude 15° or so, this period increases abruptly to about 9 hours and 55 minutes. Regions near the equator are whirling round at a speed of over 90,000 km per hour! Not surprisingly, the centrifugal force has caused the globe to bulge outwards, and the disk is noticeably elliptical almost as soon as magnification reveals it at all. This rapid spin has caused the white clouds to be swept into *zones* circling the planet, and these, together with the darker *belts* in between, are reasonably permanent, having been given the names shown in the diagram. The belts are a mass of turbulent detail, particularly in the equatorial and temperate regions; while some long-lived white "ovals," like the Red Spot but smaller and lacking its tint, may also be seen.

The Jovian markings Do not expect to see all this detail at first glance, even if you are using a telescope of 200 or 300 mm aperture. Quite possibly you will see nothing but a faintly striped disk and some of the four bright moons (one or more of these may be hidden by the planet, or in passage across its face, at the time of observation). However, provided the telescope is a good one, the detail will come. The two equatorial belts, and the South Temperate Belt (STB) are usually the most prominent. Use a magnification of ×150 or ×200, and settle down to study the disk carefully. Is the planet's limb sharp and steady, or swirling as if under water? If the latter, then the seeing is poor and you might as well abandon observation until another opportunity presents itself. But if the disk seems to puff up into a bubble, and then contract briefly before blurring and exploding again, the fault is not in the atmosphere but in the telescope tube, and it should be left outside for half an hour to reach some sort of temperature equilibrium. This is the typical appearance given by a closed-tube catadioptric telescope that has been brought out from a warm

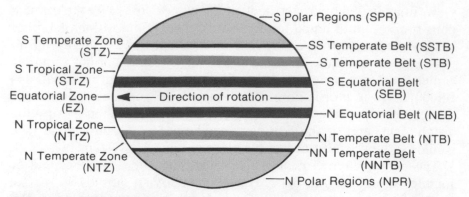

S Polar Regions (SPR)

S Temperate Zone (STZ)—

S Tropical Zone— (STrZ)

Equatorial Zone— (EZ)

N Tropical Zone— (NTrZ)

N Temperate Zone (NTZ)

—SS Temperate Belt (SSTB)

—S Temperate Belt (STB)

—S Equatorial Belt (SEB)

—N Equatorial Belt (NEB)

—N Temperate Belt (NTB)

—NN Temperate Belt (NNTB)

—N Polar Regions (NPR)

◄— Direction of rotation ——

The belts and zones of Jupiter   These are shown schematically, and when Jupiter is active they may overlap and be partly lost in the confusion of detail. Normally, the NEB and SEB are the most prominent belts, and the STB is often dark and active. Fine belts are sometimes seen in the polar regions.

room. As a general rule, out-of-focus unsteadiness lies near at hand and can be remedied; sharply focused ripples may be kilometers away. Note that the initial judgment is made by looking at the limb, not at surface detail, because the limb offers a more contrasty and easily registered standard.

A calm, or reasonably calm, limb is the signal to persevere further. Gaze steadily at the equatorial belts, and irregularities may be seen, either of form or of intensity, or even both. Projections often extend from the belts into the bright equatorial zone (EZ). Gradually, other projections and indentations will be made out—suddenly sharp and unmistakable, as the air steadies itself, then blurred and gone; and your eye may wonder if it was really deceived, until another instant of steady air reveals the feature again. Jovian observation, like all planetary work, involves training the eye and brain to snatch these precious seconds (or fractions of a second) of distinct vision, and to memorize the features that are seen long enough for them to be transferred to paper. Of course, some nights are better than others, but on no night are the features of Jupiter or any other planet simply spread out like a map, to be copied; you will have to discriminate between the good moments and the poorer ones, and, with more experience, you will be able to profit more from the good ones.

Making a drawing Don't attempt to draw a circular Jupiter! Use an elliptical blank, with an equatorial diameter 6 percent greater than the polar one. The "international" standard is 50 mm by 47 mm. Some active societies produce blacked-in outlines for the purpose.

Unfortunately, these are often printed on poor-quality paper, making it difficult to produce a satisfactory drawing. If you want to have your own made up, make an oversize master for a printer to reduce photographically. Set two pins 50 mm apart in a sheet of paper placed on a board, use a loop of cotton 191 mm long, and run a fine pen around inside the loop: this should produce an ellipse with equatorial and polar diameters of 140 mm and 131 mm respectively.

When making a drawing of Jupiter, remember that its rotation causes objects to appear on the disk at the eastern limb and to disappear at the western limb some five hours later. In practice, however, few details are visible within about an hour of limb passage, so that the useful visibility of a given feature is only about three hours. Therefore, if a drawing takes even half an hour to make, the features will have shifted very noticeably by the time it is finished. The best that can be done is to sketch in the positions of the main "landmarks," and to take this moment as the basic time of the drawing. Then start filling in the details, beginning at the western limb so that they are caught before going out of view. Always note the true beginning and ending time of the observation. A long session may result in two or three separate drawings having to be made, to show all the fresh detail that is appearing!

An alternative approach, useful when a particular latitude is being studied, is the "strip sketch." Features are drawn as they cross the meridian, and the final result looks like a horizontal strip cut from a map. Over the course of two consecutive nights, it may be possible to produce a complete 360° representation of the particular band being studied—but the Jovian features can change so quickly in appearance that there are almost certain to have been some alterations in the meantime. A scale of 15–20 mm per hour of surface advance will be suitable, depending upon the amount of detail that can be made out.

Transit work Drawing features as they pass the planet's meridian is a half-way step to the *transit* method used by all serious observers. Remember that Jupiter rotates so quickly that the clouds are advancing across the disk, from East to West, at a rate of 1° of Jovian longitude every 100 seconds. Remember, too, that the human eye is a good judge of symmetry.* If you choose a night of good seeing, and

---

* Perhaps it should be called a *systematic* judge of symmetry. Some observers tend to judge transit times a little early, others a little late. This does not matter much because the error is systematic and can be allowed for.

view the planet with an aperture of as little as 75 mm and a magnification of $\times 100$ or $\times 150$, the features will appear to have shifted noticeably westward in as little as five minutes. So, if the time when a feature is judged to be on the central meridian is noted, its longitude can be derived with a probable error of perhaps 2° or 3°. To reduce the time to Jovian longitude, use the tables published in many astronomical almanacs.

There are, as we have seen, two principal rotations times on Jupiter, and two longitude systems therefore have to be used: System I for objects in the so-called equatorial current, that extends from the northern part of the South Equatorial Belt (SEB_n) to the southern part of the North Equatorial Belt (NEB_s), and System II for features in higher latitudes. A single observation of a feature is of little use, but if you can reidentify it a few nights after the first sighting and obtain another longitude timing, you have in your hands the means of calculating its own rotation period. Almost certainly, you will find that its longitude has changed slightly between the two observations. It may be moving faster than the general system in which it lies (longitude decreasing), or it may be moving more slowly (longitude increasing). Try to recover it on a further night, and see if the longitude change is constant. Compare it with other nearby features, whether spots or projections, notches or whorls. Are they all keeping together, or are they separating? Are other features gradually approaching because of their own drifts? Out of these times and longitudes begin to emerge something of the Jovian drama: you are measuring the effect of hurricanes undreamt of on the earthly scale, and without such careful, repeated measurements they would pass unnoticed.

The first transit measurements were made about a century ago. The need for continuous work is as great as ever. No spacecraft can match such long-term coverage, and the very variety of what the planet offers means that something new will be observed on every reasonable night. Serious observation of Jupiter does, however, demand long sessions at the telescope, and you will not keep it up for very long if you have to twist your neck to see through the eyepiece: so observer comfort is of critical importance.

The following table may be useful. It indicates when a particular longitude will return to the meridian, so that the region can be reexamined, supposing that the first observation was made at 10 pm (22.00 hours) on Day 0.

Jovian latitude	Days elapsed						
	1	2	3	4	5	6	7
System I	17.41	03.30		04.44	00.24	05.56	01.37
		23.12	18.53		20.06		21.18
System II	17.53	03.49		05.31	01.24		03.01
		23.42	19.34		21.16		22.58

Individual features The most famous marking on Jupiter (in fact, probably the most famous marking in the whole solar system), the Great Red Spot, lies in the South Tropical Zone (STrZ), nudging the SEB_s. Although it has been observed fairly continuously for over three centuries, it certainly is not permanent in a "fixed" sense because it shows large and erratic drifts in longitude. Space-probe observations have indicated that the material within it is rotating inside its confines with a period of a few days, and this supports the view that it is the site of a vortex, where atmospheric masses from below are swirling up to the highest cloud tops. The reddish hue popularly associated with the spot is no doubt due to some chemical impurity. However, the feature is certainly not always red, and it is often difficult to make out at all, fading from view for months or even years, and then recovering. It was prominent in 1879, when it first received its name, and during the present century it has had several revivals, notably around 1960, while in 1971 it was vivid. Recently, it has been less obvious. This is intended as a warning to the enthusiast who automatically asks when the Red Spot will come on to the disk—it may be already there, but undetectable with a small instrument unless the indentation it makes in the SEB (the Red Spot Hollow) can be made out.

Some other oval formations, existing at higher latitudes in the Southern Hemisphere, may also be long-lasting. Although much smaller than the Red Spot's dimensions of about 39,000 km by 14,000 km, these bright white objects have been recorded for many years. In general, the Southern Hemisphere of Jupiter shows more violent activity than does the Northern. However, it can be safely said that on any night of reasonable seeing, a telescope with an aperture of 150 mm or more will keep a practised eye busy tracking the detail as it marches past Jupiter's meridian. If a feature is small and symmetrical, you will time its central passage only, but if it is slanted, or irregular, it may be best to time both the preceding and following ends, as well as the center. These are normally abbreviated *p* and *f;* for example, the *p* end of the Red Spot will transit the central meridian somewhat less than an hour before the *f* end.

Jupiter's moons The history and promise of Jovian observation is so rich that it deserves a long chapter to itself, but we must conclude by saying a few words about satellites. The four "Galilean" moons— Io, Europa, Ganymede, and Callisto—were discovered by Galileo in the winter of 1609–10, using his very primitive "optic tube," with an aperture of about 30 mm. They are, therefore, easy to see in any instrument. Their orbital planes all coincide with the plane of Jupiter's equator, which in turn is almost edge-on to the Earth at all times. This means that the satellites are always distributed in a narrow band to East and West of the planet. The maximum possible tilt of Jupiter's axis towards the Earth is only 3°, and when this next happens, in 1988–89, the outermost moon, Callisto, will be able to pass North or South of the planet's poles at its conjunctions. The other three, however, always pass in front of Jupiter (a *transit*) or behind it (an *occultation*) during each revolution. Around the time of occultation they must also pass into eclipse in Jupiter's shadow.

Eclipses, in particular, are worth observing: one moment the satellite is shining normally to the East or West of the planet, and three or four minutes later it has faded into invisibility! Using a publication such as the *Observer's Handbook,* watch some predicted eclipses and time the moment when the satellite either disappears into eclipse (abbreviated EcD), or reappears from eclipse (EcR). Normally, a satellite can be followed deeper into the shadow at a disappearance than it will be detected at a reappearance because you know where to look, and observers with larger instruments will also have the benefit of greater light-grasp. Errors of several minutes have been known to occur in the predictions, and these phenomena are always interesting to watch. It is also a challenge to follow a moon onto the planet's disk during a transit, while their shadows can often be seen as small black specks passing slowly over the clouds. Truly, with so much to see on any night, Jupiter deserves the title of the "amateur's planet."

SATURN

Two objects in the sky never fail to thrill the newcomer to astronomy who observes them for the first time through even a modest telescope. One is the Moon; the other is Saturn. Floating against the blackness of space, sprinkled around with its faint, starlike moons, this unbelievable wonder always comes as a fresh surprise, as clear and as clean as a newly minted coin. Surprisingly, too, it can be seen with a very low power; even ×30 will reveal this lovely gem. You will find that a common reaction of people gazing at Saturn for the first time is "I never expected it to be *that* clear!"

The rings of Saturn The reason why Saturn looks so sharp and neat lies in the rings: they supply the eye with something really sharp and crisp on which to focus. The fact that the globe itself is less well defined (because, like the globe of Jupiter, the limb tends to fade away against the sky) passes unnoticed. The rings of Saturn are, in the optical sense—perhaps in the physical sense too—one of the sharpest things in the solar system. They are thinner than any razor blade. The *Voyager* results suggest an overall thickness of a kilometer or so, compared with an outer diameter of 272,300 km: this is in the same proportion as a sheet of newspaper 22 meters across.

For many years, it was believed that Saturn's rings are unique. We now know that both Jupiter and Uranus have faint, ring-like strings of fragments orbiting around them, and Jupiter also possesses a hazy band of sulphur particles. But these are excessively dim, whereas the rings of Saturn are actually brighter than the planet itself. They consist of highly reflective icy particles, probably a meter or so across, and if the system could ever be turned square-on to our line of sight, so that they were seen in plane view, the planet would shine as brightly as Jupiter. We can never obtain such a view, but we are fortunate that the axis of Saturn is tipped from its orbit vertex by 26½°, so that we do at times have a reasonable view, because they rotate in the plane of the planet's equator. At such times, the northern and southern extremities of the rings can just be detected beyond the planet's poles. This opened-out aspect last occurred in 1973, and will happen again in 1988, while the edge-on phase, when they cannot be seen for a few nights because of their extreme thinness and faintness, was last passed in 1979–80 and is due again in 1995. The diagram shows why Saturn passes through these different phases of ring-presentation. During the next decade, the North Pole of the planet and the northern face of the rings will be on view.

The structure of the rings Everyone is so familiar with the wonderful photographs sent back by *Pioneer* and *Voyager* that it is easy to doubt whether Earth-based observation is of much use. The three

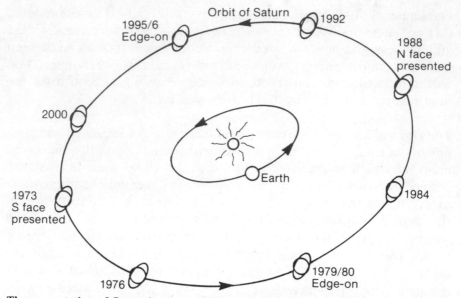

The presentation of Saturn's rings.   The axis of Saturn and its ring system always points in the same direction. Therefore, during its orbit of the Sun, the aspect as seen from the Earth goes through a steady change. The Earth's orbit is so small, compared with that of Saturn, that its own motion has little effect on the rings' presentation except when they are near the edge-on position.

well-known annuli into which the system has been divided for more than a century—outer ring A, of medium brightness, brighter ring B, and dim inner ring C (the Crepe ring) have now been seen by spacecraft to consist of narrow strands too numerous to count properly. The wide gap known as Cassini's division, which separates rings A and B and can be seen as a dark line with apertures of about 60 mm when the system is well presented, is now known not to be completely empty, but to contain a few thin strands of particles, while some other faint divisions, suspected by observers with large telescopes, have either been confirmed or shown to be optical illusions.

Is there any point in trying to observe the rings for changes of brightness intensity, or for new divisions? The answer must depend upon whether the rings themselves are in a state of change. If not, then *Voyager* is likely to remain the last word. But it has been suggested in some quarters that material from the outer parts of the rings may be spiralling inwards, finally being shed from the Crepe ring on to the planet's cloudy surface. If this is true, then subtle changes of intensity could be taking place—although whether such changes could be detected from the Earth is another matter!

It does seem extraordinary that Saturn was intensively studied by as great an observer as William Herschel, using large apertures, and yet the Crepe ring went undiscovered until it was found independently by two observers within a fortnight of each other in the year 1850. Had it undergone a sudden brightening around that time? If so, could its brightness be fluctuating now? The brightness of the different rings in the system is evidently related to the density of the orbiting particles rather than to their individual reflecting powers, so that a sudden release of material from one orbit to another could alter the relative intensity of different zones. This is speculation, of course, and may or may not be considered sufficient to justify a monitoring program.

In good conditions, the Crepe ring can be seen with an aperture of 150 mm, but 250–300 mm telescopes are really much more suitable for regular work on Saturn, and powers of ×250– ×350 are necessary for a good view of the disk. However, as we have already said, the beauty of the planet can be appreciated with quite modest equipment, and you don't need any more justification than this to turn to it time and time again.

Drawing Saturn Whether or not the rings are permanent, the globe markings are very definitely in a state of change. Its physical make-up is similar to that of Jupiter, although there is much less ammonia in the outer clouds, and its equatorial rotation time of 10 hours 14 minutes is still very fast. Unlike its giant companion, however, higher latitudes do not exhibit a fairly constant "System II" rate, but seem to rotate steadily more slowly as the poles are approached, the longest recorded rotation period for high latitudes being about 10 hours 45 minutes.

It is slightly smaller than Jupiter, and much less dense—in fact, its mean density is less than that of water—the rigidity of the globe being so low that it is even more flattened, despite the slower rate of rotation. The polar diameter has been accurately measured as 0.90 of the equatorial diameter. However, the main problem encountered in drawing Saturn is to reproduce the outline of the ring system because this changes appreciably even from month to month as the Earth swings past Saturn around the time of opposition, and of course there is a very noticeable opening-out or narrowing as the years pass. Some national (or even local) societies help their members by producing stencils, or printed outlines, corresponding to different presentations of the planet and its rings. If you can obtain the values of the major and minor axes (or the long and short widths) of the ring outline from an almanac, and prepare an appropriate ellipse, the following features can be sketched in, assuming that on the drawing the major axis of the outer edge of the ring system is 100 mm long:

Major axis of Cassini's division	87 mm
inner edge of ring B	66 mm
inner edge of Crepe ring	55 mm
Saturn's equator	45 mm

On the basis of a 100-mm major axis for the drawing, the minor axis of the ring outline will be approximately as follows at forthcoming oppositions (note that it alters somewhat during each apparition, because of the changing angle from which the planet is being viewed, but the outline can be adjusted at the telescope if necessary; this change is particularly noticeable near the edge-on phase, and the values for 1994–97 are subject to large change during the apparition).

Opposition date	Ring outline, minor axis (Major axis 100 mm)
1984 May 3	31 mm
1985 May 15	37
1986 May 27	42
1987 Jun 9	44
1988 Jun 20	45
1989 Jul 2	42
1990 Jul 14	38
1991 Jul 26	35
1992 Aug 7	31
1993 Aug 19	25
1994 Sep 1	12 approx.
1995 Sep 14	edge-on
1996 Sep 26	9 approx.
1997 Oct 10	16 approx.
1998 Oct 23	26
1999 Nov 6	35

It is worth noting that the orbit of Saturn is appreciably eccentric, and it is at perihelion when opposition occurs in the constellation Gemini, in December or January. At this time the disk diameter is about 12 percent larger (equatorial width 20.5″ arc) than at an aphelic opposition in Sagittarius, as in 1988.

Markings on Saturn Because its disk is almost always less than half the diameter of Jupiter's, markings are much harder to make out in any case; but we now know that the fault is not entirely that of distance, for Saturn really does show much less cloud activity. At first glance, all that can be made out is the bright equatorial zone with, perhaps, a faint equatorial belt (apart from around the edge-on phase, one of the equatorial belts tends to be hidden by the ring system). The overall tone of the globe, certainly compared with the bright ring B, is a dusky yellow but on rare occasions large white spots have been observed. One of the most prominent ever seen was discovered in 1933 by the well-known film comedian Will Hay, using a 150-mm refractor. When first seen, on August 3, it appeared as a large, slightly elliptical bright patch in the equatorial zone, of about the same relative size as Jupiter's Red Spot, although not so elongated. Within three weeks or so, the brightness had extended along the zone, so that it embraced about half the planet and could no longer be called a "spot" at all.

It seems probable that no other recorded marking has been as bright as Hay's spot, although prominent white features were recorded in 1976 and 1978. The fact that such outbreaks are completely unpredictable means that the disk should always be examined most carefully, even if you are not an habitual Saturn observer. It is a drawback that the beautiful rings distract attention away from its small disk; so, if the seeing is reasonably good, spend some minutes examining the equatorial zone and any belts that may be visible. A magnification of about × 250 is necessary for a good view, although a really prominent marking could be seen with less.

Bearing in mind that the equatorial region has a rotation period of 10 hours 14 minutes, you can expect to reobserve the same longitude of this region of the disk (generally the most likely to produce a detectable feature) at the following intervals, assuming, as we did with Jupiter, that the original observation was made at 10 pm (22.00 hours) on Day 0:

			Days elapsed			
1	**2**	**3**	**4**	**5**	**6**	**7**
18.28	04.42	01.11		04.22	00.50	
		21.39			21.19	

If you do ever succeed in detecting any sort of definite marking on Saturn, the first priority is to make a timing of its transit across the planet's meridian, and the second is to inform an experienced planetary observer who can check your observation and relay the news to others for further examination and study.

Saturn's moons Saturn's retinue of nine telescopic satellites and at least eight lumps of rock a few kilometers across (the latter discovered by space probes) cover a wide range of brightness. There are not, as with Jupiter, a few leaders and the rest nowhere. The largest moon, Titan, is bigger than Mercury and can readily be seen with binoculars at magnitude 8½. Rhea, Tethys, and Dione (all brighter than magnitude 10½) can be seen with an aperture of 60 mm under good conditions, or with 150 mm from almost any site. Iapetus is curious: one hemisphere is very much darker than the other, resulting in a brightness variation from near that of Rhea, when near its western elongation from Saturn, to about magnitude 12 near eastern elongation (its brightness is linked to its orbital position because, in common with all the other major satellites of Jupiter and Saturn, it keeps the same hemisphere turned inwards towards its master). The other four moons require apertures of about 200 mm upwards, and distant Phoebe, like the outer moonlets of Jupiter, has probably never been seen visually at all.

Saturn's moons all revolve in the plane of the equator, and so the outlines of their orbits match the outline of the ring system. Apart from near the ''edge-on'' phase, therefore, they can be scattered well North and South of the planet, as well as East and West, and are difficulty to identify without an almanac, particularly when the planet happens to lie in front of a rich field of stars.

URANUS

Uranus was discovered in 1781 by William Herschel, who was sweeping in a star field on the border of the constellations Auriga and Gemini when he came across an object that was, with his homemade 150-mm Newtonian, immediately distinguishable from a star. He was using a magnification of × 227, and, interestingly, regarded it initially as ''a nebulous star or perhaps a comet.'' This is curious because the disk of Uranus, although very small, should appear well-defined—and we know that Herschel's telescope was a good one, because of the close double stars that he discovered with it. Perhaps the seeing was very poor on that evening, or the mirror was not in its best condition.

The chart in Appendix III gives a finder chart and details of the position of Uranus for the rest of the century. Rather than trying to home in on it directly, either using this information or its position taken from an almanac, arm your telescope with a magnification of ×75 or ×100 and sweep over the likely area, as Herschel did. If your telescope defines the stars as it should, then the disk of the planet should be obvious as it passes through the field. If it is not noticeably different from the nearby stars of about the same magnitude (5.7) then there is something wrong with the seeing, the telescope, or your eyes!

The disk of Uranus measures about 3.8″ arc across, and although it can be detected easily enough, an aperture of about 250 mm is the minimum with which enough light can be collected to give the hope of observing surface markings, and a magnification of at least ×300 will be necessary. Excellent seeing is essential, which implies that the planet will be high in the sky. When Herschel made his discovery it was exactly at the north point of the ecliptic (the position the Sun occupies at midsummer), and so was extremely favorably placed for observation from the British Isles. Since that time it has circled the Sun about 2½ times—its year corresponds to eighty-four of ours—and it currently lies in Ophiuchus, in the southern part of the zodiac. It is therefore well placed for observers in the Southern Hemisphere. However, most amateurs will be content simply to identify this remote planet, and perhaps trying to see it with the naked eye if conditions allow. If you have only a very small telescope, or binoculars, and are trying to follow its slow drift in front of the stars, remember that the stationary points at the beginning and end of its retrograde motion occur about seventy-five days before and after opposition.

Some years ago, there was considerable interest in checking the magnitude of Uranus against that of nearby stars, and some observers have suspected slow variations of brightness over a period of several years. These are unconfirmed, but magnitude estimates are well worth making, using the techniques described in Chapter 11. The following comparison stars will lie near its path during the next few years:

	R.A.	(2000)	Dec.		Mag.
24 Oph	16ʰ	57ᵐ	−23°	09′	6.2
26 Oph	17	00	−24	59	5.8
39 Oph	17	18	−24	17	5.2
43 Oph	17	23	−28	09	5.3

Although its diameter is about four times that of the Earth, Uranus is so remote that no marking definite enough to give a clue to its rotation period has ever been observed. The length of its day, derived by indirect methods, is surprisingly vague: somewhere between sixteen and twenty-eight hours! Its axial tilt is the oddest feature of the planet, however, because it rotates practically on its side instead of more or less "vertically," as is the case with the other planets of the solar system. Twice in the Uranian year, one of its poles points to within a few degrees of the Sun, which means that much of the opposite hemisphere is shrouded in permanent night for years at a time. The next such polar presentation will occur in 1985, while the plane of its equator will pass through the Sun and the Earth's orbit in the year 2007. At such times, in the past, observers with apertures of about 300 mm or more have recorded faint equatorial belts on its minute bluish disk.

The ring system, discovered in 1977, is far too dim to be observed with any existing telescope. However, if the planet is high in the sky, its brightest moons, Titania and Oberon, may be seen with moderate apertures; both have been recorded with instruments of only 150 mm in aperture, and should be readily visible with anything much larger if the bright planet is hidden behind a hair in the field of view.*

NEPTUNE

Like its remote planetary neighbor, Neptune now lies south of the celestial equator, in the constellation Sagittarius. At magnitude 7.7 it can easily be seen in binoculars provided it is reasonably high in a transparent sky, and a good star atlas such as *Sky Atlas 2000* will enable it to be identified straight away; identification charts are also supplied in the *Handbook* of the British Astronomical Association. Appendix III gives details of its position for the rest of the century, and even if there is some doubt about its identity, careful observation over two or three nights around the time of opposition will reveal its motion relative to nearby stars.

With a disk diameter of 2.5″ arc, a magnification of about ×75 is necessary to distinguish it clearly from a star, and an aperture of 75 mm or more will be necessary if this image is to be bright enough for easy visibility. If you have a telescope of 150-mm aperture or more, try experimenting with cardboard stops of different aperture (eccen-

* See page 156.

tric stops in the case of a Newtonian or Cassegrain, so as to cut out the central obstruction), and see at what point it becomes difficult or impossible to distinguish Neptune from a star of similar brightness. You may also care to estimate its magnitude, using some of the following comparison stars:

SAO No.*	R.A.	(2000)	Dec.	Mag.
187185	18ʰ 43ᵐ.3	−22°	25′	7.7
187358	18 51.1	−20	18	7.4
187618	19 03.6	−20	08	7.8
187789	19 11.3	−20	21	7.9
162859	19 39.8	−19	14	7.8
163046	19 51.6	−19	13	7.8
163331	20 12.2	−18	09	7.8
163693	20 34.5	−19	24	7.7

* The star's identification in the Smithsonian Astrophysical Observatory *Catalogue* (1966). See page 196.

Adventurous amateurs may care to search for the larger of Neptune's two known satellites, Triton. This is probably larger than the planet Pluto, and shines in the sky at about magnitude 13½. To justify the search, at the present time, you should live in the southern part of the United States (or, still better, in the Southern Hemisphere!), and have a telescope of some 400 mm aperture. You will also need to refer to the *Astronomical Ephemeris,* which gives the dates and times of its greatest elongations from the planet. These are never more than 17″ arc, so that in a way you are trying to see the very faint companion of an unequal double star. It will be almost essential to hide the bright disk of Neptune from view, perhaps using a hair set in the focus of the eyepiece, as described on page 156.

PLUTO

Other, remoter worlds may still await discovery, but at the present time the strange double planet Pluto, with a diameter of only 3000 km and a moon, Charon, of almost half its diameter circling a mere 20,000 km away, is believed to mark the frontier of the planetary system. Or, to be more accurate, it did until the year 1979, when its markedly elliptical orbit—the most eccentric in the solar system—carried it closer than Neptune to the Sun. Having passed perihelion in 1989, it will once more become the remotest planet in 1999, heading for its aphelion, fifty times the Earth's distance from the Sun, in 2113. The vastness of its orbit is indicated by the fact that 248 years are required for it to accomplish one circuit of the Sun, traveling at an average speed of 5½ km per second!

In a transparent sky, Pluto can be detected with a telescope of about 200 mm in aperture, and some keen eyes have caught it with less because at the present time its magnitude is 13.9. The principal difficulty is in distinguishing it from the numerous neighboring stars of about the same brightness, but because it currently lies in a relatively star-poor area of sky, in northeastern Virgo (passing into the Serpens–Libra area in the early 1990s), identification will be an easier task than at the commencement of the 21st century, when it will enter the southern Milky Way. The predictions published annually, if plotted on a fairly detailed chart such as *Atlas Eclipticalis,* will permit its certain location within a field of perhaps half the Moon's apparent diameter, and its rate of motion, around the time of opposition, is over 1' arc in twenty-four hours. If the suspect "stars" are carefully drawn, reobservation on the following night could establish the planet's identity. As an alternative, you could try plotting its position directly on the photographic *Atlas Stellarum* chart, that shows stars as faint as the 14th magnitude, or use a ready-made annual chart, showing its position in the star field, as published regularly in the *Handbook* of the British Astronomical Association.

CHAPTER 7

How to Observe Comets

The word *comet* can arouse a very powerful mental image, with numerous and varied associations. Perhaps we think of great histori- cal comets, with tails that reportedly stretched across the sky. Or we may connect them with the sinister developments that they were popularly supposed to presage. They can be associated with the far depths of dim starlit space as well as with the scorching heat of the inter-Mercurial regions. Perhaps, most of all, they symbolize the un- expected. A comet seems to be a law unto itself.

The vast majority of comets appear unexpectedly because in most cases we do not know of their existence until they appear in the sky; and the huge number of comets that must belong to the solar system is proved by the few that we do see. Consider some of the brightest comets of the postwar period: Arend-Roland (1956); Mrkos (1957); Seki-Lines (1962); Ikeya-Seki (1965); Bennett (1970); and West (1976). None of these had ever been observed before because their orbital periods are on average to be measured in terms of hundreds of thousands of years. Therefore, if comets of such immensely long periods are coming into view at this sort of rate (and other, much fainter ones, are discovered every year), then they must be very numerous indeed, creeping through space in regions where the Sun's pull is little stronger than that of nearby stars, but finally submitting to its insistent gravity and setting off again on the long, long, acceler- ating path that eventually sears their ice to vapor, boils off clouds of dusty material, and then hurls them back on the rebound for more uncountable ages of nothing but starlight.

The Sun's huge family of comets is undoubtedly being reduced. If the universe contained nothing but the Sun and a comet, it would repeat its path almost endlessly. A comet, however, has a very low mass, no more than that of an asteroid a few kilometers across, and even a modest planet has sufficient gravitational attraction to perturb it from its original path in the case of a close encounter. A passage near the giant planet Jupiter can alter its orbit completely, and it is possible that these comets of immensely long period were formed in the true planetary region and subsequently perturbed by Jupiter and the other giant planets into their present huge orbits. It is also possible that some are adopted children of the Sun, picked up by our star during its passage through a more densely populated region of one of our galaxy's spiral arms eons ago. After all, if our solar system is losing comets, other stars may be collecting them, for they must go somewhere. We now observe comets that have periods ranging from only 3.3 years (Encke's) to millions of years. A fair number of those known have periods of about seven years, which allows their aphelia to lie closer to the Sun than the orbit of Jupiter, and means that they are safe from a potentially disastrous close approach and gravitational entanglement with this huge planet.

The nature of a comet A comet, then, is a small body of crumbly material, no more than a few kilometers across, cemented together with frozen liquids, probably mainly water. Almost certainly, it represents original, largely "unprocessed" planetary material. In other words, unlike the asteroids of the S-type, they have not been fired, melted, and tempered in the hot depths of a planet. The solid nucleus of a comet is, therefore, relatively fragile. The tail, which is the most dramatic feature of a bright comet, is caused by the vaporization of some of the ice and the release of the dusty material cemented on to the surface of the nucleus. The former process produces a gas tail, the latter a dust tail, and both are streamed away from the Sun, although in slightly different directions. The tail of a comet is unbelievably tenuous, and we see it only because it is hundreds of thousands of kilometers wide and perhaps millions of kilometers long; this huge appendage dims starlight much less effectively than does a cloud of smoke from a bonfire!

By no means do all comets exhibit tails. The so-called *short-period* ones—those with periods of less than a century or so—have returned so frequently to the Sun during their history that their tail-forming material has either been considerably reduced, or else exhausted altogether, and they may exhibit only a spherical hazy hood known as the *coma*. The historical evidence suggests that even Halley's Comet (period 76 years) has become noticeably less bright over the centuries, and the really brilliant comets are the total newcomers that have not previously returned to the vicinity of the Sun within historical memory.

Neither should it be assumed that all comets follow very elongated orbits that carry them close to the Sun at perihelion. Some comets have perihelion distances greater than that of the Earth from the Sun, so that they come to opposition like a superior planet! These objects, however, tend to be very faint. It seems likely that the solar system regions out to Pluto contain numerous "ghosts," comets that never swing close enough to the sun to shine brightly, and will remain undiscovered unless, perhaps, detected by a passing spacecraft.

Comet rates How many comets, per year, could you expect to see using a telescope of about 200 mm in aperture? On average, about a dozen comets are observed to pass perihelion each year; most of these are returns of known comets, but some are always new discoveries. Usually, but not always, the newly discovered objects are brighter than the returning ones. In 1982, for example, six comets were picked up: four of them had been observed on previous returns to the Sun, and the other two were new discoveries. Of these six, one became visible with the naked eye as a 4th-magnitude object, three reached about magnitude 9 and could be seen with very small instruments, and the two faintest reached magnitude 10 or 11, and were detectable with apertures of about 200 mm. Therefore the year 1982, despite its meager crop, offered the serious comet observer a regular supply of objects to look at, and if you have a telescope of even 150 mm in aperture you can confidently expect to have at least two or three comets to look at in any year.

How comets are named During the course of its career, a comet receives several labels. Most important, perhaps, is the name of its discoverer. For example, the bright comet of 1982 was found by a New Zealand amateur astronomer, Rodney Austin, and is therefore known as Comet Austin, while a fainter object, discovered in May 1983 by several Japanese amateurs, has been named Comet Sugano-Saigusa-Fujikawa (even if other observers also make an independent discovery, three is the maximum number of names that a comet may bear).

Comet Austin, 1982g This prominent comet was easily visible with the naked eye and was notable for its very narrow tail (*R. W. Arbour, 400-mm Newtonian*)

In addition, each comet is given a year designation, followed by an italic letter indicating the order, in that year, in which it was first observed. Comet Austin was the seventh comet to be detected in 1982, and is known as 1982*g*, while Comet Sugano-Saigusa-Fujikawa, the fifth comet of 1983, is 1983*e*. These letters are also applied to returning expected comets.

Eventually, this temporary letter designation is replaced by a permanent roman number, indicating the order in which a comet passed perihelion in a given year. For example, one of the comets observed in 1982 was Comet d'Arrest, 1851 II. This means that the comet was first discovered by d'Arrest, and that it was the second comet to pass perihelion in the year 1851. However, at its recent apparition it acquired the temporary label 1982*e,* as it was the fifth to be picked up in that year. Comet d'Arrest has a period of only 6½ years, and because it has been observed on at least one previous occasion its name is sometimes preceded by "P/" (i.e., P/d'Arrest) to indicate that it is periodic.

Comet notification If you want to observe comets, you must know where in the sky to find them. The expected returns of periodic comets are listed in some annual handbooks because their orbits can be calculated years in advance. The *Handbook* of the British Astronomical Association is most useful for this purpose; for example, the 1982 issue listed no less than thirteen comets, with their right ascension and declination, as well as distance from the Earth and Sun and their predicted magnitude, given for every tenth day. Using this information, the apparent path of the comet can be plotted on a star chart, and its night-to-night situation can be estimated with sufficient precision to bring the required area within the telescopic field. The *Observer's Handbook* is also useful.

It is worth noting that the predicted magnitudes in these ephemerides are often very pessimistic. This is because they usually refer to the brightness of the tiny nucleus rather then the hazy coma. If you see the magnitude of a comet given as "M_2" or "m_2," this is the nuclear magnitude. The total magnitude of nucleus, coma, and tail (if any) is given by "M_1" or "m_1," and may be several magnitudes brighter. P/Churyumov-Gerasimenko 1969 IV, which was observed by amateurs towards the end of 1982 at a total magnitude of about 10.2, was predicted to be magnitude 12.2 (M_2) at the time, while P/d'Arrest 1851 II reached a total magnitude of 8.5 at the beginning of

November 1982, although its predicted M_2 at the time was 17.0! More will be said about measuring the magnitude of a comet in due course. Meanwhile, remember not only that "magnitude" can refer to one of two things, but also be prepared for a comet's coma and tail to develop in an unpredictable way, so that it may be much brighter, or much fainter, than expected.

You will, therefore, need to have access to an astronomical yearbook or almanac in order to know all about expected comet returns (but take note that the *Astronomical Ephemeris* does not include comet information). However, the brightest objects are usually the new discoveries, and to find out about these you will have to subscribe to an information service of some kind. The monthly *Sky & Telescope* has an invaluable "Comet Digest" feature which gives current ephemerides for both periodic and new comets, but its "new comet" information is, unavoidably, some weeks out of date, although it will be able to give you current positions. Some societies and organizations offer a telephone "early warning" system, with one member subscribing to the telegram or telex service run by the Smithsonian Astrophysical Observatory at Cambridge, Mass. With this facility, you can hear about a new discovery in a few hours.

Where comets are to be found It is often assumed that a comet can be seen only in or near the twilight sky: this is because their greatest brightness is associated with perihelion passage, when they are at their closest to the Sun. It is true, in most cases, that a comet is at its brightest when seen in this direction, but some have been easily visible when more or less opposite the Sun in the sky, two recent examples being Comet Panther, discovered on Christmas Day 1980, and Comet Iras-Araki-Alcock, that passed very close to the Earth in May 1983. It is true to say, however, that most discoveries of bright comets are made in the Sun's half of the celestial sphere, partly because this is where observers tend to hunt for them: Comet Austin, found in the summer of 1982, is a good example of a bright comet discovered and best seen not far from the twilight region. Therefore it is very important to have good views towards the horizon in the eastern and western arcs of the sky, for observing before dawn or at dusk. Summer comets are often seen underneath the celestial poles as well because this region (particularly as seen by observers in higher latitudes) lies fairly near the Sun at this season of the year.

The ephemeris for a new comet usually lists its expected right ascension and declination at daily or five-day intervals, depending upon the rapidity with which it appears to be moving in front of the stars. It may also state the elongation of the comet from the Sun, in degrees, which will help you to see if it is approaching the Sun (and therefore becoming harder to observe), or moving away from the Sun into a darker sky. The comet will also, of course, be changing in declination, and its motion may be carrying it higher into the sky from night to night, and therefore improving the view, or alternatively it may be descending towards the horizon. Comets do not keep to the general plane of the solar system, as indicated by the ecliptic; some have orbits that are inclined practically at right angles to the ecliptic, and they may, therefore, pass near one of the celestial poles.

At any rate, the first task on receiving new cometary information is to plot its position on a star chart and to discover which constellation houses it. Then determine the position of that constellation relative to the Sun. You will then know whether the hunt will be a dusk, midnight, or dawn operation, and whether conditions may improve or worsen as the nights pass.

Observing comets A "typical" telescopic comet looks like a hazy patch of light. This is the coma, with a diameter of anything from perhaps 1′ arc to ½° or more across. Usually there is a brightening towards one point in the coma, known as the *condensation*. Some comets have a very pronounced condensation, while others show hardly any at all. The condensation is the site of the gas- and dust-producing activity in the comet, and should not be confused with the nucleus itself, the tiny solid part, which is relatively faint: to see it, you will need a large aperture and a high magnification. Only the brighter comets will reveal anything in the way of a tail, although the coma may appear elongated, suggesting where a faint tail may begin. A comet's tail always points more or less away from the Sun.*

The subject of magnification is important. Because a star appears as practically a point of light, magnification does not seem to enlarge it very much unless a very high power is used. (This is a half-truth because poor seeing conditions may produce a blurred and enlarged star disk that *can* be magnified noticeably: the better the seeing, the smaller a star appears to be.) But an extended object, such as a planet or comet, appears to enlarge in step with the magnification used.

---

* Due to the angle at which the tail may be presented to the Earth, however, it could on occasions *appear* to be directed towards the Sun.

Because the same amount of light concentrated by the telescope is being spread over a larger area of the retina, its unit brightness falls. A planet, being intrinsically bright and sharp-edged, can be enlarged considerably before its image becomes too dim to be made out properly, but most comets are faint and ill-defined to begin with. Increasing the magnification simply makes them even fainter and more ill-defined!

The larger the aperture of the telescope, the brighter the focused image and the higher the acceptable magnification. For most comet work, a power of about $\frac{A}{4}$, where A is the aperture in millimeters, is a good start: say, $\times 20$ with a 75-mm telescope, or $\times 40$ with an aperture of 150 mm. Very small and well-condensed comets can be seen with higher powers than these—so, of course, can bright ones, although it does not follow that more detail will be made out. An incidental advantage of using a low magnification is the ease with which the comet can be brought into the field of view.

Observing projects Having acquired your comet, what are you to do with it? First, there is the satisfaction of having made it out at all. Most telescopic comets are far from easy to locate: they are small and dim, the position may be uncertain, the altitude may be low, and twilight may interfere. The observation may only be a glimpse out of the corner of your eye. It may not even be seen at all because comets can fade by a significant amount almost overnight. So you should first of all savor the thrill of seeing something so rare and fleeting. Having done that, however, you will want to make some notes, perhaps along the following lines:

(a) *Its appearance* How big is the coma? If the comet is a large one, the diameter of the field of view can be used as a scale. If it is small, estimate its diameter in terms of the distance between two field stars that can be identified on a suitable atlas, and their angular separation measured. The *Borealis, Eclipticalis,* and *Australis* series of atlases is excellent for this purpose because they show stars as faint as the 9th magnitude (Appendix I).

Even this apparently straightforward observation is not as easy as it may sound. Just where does the coma end and the dark sky begin? There is, in fact, no "edge" as such: the comet simply fades away until its brightness is imperceptible. You must make a judgment where this point occurs. Try using averted vision, directing your stare elsewhere but concentrating on the comet. This brings the feeble light on to that part of the retina which is most sensitive to faint illumination, and the comet will probably appear to be larger than when you gaze straight at it. Averted vision is a most useful technique, but not easy to master, so strong is the instinct to look at the object of your attention rather than away from it!

Another way of enhancing contrast is to move the telescope from side to side in short but fairly rapid arcs, so that the comet seems to dance in the field of view: does this motion make it appear larger? Movement of an object across the retina does make it more noticeable. Change the magnification, or use the finder, or try a pair of binoculars if the comet is bright. How does its size appear to change when gauged by the surrounding stars? It often happens that a moderately bright comet appears to be larger when viewed through a small low-power instrument than a large medium-power one, the much smaller image resulting in a more intense impression on the retina. The use of averted vision, and the side-to-side motion of the telescope, are both useful dodges when trying to locate an excessively faint object.

Having decided upon its size, there are other questions to ask about its appearance:

1. How condensed is the brightest part of the coma? Does it come up to a bright disk or even a point of light, or is it a much more gentle and vague increase of brightness? Does the brightness increase steadily from edge to center, or does it show a sudden increase in intensity?
2. What shape is the outline of the coma? Is it circular, with a central condensation? Or is the condensation away from the center? If so, in which direction does the displacement lie? Is the coma pear-shaped, or irregular? Always make an estimate of the position angle of any asymmetry, remembering that the East–West direction can be established by swinging the telescope around the solar axis, and the North–South direction by nodding it in declination. Position angle is measured from North (0° or 360°) round by East (90°), South (180°), and West (270°).
3. Are there any streamers or hoods visible in the coma? Some comets are active and throw off clouds of material in just a few hours. In the case of some Sun-grazing comets (those passing well within the orbit of Mercury at perihelion), the nucleus has been known to disintegrate into a number of separate objects.
4. Finally, if a tail of any sort is visible, estimate its length and any condensations and streamers in it. In the case of a bright naked-eye comet, binoculars may give a more impressive view than an astronomical telescope can do, because of their much wider field of view.

(b) *Its position* In theory, if you are using an equatorial telescope with circles, the position of the comet on the celestial sphere can be written down directly. But no telescope is so accurately set up, or has such precisely divided circles, that a useful position could be obtained in this way. You need to use the stars in the field of view as reference points, identifying them from an atlas and then looking up their positions in a catalogue such as the very useful *Sky Catalogue 2000.0,* that covers the whole sky down to magnitude 8.0. Remember that the eye is extremely sensitive to alignments (or rather to misalignments) of points, and you may be able to watch a comet slowly drifting along its orbit until it lies precisely on the line joining two stars. Estimate also its relative distance between the two stars (say, ⅔ of the way from star A to star B), and you have a "fix." If you take several fixes, using different pairs of stars, an accuracy of a minute of arc is certainly possible, and perhaps better if the stars are favorably placed relative to the comet. Of course, this sort of precision is crude by comparison with the results achieved by professionals using proper cameras and telescopes, but it is always interesting to compare your own results with the ephemeris position—and ephemerides are not always right!

Sometimes, a comet passes almost centrally over a star. Being so rarefied, it is unlikely to dim the star appreciably, and a position can be derived on the basis of the single observation.

(c) *Its magnitude* Estimating the brightness of a comet is probably the most useful thing you can do with it because you are getting to the root of one of the greatest puzzles about comets: why they change in brightness as they do. The brightness of a planet, at any point in its orbit, is almost perfectly predictable because it depends only upon the distances of the planet from the Sun and the Earth, and to a small extent upon the angle Sun-planet-Earth. A comet, however, is not a solid body of constant size. Its coma grows, and the larger it grows the more surface there is to reflect light; therefore, the brighter the comet appears. However, the law of coma-growth is not general to all comets. In addition, the excited atoms in the head of the comet also emit light, and the manner in which the comet is self-luminous will, obviously, affect its brightness.

Long-period comets (i.e., newly discovered ones that have approached the Sun from a very remote aphelion) usually brighten up by a greater amount, as they approach, than do short-period ones; this is not surprising, in view of the large proportion of volatile ices that they must still contain (these volatiles turn to vapor and gas, releasing dust; most "new" comets exhibit both gas and dust tails, whereas "old" comets usually exhibit only a feeble dusty tail). However, the rule is not always observed—people have unkind memories of Comet Kohoutek of 1973, that would have become a brilliant object had it brightened as much as some samples of its type, but turned out to be several magnitudes fainter than this. Comet West (1976), however, gave an unexpectedly brilliant display. Therefore, magnitude estimates are well worth making and analyzing afterwards.

They are, however, far from easy to make. You should refer to Appendix II, that describes a way of estimating the magnitude of a variable star; but bear in mind that a comet bears no resemblance to a star. Its total light is diffused over an appreciable area, and the M_1 magnitude of a comet, which is what you are trying to measure, corresponds to the brightness it would have if all the luminosity of its coma were compressed into a starlike point. Because this cannot be done, the only method open to the amateur is to alter the focus of the eyepiece so that the stars expand into disks that match the size of the coma. You can then either compare the appearance of the defocused stars with your memory of the focused comet (and vice versa), or else compare the defocused stars with the defocused comet. The former method gives a more accurate comparison because you are leaving the comet alone; but it does depend on memorizing the brightness of one as you shift the focus for the other. Obviously the results will not be interchangeable, since the second method is likely to give a fainter magnitude value than the former, but this does not matter if a series of observations are being compared with each other.

The method is not easy. You will almost certainly find that the defocused star image does not particularly resemble the comet (it probably has a much more distinct edge, while the center, if a Newtonian or catadioptric system is being used, will have a central black hole instead of the typical condensation of a cometary image). In more practical terms, a comet passing fairly near the Earth could be moving so quickly across the celestial sphere that fresh comparison stars will have to be found from night to night, and their magnitudes obtained from a comprehensive catalogue such as *Sky Catalogue 2000.0*. If the comet is brighter than about magnitude 5½. the very useful little catalogue by Lampkin (Appendix I) can be used.

This may sound crude and open to all sorts of errors. And so it is; but it is surprising what can be done with perseverance and care. It is an excellent example of what you should be trying to do if you really want to do justice both to your instrument and to your own time and interest. Simply to look at something, and then to tick it off the list and turn to the next object, is barely to scrape the surface of what astronomy has to offer. It *can* be as passive as rushing round another country photographing as many tourist attractions as possible, but if this is all you do there is a good chance that your interest will soon wane. One star may look very like another, but in reality they are probably very different, although the amateur may lack the means of distinguishing between them. In the case of a passing comet, however, there is a wonderful chance of adding usefully to our knowledge: it has never been seen before, and it will never be seen again (at least, not in the foreseeable future), so should we pass up this unique chance of studying it and finding out as much as possible about it? Amateur comet study is really useful, because powerful professional equipment tends to concentrate on positional work and spectral analysis, rather than estimates of its brightness. For a few weeks, your observations and those of a few other enthusiasts could make an unrepeatable contribution to knowledge of this particular object.

Comet-hunting You will probably not want to embark on such a program until you have convinced yourself that you have the enthusiasm and stamina to keep it up. Both will be needed in abundance: the average times that have been logged by successful observers easily exceed 100 hours per comet, and are often much more. The most successful comet-hunter of recent years, William Bradfield, observes from near Adelaide in South Australia, and found his first eleven comets in 1600 hours of actual sweeping, spread over twelve years. His rate of one comet every 150 hours or so is much better than that achieved by other successful contemporary observers; Eric Alcock, of Peterborough, England, has averaged over 300 hours for each of his five comets and spent about 600 hours sweeping before the first was found, in 1959. Another English amateur, Roy Panther, also searched for 600 hours before making his first discovery on Christmas Day, 1980. The most successful comet-hunter in the United States during the past half-century was Leslie Peltier, of Ohio, who discovered twelve comets between 1925 and 1954.

Anyone taking up comet-hunting must expect to spend literally years sweeping the sky at inhospitable hours. The regions of dark sky near the Sun are the most profitable to sweep, carefully scanning the stars in steady horizontal or vertical arcs, making sure that each sweep overlaps the previous one so that none of the selected area is missed. Most amateurs have used altazimuth mountings for this work, finding them more versatile because a low magnification is always used—rarely higher than × 50, and usually around × 30.

In some ways, telescope type and quality are less important than for other work. But it must be admitted that a catadioptric telescope is not particularly suitable for very low-power work because there is an uncomfortably large "blind spot" near the center of the exit pupil, caused by the outline of the large secondary. Newtonians are potentially more suitable in this respect, since the central obstruction is, or should be, smaller. But Peltier and Bradfield made their discoveries with 150-mm short-focus refractors, and Alcock discovered four of his comets with 25 × 105 military binoculars, picking up his fifth while hunting for novae (exploding stars) with 15 × 80 glasses. Nevertheless, it is quite possible that very careful searching of a smaller area of sky, using a higher magnification and a larger aperture, could reveal much fainter comets than are habitually found by amateurs, and this is a field of work that could be undertaken with a catadioptric instrument of at least 200 mm in aperture, with a power of ×50 or ×75.

Recent comets It is usually only the bright or otherwise particularly interesting new comets that achieve any mention; this is perhaps understandable, but it gives a false impression of the amateur discovery scene. To give a better overall picture of the level and type of comet discoveries that are happening at the present time, the following table lists the new comets that have been discovered by amateur observers in the decade since 1978, with, wherever possible, details of the type of instrument used.

You will see that the apertures used to make these discoveries ranged from 85mm to 500mm, and that Newtonian reflectors, refractors, and large binoculars were all successfully employed. At discovery some comets were already naked-eye objects; others always remained only dimly visible even with a large telescope. On the evidence of this table it seems safe to say that almost any instrument holds out the hope of an eventual discovery, if it is used with persistence.

Name	Discovered	Notes
Bradfield	Feb 4, 1978	This comet was discovered, like his previous finds and all the others in this list, with his home-assembled 150mm short-focus refractor. It became as bright as mag. 5, but was too near the Sun to be easily seen at this time.

Meier	Apr 27, 1978	Discovered with a 400mm reflector, this comet was observable for a year and reached naked-eye brightness.
Haneda-Campos	Sep 1, 1978	A new faint periodic comet. The Japanese codiscoverer Toshio Haneda used an 85mm refractor.
Machholz	Sep 12, 1978	A faint comet, detected with a 250mm reflector.
Sergeant	Oct 1, 1978	A naked-eye object at the time of discovery, this comet had already started to fade from view.
Denning-Fujikawa	Oct 9, 1978	An independent recovery of a faint periodic comet, discovered a century ago by the famous observer William Denning and then lost.
Bradfield	Oct 10, 1978	Already faint when discovered, this discovery by Bradfield rapidly faded.
Bradfield	Jun 24, 1979	During its brightest period, this comet became visible with binoculars.
Meier	Sep 20, 1979	His second comet. A faint object, discovered wtih a 400mm reflector.
Bradfield	Dec 24, 1979	At discovery with his 150mm refractor this was already a naked-eye object, with a noticeable tail.
Cernis-Petrauskas	Jul 31, 1980	A faint comet, discovered with a 500mm reflector.
Meier	Nov 5, 1980	Detected with a 400mm reflector, this comet was a dim binocular object.
Bradfield	Dec 17, 1980	This became one of the brightest comets of the decade, being easily visible with the naked eye and having a tail several degrees long.
Panther	Dec 25, 1980	This comet was visible with binoculars. It was discovered with a homemade 200mm reflector.
Austin	Jun 18, 1982	Discovered with a 150mm reflector, this comet became visible with the naked eye, and had a short tail.

IRAS-Araki-Alcock	May 4, 1983	Discovered with binoculars when already a naked-eye object. Passed closer to the Earth (4.6 million kilometers) than any other comet this century.
Sugano-Saigusa-Fujikawa	May 8, 1983	This comet just reached naked-eye visibility, but quickly faded.
Cernis	Jul 19, 1983	A faint telescopic comet, discovered with a 500mm reflector.
Austin	Jul 9, 1984	Just visible with the naked eye, with a tail several degrees long.
Takamizawa	Jul 30, 1984	This new periodic comet was discovered with 120mm binoculars. It takes 7 years to orbit the Sun, and was not quite visible with the naked eye.
Meier	Sep 17, 1984	Discovered with a 400mm reflector, this comet remained a very faint object.
Levy-Rudenko	Nov 14, 1984	This comet was just visible with binoculars before fading.
Machholz	May 27, 1985	A telescopic object, discovered with a 250mm reflector.
Machholz	May 12, 1986	A new, faint, short-period comet, which takes just over 5 years to orbit the Sun. Discovered with homemade 130mm binoculars.
Sorrells	Oct 31, 1986	A faint comet discovered on a photograph taken using a 400mm reflector.
Levy	Jan 5, 1987	This faint object was discovered with a 400mm reflector.
Nishikawa-Takamizawa-Tago	Jan 19, 1987	A binocular object, discovered independently by four Japanese observers on the same night.
Terasako	Jan 24, 1987	Discovered with 150mm binoculars, this comet never attained naked-eye visibility.
Bradfield	Aug 11, 1987	Reached naked-eye visibility in the autumn. Bradfield's 13th comet—a record for a 20th-century observer.
Rudenko	Aug 20, 1987	A binocular object, discovered with a 150mm refractor.

CHAPTER 8

A First Look at the Stars

This chapter is not just about telescopic astronomy. It surveys the naked-eye stars that form the patterns and the constellations of the night sky; but the constellations are the universal backdrop against which the dramas and delights of astronomy are played, and it is with reference to the constellations that the brighter stars in the sky are named.

It is possible to go through an astronomical life using only the circles on your telescope, and perhaps a sidereal clock, or a star of known right ascension, to set the right-ascension circle at the start of an evening's work. But to depend upon circles to this extent will make you helpless without them, should such a disaster ever befall you. Besides, circles are not necessarily the quickest and most certain way of finding an object, and they have drawbacks. They require illumination to be read; they may be misread; they may be in error. If you find objects by using star patterns, you can at least be sure that the signposts are in the right place. You have also the repeated pleasure of seeing fields of familiar stars, of perhaps, one day, discovering an exploding star or nova that has disturbed the pattern; or of seeing an unexpected meteor burn and expire in front of the stars.

Seasons of the stars You may already be familiar with the constellations that pass above your horizon in the course of the year. If so, these exhortations, and much of the information in this chapter, may be superfluous. Other readers, however, may have acquired their telescope without going through a preliminary cycle or two of the stars using a small telescope, binoculars, or even the naked eye. Because the telescopic objects listed in more detail in Chapters 9–11 refer to constellations, and indeed are part of the attraction of each constellation, it is only sensible to familiarize yourself with them from the beginning. You will need to know at what season they are to be found in the night sky, which ones rise high above the horizon, and which ones never rise at all. You will, surely, *want* to come to know them as old and familiar friends, returning and disappearing in exact step with Nature's annual march of crop and climate. And you will, in turn, have the delight of showing others the constellations, and remembering, in their learning, your own first steps around the sky.

Observation of solar-system objects, from the Sun to the comets, has an important component of movement in it. Sunspots move across the solar disk due to the Sun's rotation; the planets move into view and away again, approaching and receding from the Earth; meteors shoot into the atmosphere; day and night crawl across the lunar surface. In addition to these effects of motion, there are real changes on the surfaces of the Sun and the planets.

When we gaze across space to the realm of the stars, movement is almost totally deadened by distance. Only the reflected motion of the Earth, which draws the blinding searchlight of the Sun around the celestial sphere, changes the way the stars appear. Without the Sun and its retinue of circling bodies, the sky would appear almost changeless—almost, but not quite, for there are indeed some changes. Selected stars alter in brightness in periods ranging from hours to years, while some particularly close ones gradually shift their position with respect to their remote companions as the centuries pass. Telescopically, some stars are seen to be revolving around each other, although many years may have to pass before any noticeable change is seen. However, only an alert astronomer would notice any of these exceptions, and most of the stars and the so-called "deep-sky" objects (star clusters, nebulae, and galaxies) appear as emblems of constancy rather than of change.

This is part of their appeal. It is interesting to compare our own observations with those of others, perhaps made a century or more ago, while feeling almost positive that the object has not changed its appearance. Perhaps it is even more interesting to compare one's own observations, made at different times, with each other! At the same time, observing the variable objects, principally the stars that change in brightness, has the fascination of the unexpected. So there are two facets to deep-sky work: the invariable and the changing, or the constant and the inconstant.

The brightness of the stars Let us begin with the obvious but far from trivial observation that the stars in the sky are not all of the same brightness or *magnitude*. Why not?

There are two possible reasons. One has to do with luminosity, and the other with distance. Suppose that the stars were all equally distant from the Earth. Then the greater their luminosity or real brightness, the brighter they would appear. We could infer, for example, that the brilliant star Sirius, in the constellation Canis Major (the Great Dog) is more luminous than the star Procyon in Canis Minor (the Little Dog), because Procyon looks fainter than Sirius. In other words, the true luminosity or *absolute magnitude* of a star is indicated by its brightness in the sky or its *apparent magnitude*. This, certainly, would account for the appearance of the sky, and in ancient times it was believed, by some thinkers, to be the true interpretation.

An alternative explanation would assume that all the stars have the same luminosity or absolute magnitude, but that their distances from the Earth are very different. The more remote a star, the fainter its apparent magnitude will appear to be. On this theory it would follow that Sirius is closer than Procyon, and we could measure the ratio of their distances by measuring their difference in apparent brightness, using the well-known law that the brightness of an object decreases inversely as the square of its distance from the observer (twice the distance, one-quarter the brightness; three times the distance, one-ninth the brightness, and so on).

However, neither single reason explains the great variety of apparent magnitudes seen in the sky. Certainly, stars are at very different distances from us: the closest known is only 4.3 light-years away (one light-year corresponds to the distance traveled through space by light in one year, 9½ million million km), while the remotest detectable stars in the Galaxy are tens of thousands of light-years away. This, however, is only part of the story, for the stars' absolute magnitudes also vary enormously. Some stars are about 15 magnitudes, or a million times, less luminous than the Sun, while others are close to one million times more luminous! Neither must we forget that some *variable stars* can themselves change in luminosity by thousands of times.

Stars and evolution Why are some stars so much more luminous than others? The reason is not so much that they are larger—although they are—but that they are more massive. More hydrogen was in-

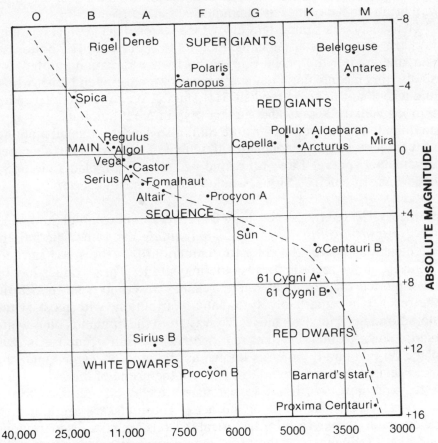

SPECTRAL CLASS

The Hertzsprung-Russell diagram This shows the more important classes into which stars have been grouped, ranging from the hottest and brightest (upper left) to the coolest and dimmest (lower right). Some naked-eye or telescopic stars are shown as examples. The diagram as presented here gives the misleading impression that giants are as numerous as main-sequence objects, but this is only because giants, being very luminous, are conspicuous; the vast majority of known stars lie on the main sequence.

volved in forming the protostar; the collapse was more violent as a result; more energy was developed at the center, where the atoms were crushed together; and the core of the star became hotter as a result. The law of starlight states, in round terms, that the more massive a star is, the more luminous it will be. This is the famous mass–luminosity law. Luminosity, or absolute magnitude, is very sensitive to mass. A star of ten times the Sun's mass, for example, will be about 4000 times as luminous.

There is another important law which states that the kind of radiation sent out by a star depends upon its temperature. The hotter the star, the greater the proportion of short-wave radiation emitted. Visually, this means that a hot star appears white or even bluish-white, and a cool star appears reddish. Hot stars also emit ultraviolet radiation, as well as X-rays and gamma-rays in great amounts. A star of medium temperature, such as the Sun, also sends out small amounts of these lethal rays—enough to kill, in fact, if they were not absorbed by the atmosphere. Life of the kind we know, on a planet circling a hot white star such as Sirius, would be out of the question.

Star types These laws are enshrined in a very famous graph known as the *Hertzsprung-Russell* or *color–magnitude* diagram. One axis of the display indicates the color (or temperature) of the star's surface, while the other records its absolute magnitude. These values having been plotted for stars that can be reliably observed from the Earth, the result is a remarkably systematic distribution, with most of the stars forming a narrow band all the way from the luminous blue-white *giants* down to the very dim red *dwarfs*. The Sun occupies its own place on this band, which is known as the *main sequence;* and the main sequence seems to indicate stars in the prime of life.

It is, perhaps, worth noting that stars are classed according to the temperature of their radiating surfaces. These classes are indicated by capital letters, which for historical reasons are now discontinuous and out of sequence. The hottest, blue-white stars belong to group W or O (surface temperatures of about 25,000° C or more). Groups B and A indicate temperatures of about 11,000°-plus and 7,500°-plus respectively, and A stars appear white to normal eyes. Then come the yellowish F and the yellower G types (about 7,000° and 5,500°, an example of the latter being the Sun), and the even cooler K and M types (about 4,000° and 3,250°), which are deep yellow and reddish. Some other types are even cooler than the M stars, but examples of these R, N, and S classes are rather rare, although they are made distinctive by their deep red tint.

Stars do not remain on the main sequence until they burn away. As they convert their hydrogen into helium—the process that gives out the energy to keep themselves hot—the waste helium core enlarges until it reaches a certain unstable size. The hydrogen–helium process, that takes place at the surface of the core, begins to lose its energy to the star's surface too quickly, and the force of radiation causes the upper layers of the star to puff outwards. If this did not happen, the star would explode. As they puff out, these outer layers cool and redden in accordance with the universal radiation law mentioned above. The star is now in the *red-giant* stage, and lies not on the main sequence but in the cooler regions to the right of the Hertz-sprung-Russell diagram.

Eventually, the star must run out of fuel. The outer layers disappear, the hot core is exposed to space, and the star dims and whitens, moving back across the main sequence to the left-hand side of the diagram, becoming a *white dwarf:* very faint (because of its small surface area) and very hot. It has now shrunk to planetary size, and must end its radiating life as a black dwarf, a dark, cold cinder.

This is the normal progress of evolution. Some stars may explode as *supernovae,* and others could become black holes, but these are exceptional events. The evolutionary sequence outlined above is, of course, far too slow to be followed in an individual star, but astronomers have discovered sufficient samples of stars in different states to infer a pattern, just as a walk through a wood can suggest how a tree must grow by revealing sprouting nuts, saplings, full-grown specimens, and leafless skeletons.

Most of the stars visible with the naked eye or with a telescope belong to the main sequence. There are also many red giants to be seen because they are very luminous and can therefore be seen even if far away. However, white dwarfs, even nearby ones, are extremely faint and difficult to detect, even with moderate telescopes.

How stars are catalogued Some of the brightest stars in the sky have acquired their own names. This is particularly the case with the stars visible from northern temperate regions because a long tradition of folklore is associated with them. Some of the more common names are used in this book. A more systematic way of listing them was developed by Johann Bayer in his pioneering catalogue of naked-eye stars, dated 1603: he selected the brighter stars in each constellation and labeled them with the Greek letters α, β, γ, etc., in approximate order of descending brightness. This letter is then followed by the genitive case of the constellation name (for example, the bright star called Regulus, in the constellation Leo, is α Leonis). Nowadays, constellation names are often written in the standard three-letter abbreviation given in the table on pages 198–99, so that, for example, the star Dubhe in Ursa Major would be written as α UMa rather than α Ursae Majoris. It is useful to know the Greek alphabet, and it is set out here.

THE GREEK ALPHABET

Alpha	A	α	Iota	I	ι	Rho	P	ρ		
Beta	B	β	Kappa	K	κ	Sigma	Σ	σ		
Gamma	Γ	γ	Lambda	Λ	λ	Tau	T	τ		
Delta	Δ	δ	Mu	M	μ	Upsilon	Y	υ		
Epsilon	E	ε	Nu	N	ν	Phi	Φ	φ		
Zeta	Z	ζ	Xi	Ξ	ξ	Chi	X	χ		
Eta	H	η	Omicron	O	o	Psi	Ψ	ψ		
Theta	Θ	θ	Pi	Π	π	Omega	Ω	ω		

(The Bayer nomenclature always uses lower-case letters)

In 1725, an important catalogue of all the naked-eye stars visible from the Greenwich Observatory, England, was published. Compiled by John Flamsteed, the first Astronomer Royal, it numbers the stars by constellation, beginning with the westernmost star (in other words, the one whose right ascension is the smallest), and working eastwards. These *Flamsteed Numbers* are still in use. Some of the fainter stars in the southern constellations, invisible from Greenwich, have received roman-letter appellations. Various telescopic catalogues have also been compiled over the years, and a single star can appear in different lists under a very different guise! Two comprehensive catalogues, going down to the 9th magnitude and prepared in the 19th century, are still often referred to: these are the *Bonner Durchmusterung* (BD), and the *Henry Draper Catalogue* (HD). BD numbers are falling out of use now, and a star is likely to be referred to by its HD number or (increasingly frequently) by its SAO number as given in the Smithsonian Astrophysical Observatory Catalogue of 1966. SAO numbers are used, for example, in the table on page 173, because the stars are too faint to have received either Bayer letters or Flamsteed numbers.

Star positions and precession In Chapter 2, it was shown how the celestial sphere is divided into circles of right ascension and declination, and how the position of any celestial object can be referred to this network or grid of reference lines.

This would be a once-and-for-all task, as is the determination of latitude and longitude of an object on the Earth's surface, if the Earth's axis remained pointing towards the same points on the celestial sphere. However, the celestial poles are very slowly describing a circle in the sky about 45° across, and since the coordinates of right ascension and declination must be referred to the celestial poles, the coordinates of the stars are steadily changing. For example, if you had wanted to point your telescope to the bright star Spica in Virgo (α Vir) in the year 1950, the R.A. circle would have read 13ʰ 23ᵐ and the Dec. circle − 10°.9. In the year 2000, to bring Spica to the center of the field of view, the readings will have to be 13ʰ 25ᵐ and − 11°.2 respectively. This change amounts to over half a degree: sufficient to put the star outside the field of a medium-power eyepiece. The phenomenon, known as *precession,* is caused by a 25,800-year "wobble" of the Earth's axis.

Therefore, if you are using circles to find an object, you need to ensure that the "epoch" of the coordinates being used to set the circles is near the year of observation. Many comet ephemerides, for example, refer to the epoch 1950, which is now some thirty-five years out of date. Of course, your circles may not be sufficiently accurate to warrant correcting for an error of 20′ arc or so; but if they are not, then perhaps they are not of a great deal of use for finding a faint object without reference to the surrounding stars!

Similarly, you must ensure that the position of an object being plotted on a star chart is measured with reference to the same epoch as the chart itself. If it is not, then the object will be plotted in the wrong place relative to the surrounding stars. The old *Atlas Coeli,* the predecessor of *Sky Atlas 2000.0* (epoch 2000), was drawn for epoch 1950, as is the very useful, small-scale *Norton's Star Atlas,* a copy of which should be on every amateur's shelf. Positions in this book have been computed for epoch 2000, but you can easily make a correction for any epoch by using the following formulas, which indicate the amount of precession to be added to a star's position for each whole year that passes (α = right ascension, δ = declination):

$$\Delta \alpha = 3.07 + 1.34 \sin \alpha \tan \alpha \text{ seconds of time}$$
$$\Delta \delta = 20.0 \cos \delta \text{ '' arc}$$

These amounts will have to be deducted if correcting to an earlier epoch.

And	Andromeda
Ant	Antlia, *The Air Pump*
Aps	Apus, *The Bird of Paradise*
Aqr	Aquarius, *The Water Bearer*
Aql	Aquila, *The Eagle*
Ara	Ara, *The Altar*
Ari	Aries, *The Ram*
Aur	Auriga, *The Charioteer*
Boo	Boötes, *The Herdsman*
Cae	Caelum, *The Chisel*
Cam	Camelopardalis, *The Giraffe*
Cnc	Cancer, *The Crab*
CVn	Canes Venatici, *The Hunting Dogs*
CMa	Canis Major, *The Great Dog**
CMi	Canis Minor, *The Little Dog**
Cap	Capricornus, *The Sea Goat*
Car	Carina, *The Keel*
Cas	Cassiopeia
Cen	Centaurus, *The Centaur*
Cep	Cepheus
Cet	Cetus, *The Whale*
Cha	Chamaeleon, *The Chamaeleon*
Cir	Circinus, *The Compasses*
Col	Columba, *The Dove*
Com	Coma Berenices, *Berenices' Hair*
CrA	Corona Australis, *The Southern Crown*
CrB	Corona Borealis, *The Northern Crown*
Crv	Corvus, *The Crow*
Crt	Crater, *The Cup*
Cru	Crux, *The Cross*
Cyg	Cygnus, *The Swan*
Del	Delphinus, *The Dolphin*
Dor	Dorado, *The Swordfish*
Dra	Draco, *The Dragon*
Equ	Equuleus, *The Little Horse*
Eri	Eridanus, *The River*
For	Fornax, *The Furnace*
Gem	Gemini, *The Twins*
Gru	Grus, *The Crane*
Her	Hercules
Hor	Horologium, *The Clock*
Hya	Hydra, *The Water Snake*
Hyi	Hydrus, *The Water Snake*
Ind	Indus, *The Indian*

Lac	Lacerta, *The Lizard*
Leo	Leo, *The Lion*
LMi	Leo Minor, *The Little Lion**
Lep	Lepus, *The Hare*
Lib	Libra, *The Scales*
Lup	Lupus, *The Wolf*
Lyn	Lynx, *The Lynx*
Lyr	Lyra, *The Lyre*
Men	Mensa, *The Table*
Mic	Microscopium, *The Microscope*
Mon	Monoceros, *The Unicorn*
Mus	Musca, *The Fly*
Nor	Norma, *The Square*
Oct	Octans, *The Octant*
Oph	Ophiuchus, *The Serpent Bearer*
Ori	Orion
Pav	Pavo, *The Peacock*
Peg	Pegasus
Per	Perseus
Phe	Phoenix, *The Phoenix*
Pic	Pictor, *The Painter*
Psc	Pisces, *The Fishes*
PsA	Piscis Austrinus, *The Southern Fish*
Pup	Puppis, *The Poop*
Pyx	Pyxis, *The Compass*
Ret	Reticulum, *The Net*
Sge	Sagitta, *The Arrow*
Sgr	Sagittarius, *The Archer*
Sco	Scorpius, *The Scorpion*
Scl	Sculptor, *The Sculptor*
Sct	Scutum, *The Shield*
Ser	Serpens, *The Serpent* †
Sex	Sextans, *The Sextant*
Tau	Taurus, *The Bull*
Tel	Telescopium, *The Telescope*
Tri	Triangulum, *The Triangle*
TrA	Triangulum Australe, *The Southern Triangle*
Tuc	Tucana, *The Toucan*
UMa	Ursa Major, *The Great Bear**
UMi	Ursa Minor, *The Little Bear**
Vel	Vela, *The Sails*
Vir	Virgo, *The Virgin*
Vol	Volans, *The Flying Fish*
Vul	Vulpecula, *The Fox*

* Strictly speaking, *Major* and *Minor* should be interpreted as *Greater* and *Lesser,* but *Great* and *Little* are now accepted usage.

† Divided into two separate constellations, Serpens Caput (head) and Serpens Cauda (body), to the West and East of Ophiuchus.

A cycle of constellations The writer, and doubtless many other people, spent a lot of time getting to know the constellations by examining the stars and deep-sky objects to be found in them, using a small telescope. Many books have adopted the same plan, and there is much to be said for it. However, since many observers may set themselves the task of observing a certain class of object, the following two chapters are devoted to double stars and deep-sky objects respectively, with Chapter 11 (*Windows into Space*) examining some selected regions in different parts of the sky.

Because this method can be confusing to the newcomer, we also include here a listing of the order in which the constellations come into good and convenient visibility in the course of a year. This means that if you are reading this book and beginning your observing career in September, you can turn to the "September" entry in the list and discover which constellations are well placed for observation. This list gives the constellations that come into best visibility (i.e., are highest in the sky) for observers at 22.00 hours in the middle of the month in question. For each month, they are grouped into those constellations whose main area is more than 20° north of the celestial equator (*Northern*); those lying within 20° of the celestial equator (*Equatorial*); and those more than 20° south of the celestial equator (*Southern*).

January	*Northern* Auriga Camelopardalis
	Equatorial Lepus Orion
	Southern Columba Doradus Mensa Pictor
February	*Northern* Cancer Gemini Lynx
	Equatorial Canis Minor Monoceros
	Southern Canis Major Carina Puppis Volans
March	*Northern* Leo Minor
	Equatorial Hydra Leo Sextans
	Southern Antlia Pyxis Vela
April	*Northern* Ursa Major
	Equatorial Crater Corvus
	Southern Chamaeleon Crux Musca
May	*Northern* Boötes Canes Venatici Coma Berenices
	Equatorial Virgo
	Southern Centaurus
June	*Northern* Corona Borealis Draco Ursa Minor
	Equatorial Libra Serpens Caput

	Southern Apus Circinus Lupus Norma Triangulum Australe
July	*Northern* Hercules
	Equatorial Ophiuchus Serpens Cauda
	Southern Ara Corona Australis Scorpius
August	*Northern* Cygnus Lyra Vulpecula
	Equatorial Aquila Delphinus Sagitta Scutum
	Southern Pavo Sagittarius Telescopium
September	*Northern* Cepheus Lacerta
	Equatorial Capricornus Equuleus
	Southern Grus Indus Microscopium Octans Piscis Austrinus
October	*Northern* Pegasus
	Equatorial Aquarius Pisces
	Southern Phoenix Sculptor Tucana
November	*Northern* Andromeda Aries Cassiopeia Triangulum
	Equatorial Cetus
	Southern Fornax Hydrus
December	*Northern* Perseus
	Equatorial Eridanus Taurus
	Southern Caelum Horologium Reticulum

If you are observing at a local time that is different from the standard one of 22 hours (and remember that Daylight Saving Time, if in force, should be allowed for, the clocks then being one hour in advance of true time), remember that the celestial sphere presents the same appearance approximately four minutes earlier on each successive night, which works out at about two hours per month. Therefore, January constellations can be observed equally well in February, but at 20 hours instead of 22 hours; while those that were not observable until after midnight can now be caught at a more reasonable hour.

One of the most exciting and satisfying observations of all is to work this time effect in reverse, and to have a view of the constellations of the coming season. Look out eastwards in the August dawn, for example, and you will see the Pleiades, Aldebaran in Taurus, the "twins" Castor and Pollux in Gemini, and Orion himself—stars and constellations that will bedeck the winter sky at midnight—all freshly emerged from their conjunction with the Sun. You may find enjoyment in noting the dates on which specific stars are either last seen in the West as the Sun moves into their region, or are first seen in the

East as it leaves them behind. How late into May can you follow Betelgeuse, in Orion, before it vanishes in the evening twilight? Such little challenges add zest to routine observation, and may lead to some interesting competition in the local society!

In the following chapters, each object will be referred to the constellation in which it lies, and its right ascension and declination (for the epoch 2000) will also be given. There are two ways of bringing it into the field: by using the circles on an equatorial telescope, or by "star-hopping."

Locating an object using circles In theory, as we have seen, any object in the sky can be brought into the field of view by using the right ascension and declination circles, together with a sidereal clock by means of which the right ascension circle can be set to time. This method assumes (a) that the circles are accurately set (admittedly easy enough to check, by pointing the telescope to a bright star of known position and seeing what they read), and (b) that the polar axis of the telescope is well aligned. This is not so easy to achieve with a portable instrument, unless you have a special polar finder or spend some time "fiddling." Furthermore, the various components of the mounting (and the optical axis of the telescope) must be accurately at right angles to each other.

In practice, the method of *off-setting* should produce more satisfactory results. Find a star whose right ascension and declination is as near as possible to that of the object being sought, and bring it into the field of view. Lampkin's little book is very useful for this purpose, but the epoch is 1950, and precession must be applied to make it agree with the 2000 positions in this book; if you use *Sky Catalogue 2000.0,* no conversion will be needed. Once the telescope is pointing in the right direction, note the difference in both right ascension and declination between the positions of the star and the object being sought, and adjust the telescope to this difference by the circles.

This method eliminates circle-adjustment error because you are interested only in the *difference* between the coordinates. It also greatly reduces the significance of polar-axis error because the telescope may have to be turned by only a few degrees from one object to the other. In this small angle, an object should turn up safely in the field of view of a low-power eyepiece, even though the alignment of the polar axis is not as good as it might be.

Let us take an example. Suppose that you are seeking the rather

faint galaxy M 77 in Cetus, and you are using Lampkin's star catalogue to find a suitable guide star. You choose the nearby star α Cet,
whose 1950 coordinates are R.A. 02ʰ 40ᵐ.7, Dec. + 03° 02'. The
coordinates for M 77, for the epoch 2000, are given on page 301 as
02ʰ 42ᵐ.7, − 00° 01'. Using the formula on page 197, you can either
update the α Cet position to epoch 2000, or backdate the M 77 position to 1950; it does not matter which you do because the differences
in right ascension and declination will be the same in either case.

Suppose that you decide to backdate M 77 to 1950. Then you have
the two positions as follows:

	R.A. (1950)		Dec.	
M 77 Cet	02ʰ	40ᵐ.1	− 00°	15'
α Cet	02	40.7	+ 03	02
Difference		− 0.6	− 3	17

R.A. adjustment Because its right ascension is slightly *smaller* than
that of the guide star, the galaxy must lie to the *West* (be careful
about this: it is easy to turn the telescope the wrong way!). A shift of
0.6 minute is, however, a very small angle; it takes the sky four
minutes to turn through 1°, so that the difference in right ascension
here amounts to only 9' arc. This is hardly readable on most circles,
and even if neglected the object will still lie within the confines of a
low-power eyepiece giving a field of ½° or so.

Dec. adjustment The galaxy is 3° 17' south of the star. It is easier
to read the declination circle divisions if you mentally convert the
minutes of arc into fractions of a degree—in this case calling the
difference 3⅓°.

If your telescope has a motor drive, it makes no difference how
long you take to make the settings. If it is hand-driven, then remember that the sky is turning at a rate of 15' arc in every minute of time.
If you are likely to spend a minute making the two circle settings, try
to allow for this when moving the telescope in right ascension. Alternatively, look slightly to the West of the center of the field of view,
once you get your eye to the eyepiece. With a little experience you
will become adept at "guesstimating" the position of a celestial object.

Remember that the closer you are pointing the telescope to the
celestial pole, the more slowly the stars appear to travel through the
field of view. Therefore, there is no single conversion from minutes

of right ascension to degrees of sky. M 77 lies at the celestial equator, where the sky appears to spin at its fastest rate of 1° per four minutes of sidereal time (or 3 minutes 59.3 seconds of ordinary civil time). But if you were hunting an object at a declination of 70° or so (for example, the galaxy M 82 in Ursa Major, up near the North Celestial Pole), you would find that it was moving through the field of view of a fixed eyepiece at a rate of 1° in about twelve minutes. The apparent rate of drift is inversely proportional to the cosecant of the declination.

Locating an object by star-hopping This is the purely visual approach. You take no notice of right ascension and declination, except for the purposes of locating the object on a star map. You do not, therefore, need an equatorial telescope, although the method can, of course, be used with one.

Do away with the need to read circles, and you do away with a nuisance. Setting a pointer to a scale is not the best exercise for an eye that is about to gaze into the depths of space: it has to concentrate upon something relatively brightly lit to enable the lines and divisions to be clearly seen. The eye must then recover from this operation before it is ready to be in its most receptive state as a photon-gatherer! Of course, you need to use an artificial light to study a chart when locating an object by the visual method, but there are two points to reckon with: first, it is much easier to grasp the essentials of a chart with a dim light—there are no adjustments and fine judgments to be made—and, second, and far more important in the long run, you will soon find that the object can be picked up from memory, without needing any chart-study at all. Circles can never be used in this way. If you commit yourself to a "circles" approach, you are condemning yourself to a life of downward-peering dial adjustment instead of star-gazing. You may come to learn the differences between the coordinates of the guide star and the object off by heart, but you will still have to use the circles to make the adjustment.

To find M 77 by star-hopping, you must first locate Cetus itself and then the locality of the galaxy, shown in the diagram. The little group of α, γ, and δ is then picked up. With the finder, the faint Flamsteed-numbered star 84 Cet is then located, about a degree south of δ. You point the finder between δ and 84, and a little to the East, and M 77 must now lie in the field of the main telescope.

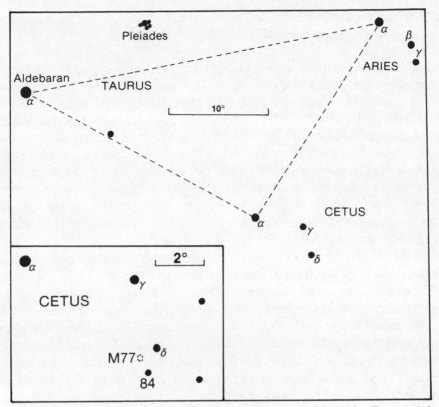

Star-hopping to M77   Identify α Ceti by some strategy such as the Taurus–Aries triangle shown here. Then proceed from α Ceti to δ Ceti, and use the little triangle δ–84–M77 to acquire the galaxy.

It may be argued that M77 is particularly easy to find, with such convenient locating stars. This is true. But every object in the sky can be located by star-hopping, even though the path may seem tortuous to begin with—it is extraordinary how the brain manages to remember quite involved journeys from one part of the sky to another, and it is satisfying to pick out a faint object in this way. You can use any sort of telescope, equatorial or altazimuth, and do the thing on any occasion. The "circles" observer is bound to just one kind of instrument. Take away his circles, and he is helpless.

This is not to say that there is never any advantage in using circles. They may be helpful in locating an asteroid or a comet whose position is changing from night to night. They are extremely valuable when trying to pick out a star or planet in a bright sky. But it is probably a good idea to look upon them as a yachtsman may look upon his engine: something to help him when in difficulties, but to be managed without whenever possible!

Finders The essential for successful star-hopping, in addition to an atlas showing guide stars, is an effective finder. The writer has in front of him the catalogue of a well-known telescope manufacturer, and the standard finder offered with their 200-mm (8-inch) Newtonian reflector has an aperture of only 30 mm, with a magnification of ×6. The limiting magnitude of this telescope, in a typical suburban sky, is probably only about 8½, and it may be very difficult to relate what you see in this instrument to what the main telescope shows. The ideal finder for most amateur instruments that are going to be used to observe faint objects would have an aperture of about 50 mm and a magnification of about ×10.

In the writer's view, an erect-image finder is a help because it relates more easily to what is seen with the naked eye in the sky. Others disagree and feel that an inverting finder, corresponding in orientation to what is seen in the main telescope, is preferable. There are arguments for both views, and most commercial finders are, in fact, inverting because they are cheaper to make. However, erect or inverting, to be really effective the finder will probably need to be considerably larger than the one supplied as standard, if your deep-sky program is to be pleasurable and not frustrating.

CHAPTER 9

Observing Double and Multiple Stars

The Sun has no stellar companion, although it does have a family of planets keeping it company. However, fully a quarter of all the stars in the Galaxy are twins, revolving around each other in stable orbits. These are known as *binary* systems, and the stars in such a system may be alike or very dissimilar, depending upon their individual stage of evolution. Massive, luminous stars evolve much more rapidly than do average stars like the Sun: Sirius, for example, is a binary, but its more massive component—which must once have shone far more brilliantly in the sky than Sirius does now—has evolved to the white-dwarf stage, and is hard to see with a small instrument. Other binaries, such as the glorious southern pair γ Virginis, consist of stars at the same evolutionary stage, looking like true twins.

Undoubtedly, binaries came into existence as a result of some fragmentation inside the original cloud of gas and dust (dare the writer suggest the term *gust* for this material?) that condensed into the stars. Some binaries are so close together that they may complete one orbit in a few hours, while others may take thousands of years to circle once. (It is worth noting that these very short-period binaries are too close to be resolved in any telescope, and are detected by indirect methods; few visual binaries have a period of less than two or three decades.) Like the planets orbiting the Sun, the stars in a double system obey Kepler's Third Law, which states that the square of the periodic time of the system is proportional to the cube of their mean separation. In other words, if the distance between two stars were doubled, they would require almost three times as long to complete one orbit.

Line-of-sight or optical doubles If you sprinkle some sand grains on a sheet of paper, you are likely to find several instances in which three of the grains, although widely scattered in themselves, happen to lie almost exactly in a straight line. Therefore, if your eye were located in the position of the first or third grain, the other two would appear to be almost touching each other. Sprinkle countless millions of stars into the millions of cubic light-years of the Sun's Galactic suburbs, and the same effect can happen, this time in three dimensions: two stars, separated by a very small angular distance on the celestial sphere, are the result of random alignment rather than signifying a physical association.

It is obvious that the greater the separation of two stars in the sky (as always in double-star work, we speak of *angular* separation by inference), the greater is the chance that the pair is an optical one, and the smaller the chance that it is a true binary, because a widely separated binary pair would have to be very near the Sun in order to appear so spread out, or so far apart from each other in space that the stars could not have preserved a permanent gravitational attachment. As a very general guide, most optical pairs are separated by more than about 30" arc.

There is no international definition of how wide a double star can be before it ceases to be called a double at all. In some cases we can speak of "naked-eye doubles," a good example being the bright pair of stars constituting α Capricorni, separated by several minutes of arc. Many bright stars have had their vicinity examined for faint companions, even though they are a considerable distance away, partly because faint stars in the vicinity of bright ones are good tests of vision and telescopic quality: for example, the northern star α Lyrae (Vega), of magnitude 0, has an 11th-magnitude companion about 80" away. So distant a companion would not be considered worth noting if the primary star were of magnitude 6 rather than magnitude 0. Very wide and faint companions are almost always optical, revealed by the fact that the proper motions of the two stars through space are very different, instead of being shared, as they are by the members of a binary or multiple system.

True physical pairs, or binaries As a general rule, all binary systems lie within about 40″ arc of each other, and the great majority are within 20″ arc. Of course, their angular distance tells us nothing about their real separation, unless their distance from the Sun is also known. However, if their orbital period is measured, and their masses are estimated, then their distance apart can be calculated. Here are some examples of possible binary systems, their separation being given in *astronomical units* (AU), corresponding to the mean distance of the Earth from the Sun:

MEAN SEPARATION (IN AU) FOR DIFFERENT BINARY SYSTEMS

Combined mass of the stars (Sun masses)	Orbital period in years		
	25	100	500
2	11	27	79
5	15	37	108
10	18	46	136

It is worth remembering that the mean distances of some of the planets from the Sun, in astronomical units, are 9.5 (Saturn), 19 (Uranus), 30 (Neptune), and from 30 to 50 for Pluto.

There are plenty of fairly wide double stars that are known to be physically connected because they share a common motion through space, but whose orbital motion around each other is so slow that their periods must be a matter of thousands of years. An example of this is the most famous double in the northern sky, ζ Ursae Majoris or *Mizar,* whose components seem hardly to have shifted in the two centuries since they were first accurately measured by William Herschel; it has been estimated that the pair will require 14,000 years to complete one orbit! Knowing that these stars lie at a distance of about 60 light-years from the Sun, and that their combined mass is about four times that of the Sun, their mean distance apart can be worked out at approximately one thousand astronomical units. It seems extraordinary that their gravitational attraction can bind them together even over this immense distance.

At the other end of the scale are stars almost touching each other, completing an orbit in a few hours. Such objects are far too close to be resolved directly, but their duplicity is proved by analyzing their light with a spectroscope. Many of these *spectroscopic* doubles seem to brighten and fade as one component passes in front of the other and blocks out its light, and we shall discuss such variable stars in Chapter 11. The most rapid directly observable binary is δ Equulei, with a period of 5.7 years, and a maximum separation of 0.5″ arc.

How a double star is measured The visual appearance of a double star must be described in terms of the magnitudes of the components, their separation in ″ arc, and the orientation of the fainter star B in relation to the brighter star A (the further components of a multiple star may be labeled C, D, etc.). This orientation is measured as *position angle*. Imagine a dial graduated into degrees, with star A at its center. The northerly direction from A corresponds to a position angle of 0° or 360°; East is 90°; South is 180°; and West is 270°. Therefore, a position angle of 135° would mean that star B was to be found to the southeast of A.

No measure of the separation and position angle (PA) of a double

Position angle   This gives the general appearance of the cardinal points as seen through an inverting telescope situated in the Northern Hemisphere. In order to define completely the relative position of the components of a double star, the distance from A to B in seconds of arc needs to be measured. In this example, the position angle of B relative to A is 135°.

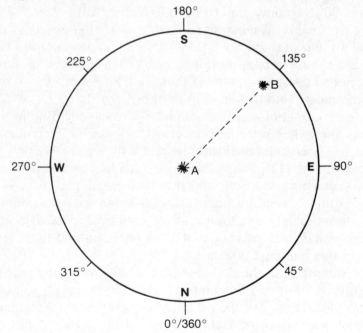

star is complete without the date or epoch of the measure. If the binary is fast-moving, it may be necessary to repeat measures at almost yearly intervals, while other stars can be remeasured at intervals of a decade or more (and some have been left unmeasured for half a century!). As an example of a series of measures of a leisurely binary here are three of Mizar, to which we have already referred:

Mizar 2.4, 4.0; 14″.3; 147° (1782)
2.4, 4.0; 14″.2; 148° (1852)
2.4, 4.0; 14″.5; 150° (1926)

Note the order in which the measure is set out: first, the magnitudes of components A and B respectively; then their separation; then the position angle of B relative to A; and finally the epoch of the measure.

You will notice two interesting things about this sequence of measures, which have been selected from a long list. The separation appears to have decreased slightly between 1782 and 1852, and then to have widened by 1926; and the position angle seems to have increased steadily. With regard to the separation, it is perfectly possible for an individual measure to have an error of a few tenths of a second of arc, so that a jump of 0.1″ arc between two individual measures is probably not significant, and the best interpretation of the three distance measures is that the distance has remained unchanged! However, position angle measures of a relatively wide double such as Mizar should be fairly accurate, and there is good evidence that the slight increase in this angle is significant. It is on this basis that the 14,000-year period has been calculated.

You are unlikely to be interested in tackling double-star measures. Few amateurs are, believing that the necessary equipment is expensive and complicated. This need not be so; but it is a fact that the really interesting stars are the close, rapid binaries, and these need large apertures to be well resolved. They are the prerogative of the big refracting telescopes built around the turn of the century, but only a handful of professional astronomers around the world are doing this sort of work today.

Resolving power and double stars To many amateurs, the word "double star" has strong associations with tests of telescopic resolving power. In Chapter 1 we pointed out how the angular diameter of the telescopic image of a star depends solely upon the aperture of the lens or mirror.* If two stars are separated on the celestial sphere by less than this resolving limit, then they cannot be seen as separate images with dark sky in between, although they may be seen as an elongated or even as a notched single image. The formula for this limit is 11.6″ arc divided by the aperture in centimeters. However, distance apart is not the only criterion of visibility. This resolving limit, often known as the *Dawes' limit* after the English observer who researched it, was originally referred to two 6th-magnitude stars viewed through a 1-inch (25-mm) aperture telescope; and the magnitude as well as the separation of the double is significant. If the stars are very bright, some of the light will inevitably flare out beyond the true diffraction disks, making them merge together; if they are too faint, they may be difficult to see properly. Either way, resolution suffers. Although it is often stated that the Dawes' limit applies to a pair of 6th-magnitude stars observed with an instrument, it is the writer's experience that telescopes in the 200- to 300-mm aperture range may show their true resolving performance more easily on doubles that are rather fainter than the 6th magnitude—perhaps the 8th magnitude in the case of a 300-mm instrument, viewed with which a pair of 6th-magnitude stars would appear uncomfortably bright. (This argument assumes, of course, that the air is of good clarity. Haze often produces steady air, ideal for examining close pairs, but the light of the stars will be dimmed.)

So the bare statement that "a telescope of such and such an aperture should resolve two stars such and such a distance apart" is practically meaningless. The stars' apparent magnitude does matter. Furthermore, the claim also assumes that the stars are of approximately equal brightness. If their brightness is very unequal, the fainter one is all too easily lost from view in the glare of the primary. The star Sirius (α Canis Majoris) is the classic example. Its white-dwarf companion is of the 7th magnitude, and currently is about 10″ arc away from the primary—but the glare of Sirius A, at magnitude −1.45, is so overwhelming that few amateurs have ever seen Sirius B.

In fact, the appearance of a close double star tells you no more about the performance of a telescope than does the image of a single star. If a single star focuses down to a tiny disk, almost a point, with no obvious flare; and if its out-of-focus images, seen with a high-power eyepiece shifted just two or three millimeters from the position

---

* See page 220 for a possible method of diminishing the size of the spurious image.

of best focus, are identical both inside and outside focus, then you can be confident that the telescope is a good one. The out-of-focus test is the most reliable of all because it is practically independent of seeing, whereas the focused image, even if formed by an excellent objective, can look woolly in bad air. (Chromatic aberration gives the defocused refractor image a border of contrasting color—reddish inside focus, and greenish-blue outside; to reduce this confusing effect, hold a yellow photographic filter over the eyepiece.)

Some "test doubles" Despite all this, you will probably want to have a look at some double stars that are near the limit of your telescope's performance. This is a valuable exercise: you will learn something about the use of high magnification and the effect of bad seeing, and you will also learn the positions in the sky of some challenging stars. So let us start off with a list of measures (see page 214) that are correct for the year 1985. Many of the pairs move so slowly that they will remain virtually unchanged for several years, but others seem to move quickly. Consult the extended list later in this chapter for more information about these and other stars.

Double-star nomenclature You will have noticed a number of curious prefixes to these stars, jumbled up with others that are familiar. For example, the fourth star on the list, α Piscium, is a star originally lettered in Bayer's catalogue of 1603, and later discovered to be a double. In such a case, the original designation is retained, although it may acquire one or more other catalogue names as well. The second star, 36 Andromedae, is Flamsteed's 36th star in Andromeda (working, it will be remembered, from West to East, in order of increasing right ascension), in his catalogue of 1725. All of the other identifications, however, refer to special double-star lists published by observers who went hunting for them. These are usually objects below the 6th magnitude, that had not previously been included in general star catalogues. The most commonly encountered symbol is that for F.G.W. Struve (Σ). Struve used what was then the largest refractor in the world, of 240-mm aperture, to examine 120,000 stars, publishing a catalogue including about 2200 discoveries (together with an examination of many known doubles), in 1837. Most of Struve's stars are fairly easy with moderate apertures, although he took it as a general rule not to include any pairs more than 32″ arc wide.

Struve's work extended down to a declination of about − 15°.

SOME CLOSE BINARY STARS (1985 predictions)

Star	Const.	R.A.	(2000)	Dec.	Magnitudes A	B	Sep.	P.A.
β 395	Cet	00ʰ 37ᵐ.2		−24° 46′	6.3	6.4	0″.51	097°
36	And	00ʰ 55ᵐ.0		+23° 38′	6.0	6.4	0″.66	278°
Σ 186	Cet	01ʰ 55ᵐ.8		+01° 51′	6.8	6.8	1″.14	055°
α	Psc	02ʰ 02ᵐ.0		+02° 46′	4.2	5.2	1″.67	277°
ι	Cas	02ʰ 29ᵐ.0		+67° 24′	4.6	6.9	2″.46	233°
7	Tau	03ʰ 34ᵐ.5		+24° 28′	6.6	6.7	0″.62	004°
Hu 445	Tau	05ʰ 01ᵐ.7		+20° 49′	8.6	8.9	0″.42	278°
12	Lyn	06ʰ 46ᵐ.3		+59° 27′	5.4	5.6	1″.69	076°
ζ	Cnc	08ʰ 12ᵐ.2		+17° 40′	5.6	6.0	0″.67	239°
I 314	Pyx	08ʰ 39ᵐ.4		−36° 36′	6.5	7.6	0″.50	233°
Σ 1338	Lyn	09ʰ 21ᵐ.0		+38° 12′	6.5	6.7	1″.06	262°
β 411	Hya	10ʰ 36ᵐ.1		−26° 40′	6.7	7.5	1″.21	320°
ξ	UMa	11ʰ 18ᵐ.2		+31° 33′	4.3	4.8	2″.33	091°
I 83	Cen	12ʰ 56ᵐ.7		−47° 41′	7.4	7.6	0″.76	223°
h 4707	Cir	14ʰ 54ᵐ.2		−66° 25′	7.6	7.9	0″.67	320°
η	CrB	15ʰ 23ᵐ.2		+30° 18′	5.6	5.9	0″.86	085°
π²	UMi	15ʰ 39ᵐ.7		+79° 58′	7.4	8.2	0″.62	026°
20	Dra	16ʰ 56ᵐ.3		+65° 02′	7.1	7.3	1″.28	068°
h 5014	CrA	18ʰ 06ᵐ.8		−43° 25′	5.7	5.7	1″.19	006°
OΣ 358	Her	18ʰ 35ᵐ.9		+16° 59′	6.8	7.0	1″.52	158°
ε²	Lyr	18ʰ 44ᵐ.3		+39° 40′	5.2	5.5	2″.35	082°
I 253	Sgr	19ʰ 19ᵐ.0		−33° 17′	7.6	7.7	0″.64	139°
ε	Equ	20ʰ 59ᵐ.1		+04° 18′	6.0	6.3	1″.04	285°

Some years after Struve's catalogue appeared, William Herschel's son, John, published the results of four years' work carried out at the Cape of Good Hope, which included 3347 new double stars. These are prefixed by the letter h, as in h 4707 Circini, but most of his pairs are wide and easy.

Following the publication of these two extremely important catalogues (and we should also take note of the doubles discovered by Struve's son, Otto, symbol OΣ), observers began to take a particular interest in discovering very close binary pairs, whose fast-moving members would provide useful observational data for the study of stellar dynamics. Perhaps the most remarkable was S. W. Burnham, a Chicago court reporter, who used a 150-mm refractor to discover a thousand new very close double stars, and was then employed at the Lick Observatory in California to proceed with his work using the great 900-mm (36-inch) refractor. Burnham's catalogue of 1290 new doubles was published in 1900, and the stars listed here are signified by the letter β, as in the first on our list. Over half of Burnham's stars were no more than 2″ arc apart when discovered; some have since widened, while others have become more difficult. Close doubles in the southern sky appeared in the catalogue of R.T.A. Innes (prefix I, as in I 314 Pyxis), dated 1899, while other catalogues have been published by observers including R. G. Aitken (A), G. W. Hough (Ho), W. J. Hussey (Hu), and T. J. J. See (See). Aitken's catalogue is particularly important, as it summarizes particulars of 17,180 double stars, and many lists refer to doubles by their Aitken catalogue (ADS) number.

It is not necessary to know any of this information when you are about to observe a double star, but it seems to the writer that doubles have strong historical associations. For they do change, albeit slowly: these points of light in the eyepiece are tiny clocks, gradually turning away the ages—not, like galaxies, so slowly as to be imperceptible, but not, like the planets of the solar system, shifting from night to night. Some of these clocks measure decades, others measure millennia. Furthermore, there is the personal link with previous observers who either discovered them or at least studied them carefully. We can directly compare our observations with theirs, whether or not the position and distance have changed: if they have, then we are aware of the grand scale of stellar motion, and if they have not, then we are staring at the original noted down by a predecessor, whether a decade or a century before.

Magnifications for double stars You will be well advised to approach the double, whether or not it is a very close pair, via a low-power eyepiece. This may even be necessary, if you are hunting a faint star with an atlas and need to identify star patterns. But the low-power view is valuable in another way, because you may have your attention drawn to other interesting pairs and groups of stars nearby. If the pair lies in the Milky Way, the field may be magnificent; so enjoy it at your leisure.

Do not overmagnify. Stars stand high magnification better than do any other celestial objects because they are in very high contrast with the sky, and do not (unless the seeing is bad) become hazy and indistinct with a powerful eyepiece. But if the pair is seen well with a lower power, then that is the best power to use. A double such as β Cygni (Albireo), with a separation of 35″ arc, can just be separated with a magnification as low as ×10, and the stars will be well seen with ×50. At ×200 they are so far apart in the field of view that the best effect is lost. However, if you were examining the star 7 Tauri (see the list on page 214), to try to resolve it with a 200-mm telescope, a magnification of about ×250 would be the minimum worthwhile power, and ×350 would give your eye a better chance.

You can derive some idea of the minimum useful magnification for resolving any particular pair by remembering that an average human eye, without any artificial aid, apart possibly from spectacles, can just resolve two stars that are about 4' arc apart. So, taking this rule of thumb as a guide, and bearing in mind that it is only approximate, we should have to magnify a double until its components appear to be about 4' arc apart, before there is much chance of resolving it on the retina. Therefore a pair 4" arc apart will need at least ×60, a pair 1" arc apart ×240, and so on. In theory, to separate 7 Tauri (1985 separation 0.62" arc) should need a power of ×400. However, the telescope-aided eye does seem to resolve rather better than 4' arc apparent separation, perhaps because the telescopic image is superior to that formed by the crystalline lens of most eyes, so you may succeed in resolving close pairs with rather lower powers than this "law" implies.

Never forget that the rule assumes that the image has been resolved in the focal plane of the telescope. No amount of magnification can divide a pair if their separation is below the theoretical limit.* You will also find that a large telescope tends to resolve a given star more easily than will a small one, even with the same magnification, because it forms a smaller image of each star, assuming that seeing conditions allow it to achieve its full resolving power.

Unequal pairs There are instances where it may be necessary to use a magnification considerably higher than the mere separation of the stars would indicate as being necessary. These will happen if the components of the double differ in brightness by a couple of magnitudes or more, and are fairly close; or if the companion is quite distant but is very faint indeed, near the limiting magnitude of the telescope.

In the first case, the problem is to remove the companion to a sufficient distance from the primary to allow it to be seen more easily. Typically, star A will be surrounded by a fluctuating corona of scat-

---

* In very special cases, resolution may be improved superficially; see page 218.

tered light, that occasionally shrinks in moments of better seeing to allow a brief glimpse of B. A high magnification, by removing B as far as possible from A on the retina, reduces the dazzling effect and permits the fainter star to be seen more clearly in the steady moments, although it may be no more distinct than with a lower magnification at other times.

Sky-darkening The second case, that of a very faint, distant companion, is perhaps more interesting. The object of using a high magnification is not so much to remove B from the vicinity of A—it may be sufficiently far away already—but to darken the background of the sky, so bringing the faint star B into better visibility.

This sky-darkening effect of magnification is an important but little-appreciated aid to sighting faint objects. Why, to take an extreme case, are stars not visible with the naked eye in the daytime? The answer is that they are indeed detectable, in the sense that their light is affecting the retina—but the light of the sky produces so strong a retinal response that the tiny extra signal from the star passes unnoticed. As the sky darkens after sunset, this basic signal begins to be noticed over the foreground "noise."

In fact, it *is* possible to see bright stars during the daytime if a telescope is used. The reason for this is not that the telescope collects more light than the eye—if this were all it did, then the star would stand out no better than it does with the naked eye because the sky's brightness would be augmented in the same proportion as that of the star, resulting in no relative gain.* What the telescope does do is to allow the area of sky containing the star to be *magnified,* and this darkens the sky relative to the star.

Suppose that the field of view includes a circle of sky 1° across, when a magnification of ×50 is used. Suppose also that a telescope of 100 mm aperture is being used. In daylight, the aperture of the human pupil is about 2mm, and therefore the telescope has fifty times its diameter or 2500 times its lightcollecting area. But the magnification of ×50 also makes the selected 1° circle of sky appear to be fifty times the diameter, so that the skylight is spread over an area 2500 times as large. In theory, the brightness of the sky background will appear the same with a 100-mm telescope at ×50 as with the naked eye.

* There is, in fact, some gain, because the telescope (a) produces a larger and more definite star image than does the naked eye; and (b) helps to shield the eye from general daylight glare.

But if the magnification is doubled, to × 100, the field of view drops to about ½°. The amount of skylight from an area only one quarter as large is now being spread across the field of view of the eyepiece. Its intensity is, therefore, reduced by four. Because a star does not expand in proportion to the magnification used—certainly not at the low and middle range of the magnification scale, at any rate—it should now have four times the contrast with its background, and be that much easier to see. The same argument can be applied to night-time observing because the sky background is never totally black, and it can be suppressed by increasing the magnification, although the argument breaks down when very high powers are used and the star image begins to appear as a disk rather than as a point, or if the seeing is so poor that the stellar image is greatly expanded.

Strategies for difficult doubles If the principal star is very bright and glaring, you may find that the faint companion or *comes* (pronounced komez) is impossible to detect, even with a high magnification. Try putting star A just outside the field of view, in such a way that the line A–B is along a radius of the field, with B towards the center. The eyepiece needs to be a reasonably good one, so that the definition towards the margin is tolerable—but you should always use the best possible eyepiece if you want to do justice to your telescope.

A better way of hiding an awkward bright primary—whether it be a member of a double star, or Uranus or Neptune when their satellites are being sought—is to use a "field bar." This is simply an obstruction inserted in the eyepiece. It needs to be in sharp focus, in the same plane as the "field stop," which is a circular diaphragm giving you a nice sharp edge to the field of view when you look through the eyepiece (see illustration on facing page). Some eyepiece types have a field stop inside, between the lenses, and not much can be done with them (these are often referred to as "negative" eye-pieces, the common Huyghenian being a good example), but most eyepieces are "positive," and the field stop will be found a little way in front of the field lens. To cement a bar to this requires care, partic-ularly if it is a high-power eyepiece with a field stop only three or four millimeters across, but it can be done with the aid of some adhesive, a magnifying glass, and a pair of tweezers!

You can, for instance, cover half the field with a scrap of aluminum foil, or perhaps a fragment of red photographic filter (which means that the bright primary can still be seen, much dimmed). The writer

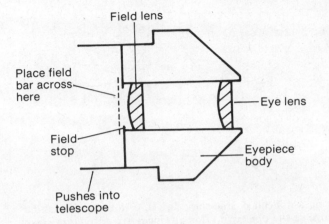

has a liking for a piece of hair stretched across a field diameter. Taking the thickness of a sample as 0.06 mm, it will subtend an angle of about 8″ arc when placed at the focus of a telescope of 1500 mm focal length, and correspondingly more or less if the focal length is smaller or greater than this value.

In the case of very close and difficult unequal pairs, that defy the use both of high magnification and a field bar, the following dodge has produced results. The telescopic image of a star, in a perfect telescope, is a tiny disk surrounded by very faint diffraction rings. These rings are caused by interference of light at the sharp edge of the telescope's object glass or mirror, and they are circular and even because the aperture is circular, with no corners or straight edges. However, if the objective is covered with a mask having six or eight straight edges (i.e., hexagonal or octagonal in shape), the diffraction rings are now converted into a corresponding number of spokes.

The difficulty in seeing a faint companion close by a bright star is not so much that it is involved in the image of the star—although it may be, if that image is smeared by an unsteady atmosphere—but that it is overlaid by the star's diffraction rings. The advantage of a polygonal mask is that the rings are now converted into rays. If you twist the mask so that the position of star B is in between two of these rays, there is a chance that it will be seen in the little cove of dark sky created by the mask.

Some reflecting telescopes have a four-vaned mounting to hold the Newtonian diagonal mirror or the Cassegrain secondary. Such a system will already be producing four faint spikes, and these can be brightened by superimposing a much wider cardboard cross over

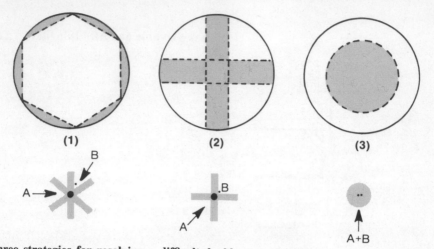

Three strategies for resolving a difficult double star In (1), a hexagonal mask has been placed around the edge of the objective. This produces six diffraction spikes around the image of bright star A, between two of which the faint companion B may be discerned. The coarse, opaque cross in (2) achieves a similar effect, but produces four spikes. Strategy (3) achieves a superficial improvement in resolving power by throwing so much diffracted light out the the star image and into the surrounding halo that the diameter of the spurious disk is reduced, possibly helping the resolution of a very close, bright, equal pair of stars.

them. In the case of a refractor, a simple expedient is to stretch two pieces of insulating tape perhaps 10 mm wide, in the form of a cross, in front of the lens. This will produce four bright diffraction spikes and four quadrants of darkness around the star. But the simple sticky-tape method has the disadvantage that the mask cannot be rotated to bring the cross and the companion into the best orientation, unless you stick the tape to a loose, rotatable ring around the object glass.

Another dodge, tried by observers of very close and *equal* pairs, is to cover the objective with an opaque disk of perhaps 50 percent or 75 percent of its full aperture, leaving only an annulus. The effect of this is to throw the majority of the light into greatly exaggerated diffraction rings, and to make the central spurious star disk appear faint and shrunken. By this means, two very close stars may be made to emerge from what otherwise would be a messy spot. But the stars must be equal in brightness, or very nearly so, for this resource to be successful, and the best diameter for the mask must be determined by experiment.

A cycle of double and multiple stars Here, then, is a list of 145 fine double and multiple stars scattered over the sky. They will not all be visible from any single site because some will be too far South, or

North, to rise above your horizon, but on any single night there will be enough to keep you occupied for some hours. The notes given against each object mention some interesting and important facts, but they cannot anticipate the most important thing of all, which is how you personally respond to them. Different individuals return again and again to their particular favorite pairs, so do not consider this list as something to be gone through and then ticked off as "done." At least, if you do that, then the doubles have not worked their magic— or perhaps you have not tried hard enough and spent long enough with each one.

Write notes about them. What is the best magnification to use? What nearby stars lie in the field? What is the best way of star-hopping to find it, if it is faint? And the thorniest question of all: how you would describe the colors.

Stars have different colors, but it does not follow from this that the tints recorded by an observer correspond with those of nature! The effect of contrast is one difficulty: if you place a white star near a red star, the white star will appear bluish. "Yellow and blue" and "orange and blue" pairs abound in the sky. Probably in most cases the "blue" star would appear white if seen on its own, for truly bluish stars (the hottest known) are rare. However, this is not the point: we are not trying to classify the stars by type or temperature, but to describe how they *appear*. If a star appears bluish, try to describe the sort of blue, and stand by your decision!

But remember too that instrumental influences must be taken into account, as well as the bias of the eye. It is sometimes difficult, for example, for individuals to agree upon whether certain shades fall into the "green" or "blue" category. This is partly a matter of personal opinion, but it may also be due to a lack of discrimination in the retinal response. This, and the telescope's own color bias, can have a huge influence upon star colors. William Herschel, as the great 19th-century observer T. W. Webb noted, was "partial to red tints." This could, possibly, have been due to the metallic mirrors in his telescopes, which tended to reflect long-wavelength light particularly well, or again it could have been for purely physiological reasons.

The matter of telescope bias becomes obvious when a star is viewed through a refracting telescope and then through a reflector of similar aperture. An achromatic lens scatters a disproportionate amount of blue and violet light into the secondary spectrum, away from the star image, which therefore acquires a yellowish tinge; a reflector, in theory, gives a slightly bluish tinge because it reflects red light the least well of all the tints to which the human eye is sensitive. On the other hand, some refractors are less "yellow" than others. We cannot draw general conclusions that have much application to specific instances, and, no matter how many times a double star has been observed before, it will never have been examined by the combination of eye and telescope that is now about to be turned upon it.

In the following list, which is divided into two parts listing stars to the North and South of the celestial equator respectively, the right ascension and declination are correct for the year 2000. Separations and position angles, if undated, may be taken as showing no significant change over many years. Otherwise, predictions for selected future epochs up to the year 2000 are given, and in some cases the orbit is represented by a diagram.

NORTHERN DOUBLE STARS

η Cas (Σ 60)
00ʰ 48ᵐ.8, + 57° 50′

AB 3.6, 7.2; 12″.2; 310°
This binary has curious and difficult colors: to the writer, A is cream-colored and B has a port-wine tint. The period is about 480 years, and the stars' true separation is equal to twice the distance of Neptune from the Sun.

β 1 Cas
00ʰ 52ᵐ.5, + 56° 37′

AB 7.9, 9.9; 1″.4; 82°
AC 7.9, 8.9; 3″.8, 133°
AD 7.9, 9.4; 8″.9; 194°
AE 7.9, 12.1; 15″.7; 332°
The stars all appear to be fixed in position.

36 And (Σ 73)
00ʰ 55ᵐ.0, + 23° 38′

AB 6.0, 6.4; 0″.66; 278° (1985)
            0″.75; 292° (1990)
            0″.84; 304° (1995)
            0″.94; 313° (2000)
The primary is yellowish, and the distance is increasing; period 165 years.

γ Ari (Σ 180)
01ʰ 53ᵐ.5, + 19° 18′

AB 4.8, 4.8; 7″.7; 360°
A splendid pair, bright and easy. Being so similar in appearance, the stars look like a binary pair; but a very slow approach of B to A is believed to be due to their independent motion through space. The stars are white.

Σ 186 Cet
01ʰ 55ᵐ.8, + 01° 51′

AB 6.8, 6.8; 1″.14; 55° (1985)
            1″.04; 57° (1990)
            0″.94; 59° (1995)
            0″.83; 62° (2000)
A yellowish pair, period 155 years. The stars are now closing.

α Psc (Σ 202)
02ʰ 02ᵐ.0, + 02° 46′

AB 4.2, 5.2; 1″.67; 277° (1985)
            1″.48; 264° (2000)
This binary pair has a period of about 700 years, and the stars are expected to be at their closest (0″.98 apart) in 2074. The primary is a hot A-type star, and would be expected to appear white. However, observers often have difficulty in describing the colors of this pair, and A has been called numerous different tints, from greenish-yellow to brown!

γ And (Σ 205)
02ʰ 03ᵐ.5, + 42° 23′

A(BC) 2.3, 5.1; 9″.8; 64°
BC 5.5, 6.3; 0″.45; 105° (1985)
A magnificent double star. The wide pair, A and BC, undoubtedly form a binary system, but the motion is so slow that the pair have remained in practically the same position for over two centuries, shining with glorious colors of deep yellow and blue. The close pair BC forms a binary with a period of fifty-five years. At the present time it should be resolvable with apertures of about 250 mm, but it is rapidly closing, and by the year 2000 star C will be sweeping past B at a distance of only 0″.05, passing through 180° of position angle in just a couple of years!

ι Tri (Σ 227)
02ʰ 12ᵐ.4 + 30° 18′

AB 5.2, 6.6; 3″.9; 72°
Yellow, bluish. A binary pair, but the motion is so slow that the period must be thousands of years. A superb, delicate object.

ι Cas (Σ 262)
02ʰ 29ᵐ.0, + 67° 24′

AB 4.7, 7.0; 2″.5; 233°
AC 4.7, 8.3; 7″.3; 115°
A glorious triple star with colors of yel-

Circumpolar star trails   A 20-minute exposure using a 35-mm focus lens. The Pole Star is at the center, and the two other bright stars in Ursa Minor are near the top of the picture. (*N. W. Scott*)

low-white, bluish, and deeper blue. The close pair AB constitute a binary system with a period approaching a thousand years. The third star, C, belongs to the same system, but its orbital period must be immense. It is also worth pointing out that Star A is a spectroscopic binary, too close to be resolved in any telescope, with a period of about fifty years.

α UMi (Σ 93) *Polaris*
02ʰ 31ᵐ.6, + 89° 16′

AB 2.1, 9.1; 18″.4; 220°
AC 2.1, 13.1; 44″.7; 83° (1890)
AD 2.1, 12.1; 82″.7; 172° (1890)
An optical pair, practically fixed. In good conditions, this double can readily be seen with a 60-mm refractor, but in unsteady seeing the companion may be invisible with twice the aperture. Observers with larger apertures could tackle the rarely reported companions C and D, discovered by Burnham with the 470-mm refractor at Chicago, but apparently not measured since. The distance of AC may have increased by several seconds of arc.

γ Cet (Σ 299)
02ʰ 43ᵐ.3, + 03° 14′

AB 3.6, 7.4; 2″.8; 295°
White or yellowish primary; there is some disagreement over the color of the companion. This is a slow-moving binary pair of very long period. Delicate, but well seen with a 90-mm refractor.

20 Per (Σ 318)
02ʰ 53ᵐ.7, + 38° 20′

(AB)C 5.3, 10.0; 14″.2; 237°
The yellowish primary is itself a very close binary (AB), with a period of about thirty years. The stars of this binary are never more than 0″.3 apart (mags. 5.6, 6.7). Star C forms a delicate test object for apertures of about 75 mm.

ε Ari (Σ 333)
02ʰ 59ᵐ.2, + 21° 20′

AB 5.2, 5.5; 1″.6; 205°
A very slow binary star, and a useful test object for smaller instruments.

7 Tau (Σ 412)
03ʰ 34ᵐ.5, + 24° 28′

AB 6.6, 6.7; 0″.62; 4° (1985)
　　　　　　0″.70; 358° (2000)
AC 6.6, 10.0; 22″.4; 54°
An interesting close binary, with a period of about 200 years. Star C will be found difficult with apertures of less than 150 mm or so.

ζ Per (Σ 464)
03ʰ 54ᵐ.1, + 31° 53′

AB 2.9, 9.5; 12″.6; 208°
AC 2.9, 11.3; 32″.8; 286°
AD 2.9, 9.5; 96″; 194°
A white star with several faint companions. A is a young and massive supergiant, some 10,000 times as luminous as the Sun, a member of a very loose cluster or *association* of stars in this part of the sky that came into existence only a few tens of millions of years ago. The companions are all optical, and make interesting light-tests.

ε Per (Σ 471)
03ʰ 57ᵐ.8, + 40° 01′

AB 2.9, 8.2; 8″.8; 10°
A delicate pair, the primary star being a clear white. Relatively fixed.

α Tau *Aldebaran*
04ʰ 35ᵐ.9, + 16° 31′

AB − 0.85, 11.3; 128″; 33° (1985)
A famous pair. The primary is a red giant about one hundred times as luminous as the Sun, and about 65 light-years away. The optical companion appears to be increasing in distance from A at the rate of about 13° arc per century, due to the proper motion of Aldebaran across the celestial sphere. It has been seen by keen eyes under good conditions with apertures of about 70 mm.

ω Aur (Σ 616)
04ʰ 59ᵐ.3, + 37° 53′

AB 5.1, 8.1; 5″.4; 359°
A neat, small pair. They evidently constitute a binary, but the motion is very slow.

Hu 445 Tau
05ʰ 01ᵐ.7, + 20° 49′

AB 8.6, 8.9; 0″.42; 278° (1985)
A test object for apertures of about 300 mm. Although the period is given as eighty-two years, there will be practically

no change of distance by the year 2000, although the position angle will increase to 300°.

λ Ori (Σ 738)
05^h 35^m.1, + 09° 56′

AB 3.4, 5.6; 4″.4; 43°
AC 3.4, 11.2; 28″.6; 184°
The primary is one of the hottest naked-eye stars in the sky, with a surface temperature of about 25,000°C. Some astronomers have suggested that it began shining only a few tens of thousands of years ago, in which case it is a true stellar infant, even though several times as massive as the Sun. Star B is also hot and bluish. The distant C has been detected with apertures of 150 mm, but a field bar will be of assistance. Although AB is believed to form a binary system, no orbital motion has been noticed.

θ Aur (OΣ 545)
05^h 59^m.7, + 37° 14′

AB 2.6, 7.2; 3″.5; 310°
AC 2.6, 10.7; 54″; 300° (1985)
AB is a most delicate and difficult object with a small aperture, worth the attempt only on a very steady night. With a 90-mm object glass, the writer has glimpsed the companion as a thickening of the diffraction ring: it might be a good candidate for the polygonal mask technique. It is evidently a very slow binary system, with companion C (optical) moving away at a rate of about 8″ arc per century.

8 Mon (Σ 900)
06^h 23^m.8, + 04° 36′

AB 4.3, 6.7; 13″.4; 27°
AC 4.3, 12.7; 94″; 254°
Both bright stars are white or bluish-white. They seem to form a binary system, although the motion has not yet allowed its very long period to be determined. Owners of telescopes in the 200-mm range may care to hunt for the optical companion C, discovered by Burnham. 8 Mon is a splendid object in any telescope, and the fine field includes two wide pairs of 8th and 9th magnitude stars.

12 Lyn (Σ 948)
06ʰ 46ᵐ.3, + 59° 27′

AB 5.4, 5.6; 1″.7; 76°
AC 5.4, 7.2; 8″.7; 308°
A beautiful triple, A appearing slightly yellowish compared with B, and well visible with an aperture of 100 mm. The binary AB has a period of several centuries, and C is also believed to be a member of the system.

α Gem (Σ 1110) *Castor*
07ʰ 34ᵐ.6, + 31° 53′

AB 1.9, 2.9; 2″.6; 83° (1985)
         3″.0; 74° (1990)
         3″.4; 67° (1995)
         3″.8; 62° (2000)
AC 1.9, 9.3v; 72″.5; 164°
This is one of the most notable binaries of the northern sky, and it has been well observed since its discovery in 1718. Surprisingly, its orbital period is still not known with certainty, although a revolution time of about 500 years seems likely. In 1969 the components passed within 1″.9 of each other. Companion C, an eclipsing binary that changes from magnitude 9.1 to 9.6 in a period of 19½ hours, is also a member of the system, and both A and B are spectroscopic binaries, so that the Castor system consists of six stars.

κ Gem (OΣ 179)
07ʰ 44ᵐ.5, +24° 24′

AB 3.7, 8.2; 7″.3; 240°
The primary is a fine yellow. This pair has not shown much change, but is probably a binary of long period. Herschel considered that B might be a planet, reflecting the light of A! A difficult object with less than 100 mm or so.

ζ Cnc (Σ 1196)
08ʰ 12ᵐ.2, +17° 40′

AB 5.6, 6.0; 0″.67; 239° (1985)
        0″.55; 185° (1990)
        0″.61; 122° (1995)
        0″.81; 83° (2000)
AC 5.6, 6.3; 6″.0; 80°
A fascinating triple star, the close pair AB having a period of only sixty years, in which time its distance varies from 0″.54

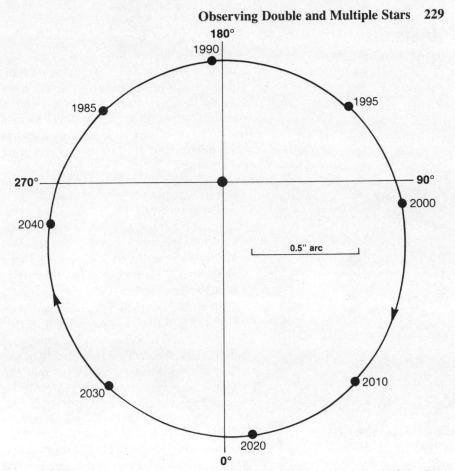

The orbit of ζ Cancri AB   The plane of the orbit is only 8° away from a perfect plane view.

(1991) to about 1″.2. With the third star C so close, this is a bright and beautiful object, AB appearing golden yellow, and C —unusually—white. Even when closest, the stars should be resolvable with an aperture of about 220 mm.

φ² Cnc (Σ 1223)
08ʰ 26ᵐ.8, + 26° 56′

AB 6.3, 6.3; 5″.1; 218°
A fine white pair. To the writer, and according to most catalogues, the stars are of equal brightness, but some observers have recorded a difference of up to half a magnitude between them. Practically fixed: probably a very long-period binary.

ι Cnc (Σ 1268)
08ʰ 46ᵐ.7, + 28° 46′

AB 4.2, 6.6; 30″.5; 307°
A beautiful object, wide and easy, with colors of yellow and blue. It can be divided with × 12 binoculars. The stars are moving through space in the same direction, and therefore, presumably, form a binary system, but no orbital motion has been detected in 1½ centuries of accurate observation!

σ² UMa (Σ 1306)
09ʰ 10ᵐ.4, + 67° 08′

AB 5.0, 8.2; 1″.8; 45°
A most interesting binary star. Observations made between its discovery in 1783, when the separation was 7″ arc, and 1887, when it had closed to 2″, suggested that star B might be traveling independently in a straight line, and that the pair formed a "near miss" optical type. Since then, however, B has swept round A at a distance of only 1″, and is now receding, but only slowly. The period is probably of the order of one thousand years, so that the greatest separation, about 11″, will be reached around the year 2500!

38 Lyn (Σ 1334)
09ʰ 18ᵐ.9, + 36° 49′

AB 3.9, 6.6; 2″.8; 229°
The primary is white. This close pair can just be divided with an aperture of 100 mm, but it is more distinct with larger instruments. The position angle is reducing at a very slow rate—about 5° per century.

Σ 1338 Lyn
09ʰ 21ᵐ.0, + 38° 12′

AB 6.5, 6.7; 1″.06; 262°
A slow binary, with a period of about 400 years. By 2000, the position angle will have increased to 289° with no change of distance.

ω Leo (Σ 1356)
09ʰ 28ᵐ.5, + 09° 04′

AB 5.9, 6.7; 0″.46; 40° (1985)
            0″.49; 59° (1990)
            0″.54; 75° (1995)
            0″.62; 88° (2000)
A yellow binary with a period of 117 years. Its separation varies from 0″.18 (as

in 1956) to about 1″.1, which it will reach around the year 2038. At present, it forms a fine test object for apertures of about 250 mm. It is interesting to note that the English observer W. H. Smyth, using a 150-mm object glass, was able to elongate ω Leonis into an "egg," and measure its position angle as well as estimate its separation, when the stars were only 0″.4 apart in 1843.

γ Leo (Σ 1424)
10ʰ 20ᵐ.0, + 19° 51′

AB 2.2, 3.5; 4″.3; 124°
One of the finest bright binary stars, now approaching its widest opening in an orbit that may take some six centuries to complete. The stars are very slow-moving. It is wide enough to be resolved with a small telescope, and the colors, as seen by the writer, are fine crocus yellow and a greenish yellow.

54 Leo (Σ 1487)
10ʰ 55ᵐ.6, + 24° 45′

AB 4.5, 6.3; 6″.5; 111°
A gem: a beautiful combination of pure white and a rich blue-gray. Presumably a binary, but practically fixed.

ξ UMa (Σ 1523)
11ʰ 18ᵐ.2 + 31° 33′

AB 4.3, 4.8; 2″.3; 91° (1985)
             1″.3; 60° (1990)
             1″.1; 317° (1995)
             1″.8; 273° (2000)
A very fast-moving binary, with a period of sixty years. The maximum separation (3″.1) occurred in 1974, and they will be at their minimum distance apart (0″.84) in 1993. This cream-colored binary was the first ever to have its orbit determined, and at present it can be divided by a very modest instrument. (See page 232.)

ν UMa (Σ 1524)
11ʰ 18ᵐ.5, + 33° 05′

AB 3.7, 10.1; 7″.2; 147°
A very delicate, unequal pair. The primary is a lovely full yellow, but the com-

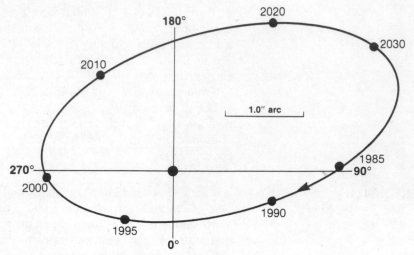

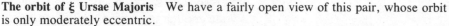

The orbit of ξ Ursae Majoris is only moderately eccentric. We have a fairly open view of this pair, whose orbit

panion may be found difficult to see with an aperture of less than 150 mm or so. The stars are relatively fixed.

ι Leo (Σ 1536)
11ʰ 23ᵐ.8, + 10° 32′

AB 4.1, 7.0; 1″.3; 142° (1985)
        1″.4; 132° (1990)
        1″.5; 124° (1995)
        1″.7; 116° (2000)
A binary star with a period of about 200 years. The stars were at their closest (0″.61) in 1942. The primary is white. At the present time, this will probably be a severe test for a 200-mm telescope.

Σ 1694 Cam
12ʰ 49ᵐ.3, + 83° 24′

AB 5.3, 5.8; 21″.6; 326°
A lovely easy white pair, relatively fixed.

α CVn (Σ 1692) *Cor Caroli*
12ʰ 56ᵐ.1, 38° 18′

AB 2.9, 5.4; 19″.6; 228°
A bright and easy fixed pair. Often described as yellow, the writer sees them as white and "warm" white. This is one of the few stars to have received a "modern" appellation: *Charles' Heart,* applied by Edmund Halley after the restoration of the British monarch, Charles II, in 1660.

ζ UMa (Σ 1744) *Mizar*
13ʰ 23ᵐ.9, + 54° 56′

AB 2.4, 4.0; 14″.5; 150°
One of the most famous pairs in the sky, with the 4th-magnitude star *Alcor* 11½′ arc away, forming an easy naked-eye pair in a fairly dark sky. A and B show little motion, but presumably do form a binary system of very long period. They suffer from being too well known, and are not often examined critically. To the writer, the pair seems white; others call it yellowish.

κ Boo (Σ 1821)
14ʰ 13ᵐ.5, + 51° 47′

AB 4.6, 6.6; 13″.5; 236°
A beautiful pair, yellowish and bluish. Practically fixed.

π Boo (Σ 1864)
14ʰ 40ᵐ.7, + 16° 26′

AB 4.9, 5.8; 5″.6; 108°
An attractive white pair, with a 10th-magnitude star about 2′ arc to the South. Practically fixed.

ζ Boo (Σ 1865)
14ʰ 41ᵐ.1, + 13° 44′

AB 4.6, 4.6; 1″.03; 304° (1985)
　　　　0″.97; 303° (1990)
　　　　0″.89; 301° (1995)
　　　　0″.80; 300° (2000)
A yellowish pair. This is one of the most remarkable binaries in the sky; the orbital eccentricity of its components is very nearly as great as that of Halley's comet. During the 123-year period, the separation varies from 1″.16 (as in 1959) to only 0″.03 (which will next happen in 2021), and during this closest approach the position angle of the two components will be changing at a rate of about 1° per day! However, the pair will then be beyond the resolving power of any visual telescope. An 11th-magnitude star lies about 100″ arc to the West. (See page 234.)

ε Boo (Σ 1877)
14ʰ 45ᵐ.0, + 27° 04′

AB 2.7, 5.1; 3″.0; 339°
This pair is famous as Struve's *Pulcherrima*—the most beautiful of all his double stars. It is a binary of very long (so far

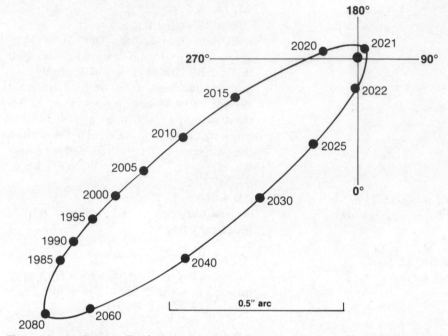

The orbit of ζ Boötis   During one revolution, the true distance between this pair of stars change by a factor of 45.

incalculable) period, and has opened by only 0″.4 since Struve's time, 150 years ago. This pair has been divided with only 40 mm. The golden primary is in spectacular contrast with the blue-green companion, although a reasonable aperture is necessary to show the tints well. A magnitude 12 star lies about 3′ arc to the West.

ξ Boo (Σ 1888)
14ʰ 51ᵐ.4, + 19° 07′

AB 4.8, 6.7; 7″.2; 329° (1985)
          6″.6; 318° (2000)

The primary is light yellow, the companion tawny or warm gray. A binary lying only 22½ light-years away from the Sun, this pair has a period of 150 years, and at the moment they are near their widest separation.

η CrB (Σ 1937)
15ʰ 23ᵐ.2, + 30° 18′

AB 5.6, 5.9; 0″.86; 8° (1985)
          1″.08; 26° (1990)

1″.02; 42° (1995)

0″.78; 63° (2000)

A close binary, consisting of two almost equal yellow stars, orbiting in a period of 41½ years. They are now approaching maximum separation, after passing minimum separation (0″.37) in 1979. A star of magnitude 12.6, 66″ arc to the North, will be an interesting light-test for small telescopes.

μ ² Boo (Σ 1938)
15ʰ 24ᵐ.5, + 37° 21′

AB 7.2, 7.8; 2″.2; 14°

This double is the fainter component of the wide binocular pair μ Boötis (magnitudes 4.5 and 7.0, separated by 108″ arc). The period of the binary is about 260 years, and at present it is near maximum separation.

δ Ser (Σ 1954)
15ʰ 34ᵐ.8, + 10° 32′

AB 4.2, 5.2; 3″.9; 179°

A very attractive pair of white stars, showing only slow relative motion, and easy with almost any telescope.

ζ CrB (Σ 1965)
15ʰ 39ᵐ.4, + 36° 38′

AB 5.1, 6.0; 6″.3; 305°

A charming white pair, practically fixed.

π ² UMi (Σ 1989)
15ʰ 39ᵐ.7, + 79° 58′

AB 7.4, 8.2; 0″.62; 26° (1985)

A very close and difficult binary, with a period of about 170 years. It is now near its widest opening, having been unresolvable even with very large telescopes when the components were at their closest, around 1902. The orbit is rather uncertain.

σ CrB (Σ 2032)
16ʰ 14ᵐ.7, + 33° 52′

AB 5.8, 6.7; 6″.8; 234° (1985)

7″.1; 237° (2000)

An attractive, easy binary, with a 9th-magnitude pair in the same high-power field, to the East. When discovered in 1781 the components were less than 2″ arc apart, and rapidly closing. Now they are slowly widening. The orbit is not yet known very accurately, but the period may be about a thousand years.

ζ Her (Σ 2084)
16ʰ 41ᵐ.4, + 31° 36′

AB 3.0, 6.5; 1″.40; 113° (1984)
1″.48; 102° (1986)
1″.54, 92° (1988)
1″.58; 83° (1990)
1″.58; 74° (1992)
1″.52; 64° (1994)
1″.37; 53° (1996)
1″.11; 39° (1998)
0″.92; 28° (1999)
0″.71; 10° (2000)

A very difficult, unequal binary in rapid motion, the period being only thirty-four years; when widest, round about 1991, an aperture of 100 mm might be able to show the companion as a bulge on the disk of A. This phase is now approaching, and the star is well worth careful attention on a really steady night, under the highest

The orbit of ζ Herculis   The orbit of this pair is not particularly eccentric, and we see it at a fairly open angle.

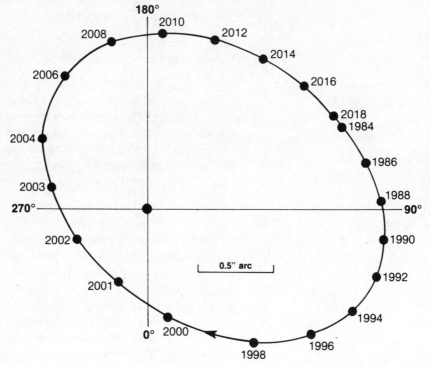

magnification you can usefully apply. The primary has a yellowish tinge. The closest approach of the two stars (0″.49) will occur in the late summer of 2001.

20 Dra (Σ 2118)
16ʰ 56ᵐ.3, + 65° 02′

AB 7.1, 7.3; 1″.3; 68° (1985)
        1″.4; 67° (2000)
A white pair, with a period of about 700 years.

μ Dra (Σ 2130)
17ʰ 05ᵐ.4, + 54° 27′

AB 5.8, 5.8; 1″.9; 33° (1985)
        1″.9; 7° (2000)
A yellowish, equal pair. This is a binary of long period; the stars are now passing through their closest phase, but should be divisible with an aperture of 75 mm, using a high power.

α Her (Σ 2140)
17ʰ 14ᵐ.6, + 14° 23′

AB 3.5, 5.4; 4″.6; 107°
A well-known double, with superb tints of gold and greenish-gray. It can be divided with a very small telescope, but the fine contrast may not be obvious. There is an 11th-magnitude companion about 80″ arc to the northeast. Although a binary, the stars appear to be fixed.

δ Her (Σ 3127)
17ʰ 15ᵐ.0, + 24° 50′

AB 3.1, 8.7; 9″.4; 270°
A cream-colored star with an unusually close optical companion. In 1780, the stars were separated by 34″ arc, and they attained minimum separation (8″.9) in 1965.

ρ Her (Σ 2161)
17ʰ 23ᵐ.7, + 37° 08′

AB 4.5, 5.5; 4″.2; 317°
A beautiful pair, white and dull white, well seen even with a small telescope. The stars form a binary system, but the orbital motion is very slow.

40 Dra (Σ 2308)
18ʰ 00ᵐ.1, + 80° 00′

AB 5.8, 6.2; 19″.0; 232°
A lovely cream-colored pair of stars, attractive with either low or high magnification: low power gives the best effect, while a high power reveals the delicate tints better. A low-power field includes a wide 9th-magnitude pair to the southeast.

95 Her (Σ 2264)
18ʰ 01ᵐ.5, + 21° 36′

AB 5.1, 5.2; 6″.3; 258°
A very neat pair, white and yellowish. A
binary, but practically fixed.

70 Oph (Σ 2272)
18ʰ 05ᵐ.4, + 02° 32′

AB 4.3, 6.3; 2″.21; 293° (1984)
             1″.92; 275° (1986)
             1″.62; 250° (1988)
             1″.55; 219° (1990)
             1″.81; 191° (1992)
             2″.3; 173° (1994)
             2″.8; 161° (1996)
             3″.4; 153° (1998)
             3″.9; 147° (2000)

A fine swift binary pair, with a period of
eighty-eight years. Because its orbit appears large, and its motion is rapid, this is
one of the best-observed of all binaries,
having been followed at all points of its
orbit for more than two revolutions. A
great deal is therefore known about this
system, which lies a mere 16.7 light-years
away. The primary, that appears yellowish, has the same mass as the Sun but only
about half its luminosity, being somewhat
cooler and redder, while its companion
has 0.7 of the Sun's mass and only 8 percent of its luminosity. The true distance
between the stars ranges from 11.7 astronomical units at *periastron* (October 1983)
to 35 units at *apastron*, which will next
occur in August 2027. At the present time,
the stars are closing up; they will be difficult with anything less than 100 mm by
1990, but almost any small telescope will
reveal them ten years later. By 2020, the
distance will have increased to 6″.8.

There are numerous faint companions
lying within 2–3′ arc of 70 Ophiuchi; observers with apertures of about 250 mm or
more could search for a magnitude 14 star
about 80″ arc to the West–northwest.

OΣ 358 Her
18ʰ 35ᵐ.9, + 16° 59′

AB 6.8, 7.0; 1″.5; 158° (1985)
             1″.3; 148° (2000)

A slow-moving, yellowish binary, with a period of about 290 years.

α Lyr *Vega*
$18^h 37^m.0, + 38° 46'$

AB 0.0, 9.5; 74″; 180° (1985)
AC 0.0, 11.0; 81″; 266° (1985)
The faint optical companion B of this brilliant star is an interesting test of vision and technique, but in good conditions it should be readily visible with an aperture of about 150 mm. It may help to block Vega with an occulting bar, or to use a high magnification and put it out of the field, because the ratio of apparent brightness of A and B is about 7000:1! Vega is only 25 light-years away, and its proper motion has almost doubled the distance to B since they were measured in 1800.

You may also care to search for a much fainter companion, C, that does not appear to have been measured for many years, but which should be found in approximately the position given above.

The orbit of 70 Ophiuchi   The true distance between the stars of this pair changes by a factor of 3 during one revolution. The orbit is seen from an angle of about 30° above the orbit plane.

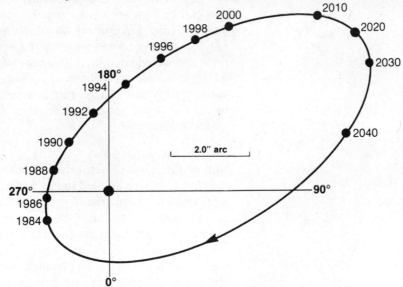

ε¹ & ε² Lyr (Σ 2382 & 2383)
18ʰ 44ᵐ.3, + 39° 40'

ε¹ AB 5.1, 6.0; 2".6; 354°
ε² AB 5.2, 5.5; 2".3; 81°
The famous *Double-double:* two very similar close pairs separated by 208" arc. Therefore, they form a severe test of naked-eye resolution, those observers who can distinctly divide the two stars without optical aid being in the minority. But a magnification as low as ×2 makes them easily visible. Both binary systems are very slow-moving, and the suggested periods of 1200 years for ε¹ and 600 years for ε² are far from certain.

Almost directly between these two pairs lies a star of about the 11th magnitude, with another, a little brighter, about 50" arc away in a southwesterly direction. These are worth hunting with apertures of about 100 mm, but in a good sky they should be readily seen with anything much larger. Other still fainter stars lie in the field.

θ Ser (Σ 2417)
18ʰ 56ᵐ.1, + 04° 12'

AB 4.5, 4.5; 22".4; 104°
A superb, almost equal pair in a fine field of stars. The stars are practically fixed, and can be well seen with any telescope. There is some uncertainty about the magnitudes of these components; to the writer they appear equal, but differences of up to a magnitude have been recorded, so that possibly one of the stars is variable; naked-eye observations have also suggested variability.

β Cyg *Albireo*
19ʰ 30ᵐ.7, + 27° 58'

AB 3.2, 5.4; 34".3; 54°
One of the most colorful double stars in the sky. It can be divided in the finder, or with ordinary binoculars, provided they are of good quality and firmly mounted. Yet for all their brightness there is no unanimity about the tints of the stars, particularly of the secondary, which some observers call a full blue while others de-

scribe as greenish. The presence of the bright primary influences color estimates, of course; try using a high magnification and hiding A behind a field bar. The blue/green decision is a notoriously difficult one because visual discrimination in this region of the spectrum is poor.

Albireo is probably a binary system, although the stars show no relative motion. The field, like many in this region of the Milky Way, is very fine.

δ Cyg (Σ 2579)
19ʰ 45ᵐ.0, + 45° 07′

AB 3.0, 7.9; 2″.2; 232° (1980)
            2″.2; 218° (2000)
AC 3.0, 12.0; 64″; 66°
This is a notoriously difficult binary; the light of A makes the companion hard to detect, even though it is no closer than many other much easier pairs, and it will probably elude any aperture smaller than 100 mm. Some observers have claimed that twilight is the best time for looking at this star. The period is very long, about 500 years, and even at maximum separation (predicted by one authority for the year 2225) the distance will be only 3″.0. Companion C will probably be a severe challenge for apertures in the 300-mm range.

ε Dra (Σ 2603)
19ʰ 48ᵐ.2, + 70° 16′

AB 4.0, 7.6; 3″.1; 20°
A delicate pair, and not too easy with apertures below 100 mm. The off-white primary gives the companion a pleasing blue appearance. Practically fixed.

π Aql (Σ 2583)
19ʰ 48ᵐ.7, + 11° 58′

AB 6.1, 6.9; 1″.4; 108°
A close, relatively fixed pair, and a good test object for an 80-mm telescope; if in doubt about its elongation, compare it with the appearance of χ Aquilae, about 2° to the West. The primary has a cream tint.

ψ Cyg (Σ 2605)
19ʰ 55ᵐ.6, + 52° 26′

AB 4.9, 7.4; 3″.2; 177°
The primary is a fine white, and there are several other pairs in a high-power field. Little relative motion.

θ Sge (Σ 2637)
20ʰ 09ᵐ.9, + 20° 55′

AB 6.4, 8.9; 12″.0; 325°
AC 6.4, 7.3; 88″; 222° (1985)
An attractive triple star, although the third star is rather wide. The golden primary lies in a magnificent field, with bright stars and groups. AB has shown little motion, but C is being left behind at a rate of about 12″ arc per century.

α Cyg *Deneb*
20ʰ 41ᵐ.5, + 45° 17′

AB 1.2, 11.7; 75″; 106°
A bright star with a very faint companion. As with other examples of its type, such as Vega, a field bar will help to reduce the glare and darken the surrounding sky sufficiently for the second star to have a chance of being detected. A 9th-magnitude star lies to the East, in a medium-power field. Deneb, despite its brilliance, is remote; it lies almost 2000 light-years from the Sun, and is a supergiant of about 60,000 times the Sun's luminosity.

γ Del (Σ 2727)
20ʰ 46ᵐ.7, + 16° 08′

AB 4.5, 5.5; 9″.5; 267°
A splendid pair, with colors of yellowish and white as seen by the writer, although there is much disagreement over the tint of the companion. The stars are very gradually approaching each other, but too slowly to give an idea of the orbit.

ε Equ (Σ 2737)
20ʰ 59ᵐ.1, + 04° 18′

AB 6.0, 6.3; 1″.04; 285° (1985)
            0″.82; 285° (2000)
(AB)C 5.5, 7.1; 10″.5; 70°
A triple system, the wide pair (AB)C of which is attractive in a very small telescope. The binary AB is a pair with a period of 101 years, the plane of whose orbit is almost edge-on to the line of sight.

When closest, in 1920, they were beyond the reach of any telescope, and they will rapidly close again after the year 2000. At the present time, they can be tackled with apertures larger than about 120 mm, and could possibly be elongated with a 75-mm refractor. A fainter pair, of about the same separation as (AB)C, lies some 10' arc to the southeast.

61 Cygni (Σ 2758)
$21^h 06^m.3, + 38° 39'$

AB 5.5, 6.4; 29".4; 147°
A wide and easy binary, with a period of many centuries, interesting not so much for its appearance, which is attractive enough, as for what this pair of stars is: among the Sun's nearest neighbors, only 11.1 light-years away, and the very first stars to have their distance accurately measured, by observing their annual parallax against more distant stars in the same field of view. This was done by Bessel in 1838. When looking at these two stars, reflect also on the fact that the distance between them is over twice the average distance of Pluto from the Sun. Both the stars are red dwarfs, about ¼₀th as luminous as the Sun, and a slight "corkscrew" effect in the path of one of the stars has prompted the suggestion that an invisible body, of planetary mass (eight times that of Jupiter, possibly) could belong to the system, orbiting once every 4.8 years. There is certainly much to reflect upon when looking at 61 Cygni!

υ Cyg
$21^h 17^m.9, + 34° 54'$

AB 4.4, 10.0; 15".1; 220°
AC 4.4, 10.0; 21".5; 182°
A white star with two faint companions, making an interesting test for modest telescopes. Both B and C are quoted as being of the same magnitude, but C, being the more distant from the bright star, will almost certainly be the easier to see.

μ Cyg (Σ 2822)
21ʰ 44ᵐ.1, + 28° 45′

AB 4.7, 6.1; 1″.7; 302° (1985)
           1″.5; 306° (1990)
           1″.4; 312° (1995)
           1″.2; 320° (2000)

A binary with a highly inclined orbit and a period of about 500 years. At present it appears as a neat white pair, well seen with an aperture of 100 mm or so, but it will require a larger telescope than this when it closes to 0″.82 in 2017. When discovered, in 1777, the components were at their greatest possible separation, about 6″.5.

Σ 2840 Cep
21ʰ 52ᵐ.0, + 55° 47′

AB 5.5, 7.3; 17″.9; 196°
A most beautiful double, clear white and pale blue, in a low-power field that includes several faint pairs. The whole effect is superb. The stars are believed to be an optical pair.

ξ Cep (Σ 2863)
22ʰ 03ᵐ.7, + 64° 37′

AB 4.6, 6.5; 8″.0; 275°
AC 4.6, 12.7; 100″; 200°
This very slow-moving pair is probably a binary of very long period. It can be resolved with almost any telescope, but is best seen with a medium power. The writer notes a splendid contrast of pale yellow and blue-green. Very few observations of C have been reported, and it would be worth a hunt with moderate apertures.

8 Lac (Σ 2922)
22ʰ 35ᵐ.8, + 39° 38′

AB 5.8, 6.6; 22″.4; 186°
AC 5.8, 10.5; 48″.8; 169°
AD 5.8, 9.3; 82″; 144°
A wide quadruple star, easy with an aperture of 100 mm. Star A is white and AB forms a finder double with a near neighbor.

σ Cas (Σ 3049)
23ʰ 58ᵐ.9, + 55° 45′

AB 5.0, 7.1; 3″.0; 326°
A neat, fixed, white and blue pair, in a fine field.

SOUTHERN DOUBLE STARS

κ¹ Scl
00ʰ 09ᵐ.4, − 28° 00′

AB 6.2, 6.3; 1″.6; 260°
A close pair of stars of almost identical brightness. They presumably form a very slow-moving binary system, and should currently be easily resolvable with an aperture of 75 mm.

β Tuc
00ʰ 31ᵐ.6, − 62° 58′

AC 4.4, 4.5; 27″.1; 169°
CD 4.6, 6.0; 0″.53; 305° (1985)
       0″.50; 276° (2000)
A magnificent fixed pair of white stars, each of which is double, although AB (4.4, 13.6; 2″.4; 151°) is far beyond the range of ordinary instruments. They are not particularly remote—only 110 light-years away—and the stars are hot main-sequence objects, about a hundred times as luminous as the Sun. Component β² (star C) is a close binary, with a period of 43 years.

β 395 Cet (82 Cet)
00ʰ 37ᵐ.2, − 24° 46′

AB 6.5, 6.5; 0″.51; 97° (1985)
       0″.61; 101° (1986)
       0″.69; 104° (1987)
       0″.74; 107° (1988)
       0″.78; 109° (1989)
       0″.78; 111° (1990)
       0″.75; 114° (1991)
       0″.69; 116° (1992)
       0″.30; 119° (1993)
       0″.46; 124° (1994)
       0″.33; 132° (1995)
       0″.17; 156° (1996)
       0″.13; 234° (1997)
       0″.28; 272° (1998)
       0″.42; 283° (1999)
       0″.51; 289° (2000)
The yellowish stars of this close pair form a binary with a period of only 25 years, and with a moderate telescope its rapid closing and opening again can be followed

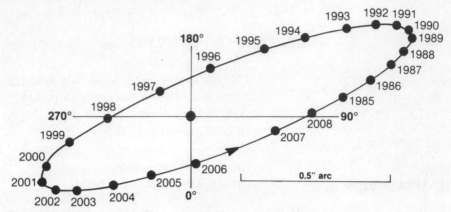

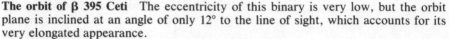

The orbit of β 395 Ceti The eccentricity of this binary is very low, but the orbit plane is inclined at an angle of only 12° to the line of sight, which accounts for its very elongated appearance.

on almost an annual basis, although even a large amateur instrument may do no more than elongate it when the components make their closest passage. At their widest, they should be divisible with an aperture of 150 mm. This binary lies only about 40 light-years from the Sun, and its large proper motion is shifting it relative to the field stars at a rate of about 1½″ arc per year, equivalent to the apparent diameter of the Moon in just over a millennium.

β Phe
01ʰ 06ᵐ.1, −46° 43′

AB 4.0, 4.2; 1″.6; 315°
AC 4.0, 11.5; 61″; 52°
A splendid yellowish pair, but the stars have not been observed for a sufficient length of time to allow the orbit to be determined, and the measure for AB is uncertain. The faint star C is an optical companion.

42 Cet (Σ 113)
01ʰ 19ᵐ.8, −00° 29′

AB 6.4, 7.4; 1″.2; 30°
A slow binary that has been observed through only a small arc of its orbit, and the separation has increased only slightly in 1½ centuries. B is a very close binary, with a separation of only 0″.1.

p Eri
01ʰ 39ᵐ.7, − 56° 12′

AB 5.8, 5.8; 11″.2; 194°
A splendid yellowish binary star. The period is uncertain, but probably approaching 500 years. Never a difficult object, the stars are now approaching their widest separation of 11″.8 in 2039. When discovered by the Australian amateur Dunlop, in 1825, they were at their closest, 3″.4 apart.

66 Cet (Σ 231)
02ʰ 12ᵐ.7, − 02° 24′

AB 5.8, 7.6; 16″.2; 232°
A pretty pair. The writer sees them as white and lilac, though others have described A as yellow. Presumably the stars form a binary system, but they are practically fixed. A 10th-magnitude pair follows in a low-power field, together with several other stars.

o Cet *Mira*
02ʰ 19ᵐ.3, − 02° 58′

AB var, 13.0; 68″; 70°
AC var, 9.3; 120″; 75°
The primary is one of the most famous variable stars in the sky, extreme magnitude range 2–9, period about 330 days. It is a close binary, with a magnitude 9.7 companion beyond the reach of ordinary instruments.

θ Eri
02ʰ 58ᵐ.3, − 40° 18′

AB 3.4, 4.5; 8″.2; 89°
A fine, bright, easy pair, relatively fixed.

α For (h 3555)
03ʰ 12ᵐ.0, − 29° 00′

AB 3.9, 6.5; 3″.1; 306°
A binary star with a period of 155 years. It is now at its widest opening, and will change little over the next twenty years. This should be readily resolvable with an aperture of 70 mm.

32 Eri (Σ 470)
03ʰ 54ᵐ.3, − 02° 57′

AB 5.0, 6.3; 6″.8; 347°
Evidently a binary, but relatively fixed. Rich yellow and blue-green: lovely colors, and an easy object.

ι Pic
04ʰ 50ᵐ.9, − 53° 28′

AB 5.6, 6.4; 12″.3; 58°
An easy fixed pair, discovered by Dunlop, the amateur who first observed p Eridani.

β Ori (Σ 668) *Rigel*
05ʰ 14ᵐ.5, − 08° 12′

AB 0.1, 8?; 9″.5; 202°
Although there is a large magnitude difference between these two stars, they are
sufficiently wide apart to be readily seen
with an aperture of 75 mm; some observers, using stops over larger apertures,
have managed to glimpse the companion
with as little as 30 mm. In fact, the writer
has the feeling that the companion is
much brighter than the magnitude 10.4
quoted in the 1963 *Index Catalogue of
Double Stars,* for it is as easy to see as the
companion to Polaris, which is magnitude
9 and much more distant from a fainter
primary. Atmospheric unsteadiness, particularly when the object never appears
very high in the sky (as is the case when
it is viewed from northern European or
Canadian latitudes), is the principal cause
of difficulty.

Rigel is one of the most brilliant known
stars, similar to Deneb in Cygnus but
rather closer, lying at a distance of about
900 light-years. The companion is, presumably, optical.

η Ori
05ʰ 24ᵐ.4, − 02° 24′

AB 3.8, 4.8; 1″.6; 78°
A very slow-moving, close pair. Both
stars are white. It is probably beyond the
ability of a 75-mm telescope to show clear
division, but such an instrument should
elongate the pair.

θ Ori (Σ 748) *Trapezium*
05ʰ 35ᵐ.3, − 05° 23′

The famous Trapezium lies within the
Great Nebula of Orion, where four stars
of magnitude 5–8 form a distinctive group
in the brightest part of the nebula. Other,
much fainter stars are also present, offering interesting tests. It is worth pointing
out that photographs of the Orion Nebula
very rarely show the Trapezium because
the stars are "burned out" in the overex-
posure necessary to show the nebulosity.
The visual observer can enjoy both at the
same time!

Instead of listing distances and position angles, a diagram of the group is given here. It shows the four principal stars A–D, together with the much fainter stars E and F, the magnitudes being as follows:

A	6.7*		D	6.8
B	7.9*		E	11.1
C	5.4		F	11.5

Both E and F have been claimed with apertures of 75 mm, although most mortals will require more than this. Because the nebular background is bright, a high magnification will help to bring them into

* Stars A and B are both variable. A fades to magnitude 7.7, and rises again, over a 19-hour period every 65 days, while B fades to magnitude 8.5 and recovers in the same 19-hour interval every 6½ days.

The θ Orionis group The very faint stars G and H have never, it is believed, been recorded with apertures of less than about one meter. Their positions are included here out of interest, the magnitude of each being estimated as 16.7.

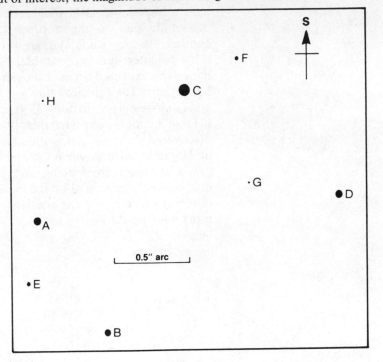

view. The 380-mm (15-inch) refractor of Harvard Observatory is reported to have revealed star E in *full daylight;* observers with equatorial mountings in good adjustment would find it interesting to try to locate θ Orionis in a blue sky.

ι Ori (Σ 752)
05ʰ 35ᵐ.4, −05° 55′

AB 2.9, 7.0; 11″.3; 141°
A delicate pair, white and grayish. The stars appear to be relatively fixed, and lie in the same low-power field as θ, being involved in nebulosity. This is a wonderful region of the sky.

σ Ori (Σ 762)
05ʰ 38ᵐ.7, −02° 35′

AC 3.7, 10.3; 11″.1; 237°
AD 3.7, 7.5; 12″.8; 84°
AE 3.7, 6.5; 41″.6; 61°
Even a small instrument shows this as a superb quadruple group, in a fine field. The wide pair AE is an easy binocular object; a high-power field includes an 8th-magnitude pair to the northwest, with another 8th-magnitude star as well. The three brighter stars appear bluish—perhaps to be expected because they are very hot objects. The principal star A is in fact a very close binary, period 125 years, and it is an extraordinary fact that Burnham discovered this pair, AB, with a refractor of 315 mm aperture, when the stars were only 0″.26 apart, by noticing that the star disk looked elliptical. For the benefit of observers with moderate and large apertures who would like to tackle this challenging star, β 1032 Ori, some positions are given here:

AB 4.1, 5.1; 0″.17; 72° (1985)
0″.20; 53° (1990)
0″.23; 38° (1995)
0″.24; 26° (2000)

ζ Ori (Σ 774)
05^h 40^m.7, −01° 57′

AB 2.0, 4.2; 2″.5; 164°
This pretty double is so easy with an 80-mm telescope that Herschel's failure to spot the companion with his powerful instruments is all the more surprising. However, there seems no doubt that B must have been present because the pair of stars are practically fixed with respect to each other. The primary is another of the very hot O-type supergiants to be found in this interstellar "cradle"; the companion has achieved notoriety through Struve's attempt at describing its tint, for which he had to devise the expression *olivacea subrubicunda*. To the writer, they appear white and gray!

52 Ori (Σ 795)
05^h 48^m.0, +06° 27′

AB 6.1, 6.1; 1″.6; 212°
Practically fixed, yellowish. A useful pair, not difficult to find, and at the resolution limit of a 75-mm telescope.

β Mon (Σ 919)
06^h 28^m.9, −07° 03′

AB 4.7, 5.2; 7″.3; 132°
BC 5.2, 6.1; 2″.8; 106°
AD 4.7, 12.2; 25″.9; 56°
Also known as 11 Monocerotis, this is a princely white triple, with an 8th-magnitude star to the northwest. To the writer, one of the most glorious sights in the sky, and far excelling some brighter and more obvious pairs. Burnham added the fourth and very difficult star D, which might be worth chasing with large apertures. The triple system is almost certainly physically connected, and lies about 700 light-years from the Sun.

α CMa *Sirius*
06^h 45^m.2, −16° 35′

AB −1.4, 8.7; 8″.7; 38° (1984)
7″.6; 31° (1986)
6″.2; 21° (1988)
4″.5; 4° (1990)
2″.9; 328° (1992)
2″.7; 259° (1994)
3″.5; 210° (1996)

4″.0; 176° (1998)
4″.6; 150° (2000)

A famous binary, with a period of fifty years. Sirius itself is the brightest star in the sky, only 8.8 light-years from the Sun and about twenty-two times as luminous. Within its glare circles the dim companion B, a white dwarf star of degenerate matter that has now collapsed into a planetary-sized but intensely hot body, with a surface temperature estimated at about 24,000°C—over twice as hot as Sirius and four times as hot as the Sun.

This companion is notoriously elusive. It was discovered in 1862 with a 470-mm refractor, at which time it was near its greatest elongation from A; and at such a time, which last occurred in 1973, observ-

The orbit of Sirius, α Canis Majoris　The orbit plane is inclined at about 45° to the line of sight. It is worth pointing out that B is the more massive of the two stars, so that it is more correct to say that A revolves around B than that B—as represented here—revolves around A. In fact, the stars of any binary system revolve around their mutual center of gravity.

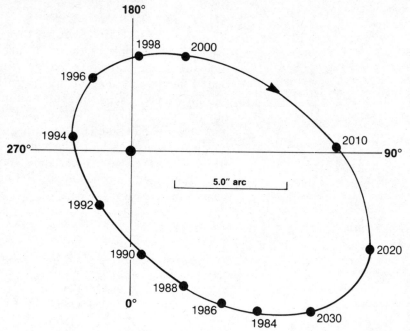

ers using apertures as small as 200 mm have reported seeing it. The use of special polygonal masks, or diffraction crosses, is mentioned on page 219. The distance is now closing rapidly, and the next closest approach will be in 1993, when, on the basis of previous close passages, no telescope in the world will be able to reveal it.

μ CMa (Σ 997)
06ʰ 56ᵐ.1, − 14° 03′

AB 5.3, 8.6; 3″.0; 340°
The primary is reddish, and the companion is faint and close. There is little relative motion.

γ Vol
07ʰ 08ᵐ.7, − 70° 30′

AB 3.8, 5.7; 13″.8; 299°
An easy pair, showing no relative motion; curiously, the components are sometimes designated γ¹ (the fainter) and γ² (the brighter), even though they cannot possibly be resolved with the naked eye. Star A is a yellowish giant star, while its companion is a hotter main-sequence object.

h 3945 CMa
07ʰ 16ᵐ.6, − 23° 19′

AB 4.8, 6.8; 26″.5; 52°
A double star noted for its colors of "fiery" red and vivid blue.

n Pup
07ʰ 34ᵐ.3, − 23° 28′

AB 5.9, 6.0; 9″.6; 114°
Also known as Hh 269, a reference to John Herschel's catalogue of William Herschel's double stars. A pair of almost equal stars, easy and attractive. Little relative motion.

κ Pup (h 273)
07ʰ 38ᵐ.8, − 26° 34′

AB 4.5, 4.7; 9″.9; 318°
A splendid pair of white stars, relatively fixed.

γ Vel
08ʰ 09ᵐ.5, − 47° 20′

AB 1.8, 4.5; 41″.2; 220°
A wide and easy pair, divisible with binoculars. Take careful note of A, for it is one of the hottest stars known; it belongs to the rare Wolf-Rayet class of stars with luminous atmospheres and "surface"

temperatures of 40,000°C or more. It is, therefore, one of the bluest stars in the sky. A and B are relatively fixed.

I 314 Pyx
08ʰ 39ᵐ.4, − 36° 36′

AB 6.5, 7.6; 0″.50; 233° (1985)
        0″.39; 232° (1987)
        0″.26; 229° (1989)
        ....
        0″.27; 253° (1998)
        0″.36; 249° (2000)

A close and difficult binary, the orbit being seen almost edge-on. At closest approach, in 1994, the separation will be less than 0″.05, and the pair will not be divisible with any instrument, but observers with large apertures may enjoy the challenge before and after this time. The period of the system is about sixty-seven years.

υ Car
09ʰ 47ᵐ.1, − 65° 04′

AB 3.0, 6.1; 5″.0; 127°
An easy, pretty double. The stars show no relative motion.

γ Sex
09ʰ 52ᵐ.5, − 08° 06′

AB 5.6, 6.1; 0″.56; 72° (1985)
        0″.63; 63° (2000)

A very close binary, discovered by the famous American telescope-maker Alvan Clark a few years before detecting the companion to Sirius in 1862. The period is seventy-six years, and they were last at their closest in 1958.

β 411 Hya
10ʰ 36ᵐ.1, − 26° 40′

AB 6.7, 7.5; 1″.2; 320° (1985)
        1″.4; 315° (2000)

A white binary pair with a period of about 210 years, now widening after their close passage in 1948.

β Hya (h 4478)
11ʰ 52ᵐ.9, − 33° 54′

AB 4.8, 5.6; 0″.7; 48° (1985)
Measures of this pair made since 1834 have suggested that this is an optical pair, but it is now considered equally likely to be a binary seen almost in the plane of the orbit. The position given here is based on the former assumption, and it will be well

worth examining, and if possible measuring, with moderate instruments.

ε Cha (h 4486)
11ʰ 59ᵐ.6, −78° 13′

AB 5.5, 6.1; 0″.9; 188° (1941)
A binary star, but its orbit has not yet been determined, and there seem to have been no very recent measures of it. This pair, like the previous one, is worth looking at; the distance may have decreased, and the position angle may have increased.

D Cen
12ʰ 14ᵐ.0, −45° 43′

AB 5.6, 6.8; 2″.9; 244°
The primary is of a deep yellow tint. This fine close pair has shown no relative motion.

α Cru *Acrux*
12ʰ 26ᵐ.6, −63° 06′

AB 1.4, 1.9; 4″.1; 114°
AC 1.4, 5.1; 90″; 202°
A magnificent pair of white stars, resolvable with a small telescope, and in very slow orbital motion. A 5th-magnitude star makes the group triple, though it is very wide.

γ Cen
12ʰ 41ᵐ.5, −48° 58′

AB 2.9, 2.9; 1″.5; 355° (1985)
           1″.4; 353° (1990)
           1″.2; 351° (1995)
           1″.0; 347° (2000)
A magnificent binary star, with a period of eighty years, moving in towards their next closest approach in the period 2013–2018. In view of their brilliance, a neutral filter may help resolve them more easily. (See page 256.)

γ Vir (Σ 1670)
12ʰ 41ᵐ.7 −01° 27′

AB 3.5, 3.5; 3″.5; 293° (1985)
           3″.0; 288° (1990)
           2″.5; 280° (1995)
           1″.9; 268° (2000)
It is interesting to realize that, by the year 2050, the effect of precession will have brought both this great binary and γ Centauri to exactly the same great circle of right ascension. Even now, if the two

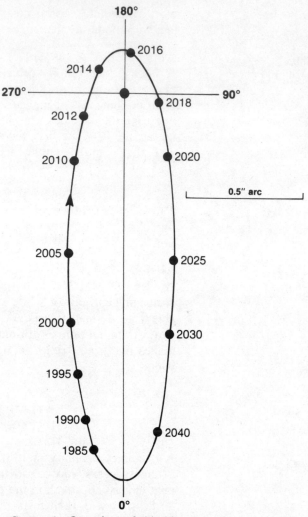

The orbit of γ Centauri   Our view of this binary is from a considerably inclined angle, only 22° from the orbit plane. The true orbit, also, is highly eccentric, the distance between A and B varying by a factor of nine.

stars are above the horizon, and you swing the telescope the necessary 47½° northwards around the declination axis after bringing γ Centauri to the center of the field of view, its splendid counterpart in Virgo should be found in the same medium-power field of view because their

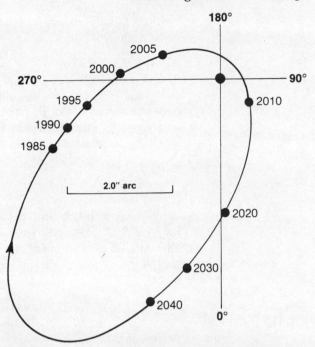

The orbit of γ Virginis   We are only 35° away from a plane view of this binary, whose true separation changes by about sixteen times in the course of a revolution. Notice the great increase in angular velocity at the time of periastron; between the years 2005 and 2010 it will move through a position-angle arc of 180°.

right ascensions differ by only 13 seconds of time. If this does not happen, then the polar axis of the mounting is in need of adjustment!

In the writer's opinion, γ Virginis is a finer object than its northern rival Castor (α Geminorum). Certainly its orbital motion is more interesting. The components take 172 years to perform one orbit of such eccentricity that the distance between these splendid yellow-white suns (lying only 32 light-years away) varies from 0″.3 to 6″.0. When they performed their last rapid close passage, in the winter of 1835–36, no telescope then existing could divide them. The pair achieved their widest separation in 1919, and are now closing rapidly as their next close approach, in 2008, draws near.

ι Cru (h 4547)
12ʰ 45ᵐ.6, − 60° 59′

AB 4.7, 8.5; 26″.8; 10°
A fine yellowish star with a wide, faint companion. There is some disagreement over the magnitude of B, one list giving it a value of 7.8, another preferring 9.5. Little change in distance.

β Mus
12ʰ 46ᵐ.1, − 68° 06′

AB 3.7, 4.0; 1″.3; 33°
A binary with a period of about 380 years. It is closing only very slowly, but the position angle will have increased to 43° by the year 2000. This pair should be resolvable with an aperture of about 90 mm, and should be a fine sight in larger instruments.

I 83 Cen
12ʰ 56ᵐ.7, − 47° 41′

AB 7.4, 7.6; 0″.76; 223° (1985)
         0″.99; 245° (2000)
A slow binary, with a period of about 300 years. These stars should be readily resolvable with instruments of 150- to 200-mm aperture.

θ Vir (Σ 1724)
13ʰ 10ᵐ.0, − 05° 32′

AB 4.4, 9.4; 7″.1; 343°
AC 4.4, 10.4; 70″; 298°
A white star with two faint companions. Curiously, John Herschel and James South, two famous early 19th-century double-star observers, found B a "very severe" test for a 75-mm aperture refractor, whereas today it can readily be seen with such an aperture. The distance cannot have changed; perhaps the star is variable? Star C can also be seen, with rather more difficulty, with a small telescope.

k Cen
13ʰ 51ᵐ.8, − 33° 00′

AB 4.7, 6.2; 7″.7; 107°
A fine pair of white stars in very slow motion. Amazingly, this pair was discovered by William Herschel, in 1783, from southern England, where its greatest altitude above the horizon could only have been

6°! Since that time, precession has carried it about 1° further South, so that a modern observer in a latitude of 51° North might care to try to emulate this remarkable feat.

φ Vir (Σ 1846)
14ʰ 28ᵐ.1, −02° 14′

AB 5.0, 9.5; 4″.8; 110°
A very delicate pair, the primary being of a pale yellow tint. It has been claimed as a test for a 75-mm instrument, but most observers would probably prefer a larger aperture than this. The stars are relatively fixed.

α Cen
14ʰ 39ᵐ.6, −60° 50′

AB 0.0, 1.4; 21″.2; 212° (1985)
19″.8; 214° (1990)
16″.3; 218° (1995)
14″.2; 222° (2000)
At a distance of only 4.4 light-years from the Sun, the α Centauri complex is our closest stellar neighbor. It is, therefore, pleasing that it should include a superb binary system. Star A is very similar to the Sun; star B is slightly dimmer, and the two orbit each other in eighty years. Their distance apart varies from 11.5 astronomical units to 36 units (i.e., from beyond Saturn to beyond Neptune, taking the Sun as one of the stars), but we view this moderately eccentric orbit from a very acute angle, and their distance apart in the sky ranges from only 1″.7 (as in 1957) to 21″.8 (as in 1980). Therefore the double is now closing up, but the stars will be resolvable in a very small telescope for the rest of the century.

The third star of the system is a red dwarf some 25,000 times less luminous than either star A or the Sun. It lies over 2° away, to the southwest, and appears of only the 11th magnitude at R.A. 14ʰ 09ᵐ.7, −62° 42′ (2000). The true distance from AB to C (Proxima Centauri) is about 30,000 astronomical units, or 750 times

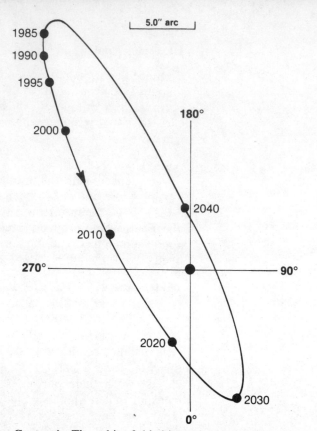

The orbit of α Centauri The orbit of this binary is not particularly eccentric, but it is viewed from an angle of only 11° above the orbit plane, which accounts for its elongated appearance.

the mean distance of Pluto from the Sun, and the orbital period of the system has been estimated at a million years. Reflecting upon the enormous angular distance between these gravitationally connected objects, and the shrunken appearance of the myriads of other stellar systems in the sky which are, perhaps, equally capacious but immensely far away, does begin to give one a sense of closeness on the one hand and remoteness on the other: of the enormous scale of space.

μ Lib (β 106)
14h 49m.3, −14° 09′

AB 5.8, 6.7; 1″.9; 5°
AE 5.8, 12.5; 27″.3; 232°
A close pair with three very faint companions, only one of which (E) is likely to be within the range of even a fair-sized telescope. To sight it, the diffraction-mask

dodge mentioned on page 219 may be put to good use. The close pair is in very slow motion, and no orbit has, as yet, been determined.

h 4707 Cir
14^h 54^m.2, −66° 25′

AB 7.6, 7.9; 0″.67; 320° (1985)
0″.72; 308° (1990)
0″.79; 298° (1995)
0″.87; 290° (2000)

The primary of this very slow binary is yellowish. The period is believed to be about 300 years.

π Lup (h 4728)
15^h 05^m.1, −47° 03′

AB 4.7, 4.8; 1″.6; 65°
A magnificent close pair of white stars. Their orbit has not yet been determined because the motion is very slow, although the distance has doubled since their discovery in 1835.

k Lup
15^h 12^m.0, −48° 45′

AB 4.1, 6.0; 26″.8; 144°
Little change: a bright and easy white pair.

μ Lup (h 4753)
15^h 18^m.5, −47° 52′

AB 5.1, 5.2; 1″.0; 130°
AC 5.1, 7.2; 23″.7; 130°
A triple star, but the principal pair AB are now very close; they certainly form a binary system, although their motion is not yet clear.

γ Cir (h 4757)
15^h 23^m.4, −59° 19′

AB 5.2, 5.5; 0″.48; 303° (1985)
0″.57; 293° (1990)
0″.65; 286° (1995)
0″.73; 281° (2000)

This bright and close pair has not been sufficiently well observed for a reliable orbit to be computed, but the above prediction is the best available, which suggests that the pair could be tackled with a 150-mm telescope by the end of the century. The period is about 180 years.

ξ Sco (Σ 1998)
16^h 04.^{m}.4, −11° 22′

AB 4.9, 4.9; 1″.04; 30° (1985)
0″.94; 34° (1987)
0″.82; 40° (1989)
0″.67; 48° (1991)

0".49; 61° (1993)
0".27; 94° (1995)
0".19; 203° (1997)
0".30; 289° (1999)
0".47; 320° (2001)

AC 4.8, 7.2; 7".9; 46°

A splendid triple star, consisting of a close binary (period forty-six years) and a third star not far away, which evidently is a physical member of the system. This object lies in a fine field, enhanced by an 8th-magnitude pair (Σ 1999) about 12" arc apart and 4' arc to the South. The stars of

The orbit of ξ Scorpii AB   The orbit of this fast-moving pair appears considerably more eccentric than it really is, due to the inclination at which we view it.

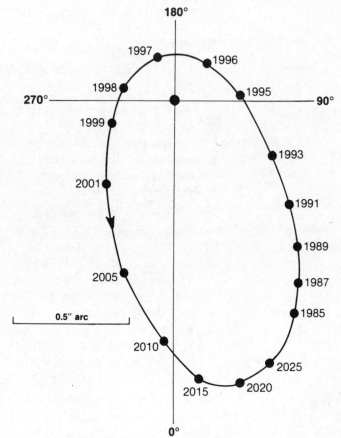

AB are now closing, and a detailed ephemeris has been given for the benefit of observers with large apertures who may care to try to divide or elongate the binary during its closest passage

β Sco
16ʰ 05ᵐ.4, − 19° 48′

(AB)C 2.9, 5.1; 13″.6; 24°
Beautiful contrast of colors distinguishes this easy pair, which to the writer appear off-white and bluish. This star, the northernmost of the three distinctive stars marking the Scorpion's pincers, lies only a degree away from the ecliptic and is frequently occulted by the Moon. The primary AB is a very close and difficult double, not resolvable with amateur instruments. Some catalogues give C a magnitude of only 6.9, which seems to be much too faint.

ν Sco
16ʰ 12ᵐ.1, − 19° 27′

AB 4.4, 6.9; 0″.9; 3°
CD 6.9, 7.9; 2″.6; 55°
A most interesting multiple group of four stars. The wide pair (AB) (CD) can be resolved in binoculars or the finder. Each component is then found to be a close telescopic double, CD being detectable with an aperture of 60 mm or so, and AB requiring at least twice this. The two pairs are 41″ arc apart, and CD is very slowly widening.

ρ Oph
16ʰ 25ᵐ.6, − 23° 27′

AB 5.2, 5.9; 3″.3; 341°
A beautiful, neat pair of yellowish stars, with little difference of brightness between them, although most catalogues indicate that A is about 0.7 magnitude brighter than B. The pair are slowly approaching each other, but no orbit has yet been determined. Prettily set between two stars of about magnitude 7½, that are visible in the finder and lie about 2½′ arc away from ρ.

α Sco *Antares*
16ʰ 29ᵐ.4, −26° 26′

AB 1.2, 5.4; 2″.5; 276° (1985)
              2″.3; 276° (2000)
A very well-known double. The pair form a binary system of very long period, perhaps a thousand years. Although the secondary star is quite bright, it tends to be obscured by the flare and brilliance surrounding the primary; and the writer's best views have been obtained in moonlight or strong twilight, which reduces the glare. The rich color of the giant primary (slightly variable) makes the companion appear a distinct green, which is, of course, an effect of contrast.

η Oph (β 1118)
18ʰ 10ᵐ.4, −15° 44′

AB 3.0, 3.5; 0″.44; 260° (1985)
              0″.50; 253° (1990)
              0″.55; 248° (1995)
              0″.60; 244° (2000)
A fine close binary, with a period of eighty-eight years. A most interesting challenge for owners of instruments in the 200- to 300-mm range.

36 Oph
17ʰ 15ᵐ.4, −26° 33′

AB 5.1, 5.1; 4″.7; 152°
A lovely pair of yellow stars, forming a binary system with a period of perhaps 500 years, and the motion is very slow. This is a beautiful sight in almost any instrument.

τ Oph (Σ 2262)
18ʰ 03ᵐ.0, −08° 11′

AB 5.3, 6.0; 1″.8; 278° (1985)
              1″.7; 283° (2000)
AC 5.3, 9.3; 100″; 127°
A close, neat pair, with colors of yellowish and white; star C belongs to the same system, while A is known to be a spectroscopic binary, so that a total of four stars are here gravitationally bound. The pair AB has a period of about 280 years, the distance varying from 0″.23 in 1832 to 2″.1 when at their widest in 1931. They are now very slowly closing.

h 5014 CrA
18ʰ 06ᵐ.8, − 43° 25′

AB 5.8, 5.8; 1″.2; 6° (1985)
0″.9; 346° (2000)
A close binary pair, now closing. The period is about 190 years.

κ CrA
18ʰ 33ᵐ.4, − 38° 44′

AB 5.9, 6.6; 21″.4; 359°
A wide pair of white stars, relatively fixed.

γ CrA (h 5084)
19ʰ 06ᵐ.4, − 37° 04′

AB 5.0, 5.1; 1″.35; 134° (1985)
1″.31; 55° (2000)
A fine binary star, with a period of 120 years. We see the orbit almost in plane view, and the separation ranges from 1″.3 to 2″.5. The stars are now at their closest, and varying little in distance, but the position angle is reducing rapidly.

I 253 Sgr
19ʰ 19ᵐ.0, − 33″ 17′

AB 7.6, 7.7; 0″.64; 139° (1985)
0″.50; 138° (1990)
0″.42; 138° (1992)
0″.30; 137° (1994)
0″.22; 137° (1995)

. . . .

A binary, period sixty years, seen from an angle of only 2° away from the plane of the orbit; this means that the stars seem to approach each other practically in a straight line, and virtually occult each other at closest passage in 1996–97. With a large aperture it will be possible to watch the two stars merging from season to season.

α Cap
20ʰ 18ᵐ.0, − 12° 33′

α ¹ AB 4.2, 9.5; 45″.5; 221°
α ² AB 3.8, 11.2; 6″.6; 180°
A wide naked-eye pair, separated by 6′ arc. Note that α¹ is the fainter and more westerly of the two. This is an easy object. α² has been successfully observed with apertures of 150 mm, but it would normally be a good test of larger instruments used with ordinary eyes.

η PsA (β 276)
22ʰ 00ᵐ.8, −28° 27′

AB 5.8, 6.8; 1″.7; 115°
A neat white and bluish pair, just divisible with an 80-mm telescope. These stars constitute a very slow binary system.

ζ Aqr
22ʰ 28ᵐ.8, −00° 02′

AB 4.4, 4.6; 1″.8; 216° (1985)
    1″.9; 207° (1990)
    2″.0; 199° (1995)
    2″.1; 191° (2000)
A very slow, beautiful, cream-colored binary system. The two stars have currently been making their closest approach in an orbit estimated to take 8½ centuries to complete, and they now form a superb sight in an instrument of 150 mm in aperture, although they can be resolved in a smaller telescope. After the year 2000 the distance will rapidly increase, and it will become a very easy object.

(*Note* The epoch dates in this chapter refer to the *beginning* of the year in question rather than to the middle. This distinction can be important in the case of fast-moving pairs.)

CHAPTER 10

Observing Clusters, Nebulae, and Galaxies

Double and multiple stars lend themselves well to the attentions of a systematic observer who likes to compare and contrast the separate personalities (as it were) amongst a vast throng of individuals. There are no two double stars alike, and yet they all have obvious features in common, whether of movement or fixity, or of magnitude difference, or of color contrast.

However, most amateurs seem to spend more time observing the other classes of "deep-sky" objects, the clusters and nebulae in our own Milky Way galaxy, and the other galaxies of the universe. This is an amorphous group indeed, not only in nature but also in appearance. It embraces brilliant star-groups like the Pleiades and tiny, glimpsed smudges against the starry background of a high-power eyepiece. This great range of sensations is almost too great to be encompassed within one chapter—or even, perhaps, one observer!

For, just as your everyday actions and choices are influenced by your personality, so, surely, is your choice of objects to observe through the telescope. It is an interesting and apparently as yet uninvestigated question: what does prompt some observers to root out elusive deep-sky objects and others to look at planets, as their main preference in astronomy? After all, amateurs are free to look at what they like, and, external circumstances excepted (such as sky conditions and telescopic equipment) the choice is theirs. What prompts this choice?

267

This is not a question that the writer or the reader is expected to answer. All that can be said, with some definiteness, is that deep-sky objects offer a challenge altogether different from that of the Sun, Moon, and planets. It may be the challenge of seeing all the objects in a list, or it may be that of detecting objects near the limit of your telescope. It could also be the more reflective satisfaction of seeing something unimaginably remote or huge.

The writer originally thought of dividing this section into its logical classes of star clusters, true nebulae, and remote galaxies. But, after some thought, it seemed best to mix them all together, and to assume that any or all of these will be fair game during your first review of the deep-sky wonders. This is not to imply that the observing technique is the same for all, because it certainly is not. But, when putting your telescope and eye through their paces, the different approaches and skills involved in getting the most out of the different objects will develop more comprehensively; and you will have the refreshing experience, if you follow the list, of coming across a fine bright object that almost seems to dazzle the eye after hunting for an elusive ghost that is sometimes there and at other times not. A continuous diet of faint galaxies could begin to pall!

Classification and catalogues It is an unfortunate legacy of the way observational astronomy developed, that the various objects to be examined in this chapter have rarely been catalogued in the orderly manner enjoyed by double stars. This is easy to understand. Some remote, dim clusters looked nebulous in the telescopes used when the first lists were drawn up, while many galaxies are telescopically indistinguishable from patches of true nebulosity. More significantly, the very existence of other galaxies in the visible universe was in doubt until the 1920s, and by that time the major catalogues had already been drawn up.

In fact, only two catalogues have popular international currency. The first appeared in 1781, and was a list of 103 nebulae and star clusters observed by the French comet-hunter Charles Messier. Messier had come across these objects more or less incidentally in the course of his comet sweeps, and his reason for noting some of them was to prevent them from being confused with the objects of his searches. However, some obviously noncometic objects (such as the Pleiades star cluster) were also included, so we may suppose that he did have an interest in compiling as comprehensive a list as possible of all the nebulae and star clusters visible from his site in Paris. His observations were carried out with various telescopes, but none of them, apparently, larger than about 100 mm in aperture. Therefore, all the objects in Messier's list are visible with modest instrumental means, and it is not surprising that this compendium is so popular with amateurs. The total now comes to 110, the additions being objects referred to in Messier's notes but not included in the original list. Messier objects are prefixed by the letter M; the Orion Nebula, for example, is M42.

The 110 Messier objects (or, more accurately, 108, because the 91st and 102nd in his list have never been satisfactorily identified) form the backbone of the list that follows in this chapter. They can be subdivided into the following groups:

Open clusters of stars	27
Globular clusters of stars	29
Nebulae	12
Asterism (coarse star-group)	1
Galaxies	39

Because all these objects were either discovered or confirmed from Paris, they all rise above the horizon for observers in midnorthern latitudes, and can also be seen from all widely inhabited regions in the Southern Hemisphere. The most southerly, M7 in Scorpius, has a declination of $-35°$, so that even this object can be reasonably well seen from the northern United States. It is not surprising that "Messier Clubs" have been formed to encourage observers to cover the whole list and gain membership!

The other widely used catalogue, which appeared in 1888, is far more comprehensive. Known as the *New General Catalogue of Nebulae and Clusters of Stars,* it includes 7,840 objects distributed over the whole sky, most of which were recorded visually using Newtonian-type reflectors. Every single one of these objects should be observable with a modern instrument of about 350 mm in aperture, assuming of course that it is well placed in the observer's sky. The writer is not aware of anyone who has succeeded in doing this, but perhaps someone, somewhere, is laboring at what literally would be the project of a lifetime. Objects listed in this catalogue are prefixed by the letters NGC; thus, the Orion Nebula has also been listed as NGC 1976, but it is usual to use a Messier number if there is one. The NGC was followed a few years later by two *Index Catalogues,* that added more objects, and these are catalogued as IC.

A selected few of the NGC objects have been added to the list

An open star cluster   The small cluster NGC 6939 in the constellation Cepheus. The stars are of about the 9th magnitude and fainter (*R. W. Arbour, 400-mm Newtonian*)

printed here, in order to spread the coverage down to the southern celestial pole. The aim has been to offer a good sample of the different types of deep-sky objects—bright and faint, near and far—that are waiting to be observed. Before starting on the list, however, it will be necessary to make a few comments on the different kinds of objects and the best ways of observing them.

Open clusters of stars   Clusters and groups of stars are very common. They range from naked-eye objects to condensed, barely resolvable hazes. Some seem to consist of well-scattered stars of similar brightness, while others have bright and faint stars massed together. Some have noticeably colored stars in their ranks. They lie in the spiral arms of our galaxy, and so are usually to be found in or near the Milky Way.

Globular star clusters   Most open star clusters are relatively young objects (up to a very few thousand million years old, and these are exceptional) because the tidal forces within the Galaxy are tending to pull the stars apart. Globular clusters are very old and very dense, and at first sight appear nebulous rather than stellar. Even with a large instrument, the majority of the stars in a globular remain unresolved, but the brighter ones may be seen sparkling in its midst. The globular clusters are scattered in a huge swarm around the nucleus of our galaxy, and some lie well away from the Milky Way. The brightest is NGC 5139, also known as ω Centauri, that shines like a hazy 4th-magnitude "comet" in the southern skies.

Nebulae The true "nebulae" covered in this section are clouds of "dust" inside the Galaxy. The word "nebulae" as used in Messier's catalogue and the NGC meant "of nebulous appearance," and many of these objects, we know now, are really galaxies of stars lying at an immense distance from our own galaxy.

Among the nebulae we find three general classes:

(a) *Bright nebulae* These are irregular in outline, and indicate regions of the Galaxy where star-formation is proceeding. These immense clouds are many light-years across, in which new stars may even now be condensing and beginning to shine. They are, in fact, the site of young star clusters. Some of these young stars will be very hot supergiants, and these emit so much short-wave radiation that they may make the womb that cradled them begin to glow, by exciting individual atoms to emit light. These are known as *emission nebulae*. The most obvious example is the Great Nebula in Orion, which is visible with the naked eye.

There are also *reflection nebulae*. If the stars are cooler, or the nebula contains dust rather than gas, their light will merely be reflected by the cloud instead of exciting it to emit light. Reflection nebulae are relatively dim and hard to see, but some are to be found in the Pleiades star cluster.

(b) *Dark nebulae* These are clouds of "dust" in their natural state, unilluminated by nearby stars. They therefore appear as a black outline against more distant objects. Dark nebulae tend not to be catalogued in the NGC, but they contribute to the irregular appearance of parts of the Milky Way, particularly in the Cygnus-Aquila region and around Crux, where black patches appear in the band of milky light.

(c) *Planetary nebulae* belong to a different class altogether, and some amateurs make a special study of them. In the later stages of its evolution, a star may expel large amounts of material. This expulsion may be relatively gentle, or it could be catastrophic, producing a supernova. The result, either way, is a star surrounded by a haze of light. Some examples are so sharp and disk-like that they earned the term "planetary" from their resemblance to a planet's disk, but others are rather irregular. NGC 2392 in Gemini is a "typical" planetary, while the Crab Nebula, M1, is a violent irregular example.

Galaxies The number of galaxies photographed with the world's largest telescopes runs literally into millions. Of these, a few thousand are detectable with moderate apertures, but most appear as nothing more than very faint blurs of light. The interest is often in detecting them at all, rather than in the beauty or wonder of the sight. But some of the nearer galaxies do show considerable detail if observed carefully and under good conditions.

Planning a deep-sky session Many double stars can be observed quite successfully in a slightly hazy or moonlit sky. If the object is to resolve a particularly close pair, haze may be a good sign, for it often indicates a steady atmosphere; while moonlight or twilight can assist the detection of a very close but rather dim companion such as that to Deneb or Antares. The case of the cluster or nebula is entirely different. The aim, almost always, is to detect *light; resolution* may well be a secondary requirement. Therefore, only nights of excellent transparency and sky darkness are really suitable for most work of this kind, as was emphasized in Chapter 3.

If the night is fairly transparent, but not as good as you would have liked, there may still be some work to be done. It will be more useful, for example, to look for small bright planetary nebulae, or to try to resolve a bright globular cluster into stars, on such a night, than to look for a remote galaxy. In other words, if the work demands a high magnification, sky darkness matters less because the magnification itself will help to darken the background. But large faint nebulae, or anything extended of low surface brightness, will appear so dim that a very low magnification is essential if the light is to be sufficiently concentrated on the retina. If the sky is already bright, then you are simply concentrating the bright background as well, and little or nothing celestial will be distinguishable.

Both altitude and season matter too. If an object never rises very high in your sky, you will make a particular effort to catch it on or very near the meridian. Furthermore, in general, the winter months offer darker skies than do the summer months, particularly in higher latitudes. It is therefore frustrating for Northern Hemisphere observers to discover that Nature has not spread her treasures impartially, and that the richest area of sky for deep-sky Galactic objects—that stretch of the Milky Way passing through Scorpius and Sagittarius—is not only well South of the celestial equator, but also comes to the meridian when it is summer in the Northern Hemisphere. If you want to have the best possible view of this part of the sky, move to southern latitudes!

Observing notes Some observers of deep-sky objects, regrettably, treat the challenge as one of getting through the largest number of objects in the shortest possible time. There is nothing to stop you from "ticking off" items in this way if you like, but you certainly will not be doing the thing justice, and you will not be helping yourself as an observer much, either. Almost anyone can locate something in the sky if its position is known and the telescope is of sufficient power. Admittedly, in the case of a very faint galaxy, this may be a triumph in itself, if you have searched for it repeatedly on previous occasions and have failed to locate it. You have persevered and won, and this is certainly cause for congratulation. Even then, however, you can pause and ask yourself "What can I see? Is it really just a barely detected hazy spot of no form whatever?" As soon as you ask yourself that question, you are implying a host of supplementary ones because anything that makes a response in the optical nervous system must have:

(a) Size—how large does it appear?
(b) Shape—is it circular, triangular, irregular, etc.?
(c) Brightness—are some parts brighter than others, and, if so, how is the brightness distributed?

To observe a nebulous object properly you must ask and answer questions such as these, not because the resulting description is necessarily of any great scientific value, but because you are forcing the eye and brain to make an effort. We have said before that the optical sense is becoming etiolated—people are so used to seeing things at a glance that, if they can't see properly, they blame their eyes, or their spectacles, but never their visual technique! Why should they do otherwise? As an astronomical observer, however, you have the advantage of realizing the unsuspected potential of the human eye, and of being able to document its sharpening-up, its improving powers of perception, as you offer it stiffer challenges. Just as the planetary observer discerns more and more detail with experience, so the deep-sky observer discovers more and more to see in fainter and fainter objects.

You will also find it fascinating to compare what you see with what others have seen, with similar or different equipment. However, a large caveat must be entered here: nebulae and clusters, far more than double stars, are affected by local sky conditions. A town observer may have as good a view of γ Virginis as a country one, but there will be no comparison between the appearance of, for example, the cluster M11 in Scutum as seen from the two sites. This is why many town observers favor compact portable instruments, that can easily be taken to a better site some miles beyond the streetlights. In general, then, the notes in the list refer to observations made by skilled eyes in good conditions.

Do not imagine, however, that all there is to see has been seen! Take the curious dark "lanes" in the globular cluster M13, in Hercules, first reported over a century ago by Lord Rosse in Ireland with his vast 180-cm (72-inch) reflector. This magnificent cluster has been observed by countless amateurs with moderate and large telescopes, but the "lanes" were not commented on again until a perceptive observer reported them just a few years ago, using a 315-mm reflector. Or what about star A in the famous Trapezium group in the Orion Nebula, one of the most closely observed objects in the sky, that was discovered less than ten years ago to be an eclipsing binary star, dimming briefly by a magnitude every sixty-five days?

There are still many discoveries and rediscoveries to be made. So try to come even to a well-known object with a fresh and unprejudiced eye, as if you were viewing it for the first time; as in a sense you are, for the sky conditions, and your own sense of enthusiasm, contribute as much to what you will see as does the telescope itself.

NORTHERN DEEP-SKY OBJECTS

M31 (NGC 224) And
$00^h 42^m.7 +41° 16'$

The Andromeda Galaxy, the nearest spiral galaxy to our own, and probably of a very similar type. Its distance is 2.2 million light-years, and it can be seen as a noticeably hazy patch with the naked eye on any reasonably clear night when it is high in the sky. Often dismissed as a disappointment by amateurs because it appears as a very large but apparently featureless haze. Even the most luminous supergiants in M31 appear of only the 16th or 17th magnitude, probably too faint to be detected visually with the largest amateur telescopes, but apertures of 200 mm or more are capable of detecting some of the brighter condensations in the spiral arms, even though the arms themselves cannot be made out. The wide wings seen around the bright nucleus are part of the central region of old red-giant stars; remember that the true limits of the arms extend for 2° or more on each side of the center.

It is worth taking some trouble with this object. Allow your eyes to become thoroughly dark-adapted (come to it after at least half an hour spent examining other celestial sights), and see how much of these elusive outer regions can be seen by

The Andromeda Galaxy, M31 Photographed with a lens of 200-mm focal length. Note the faint companion galaxy NGC 205 (also known as M110) a little above and to the right of the main nucleus (*R. W. Arbour*)

letting the telescope move fairly rapidly from side to side. The motion of a very faint object across the retina can often bring it into visibility. Also examine the two satellite galaxies, M32 (NGC 221), small and bright, with pronounced concentration towards the center, and the larger, much more diffuse NGC 205. Both of these are small elliptical galaxies containing red-giant stars.

NGC 225 Cas
00ʰ 43ᵐ.4, +61° 47′

A bright open cluster, easily detectable in the finder.

NGC 457 Cas
01ʰ 19ᵐ.0, +58° 20′

A scattered cluster of fairly bright stars containing the 5th-magnitude star φ Cas.

NGC 524 Psc
01ʰ 24ᵐ.7, +09° 32′

A galaxy, appearing as a faint condensed nebulosity. Not spectacular, but a good example of a "typical" galaxy, one of several hundred that can be swept up in a good sky with an aperture of about 150 mm.

M103 (NGC 581) Cas
01ʰ 33ᵐ.1, +60° 42′

This lies in a rich region of the Milky Way, and is really little more than a particularly crowded field, less spectacular than others nearby that Messier did not include in his catalogue.

M33 (NGC 598) Tri
01ʰ 33ᵐ.8, +30° 41′

One of the most challenging of the Messier objects. This is a spiral galaxy belonging to the Local Group, a cluster of galaxies that includes our own and the Andromeda Galaxy; it is large—about the apparent size of the Moon—but of very low surface brightness, and it may be more conspicuous in the finder, or with binoculars, than in the main instrument. A very low magnification is essential, and so is an extremely dark sky.

M74 (NGC 628) Psc
01ʰ 36ᵐ.7, +15° 47'

This is one of faintest of the Messier objects, and it is not easy to sweep up with apertures of less than about 75 mm. With a very low power it is about one field-width from the star η, the difference in coordinates of M74 being about 5ᵐ of R.A. to the East of η, and about 25' arc to the North. With the field being so readily found, this is an excellent object on which to test your telescope (or the night) for the visibility of faint objects similar to this dim galaxy.

M76 (NGC 650) Per
01ʰ 42ᵐ.2, +51° 35'

A planetary nebula, often quoted as the faintest of all the Messier objects, but in the writer's opinion easier to see than M74—however, its greater altitude in the sky as seen from northern temperate latitudes may have something to do with this. In small instruments it appears as a diffuse nebulosity with perhaps a condensation towards the center. Larger telescopes show two distinct condensations, the fainter of which is known as NGC 651. In photographs, it rather resembles the "Dumbbell" planetary nebula in Vulpecula.

NGC 663 Cas
01ʰ 46ᵐ.1, +61° 15'

A conspicuous cluster of bright stars, appearing in the finder as a condensation in the Milky Way.

NGC 869 & 884 Per
02ʰ 20ᵐ.6, +57° 08'

The famous Double Cluster in Perseus, sometimes referred to by their old designation of h & χ Persei. Two loose clusters of bright and faint stars lie almost in the same field of view, and on a dark night the sight is breathtaking. A very low power is necessary for the best effect.

M34 (NGC 1039) Per
02ʰ 42ᵐ.0, + 42° 46′

A very scattered cluster of bright stars. It seems to lack the concentration of fainter members that gives some objects such a fine "massing" effect towards the center. Visible as a hazy, partly resolved circular object in the finder.

NGC 1245 Per
03ʰ 14ᵐ.7, +47° 14′

A complete contrast to the previous object, and a sight for observers with large apertures. This is a much smaller cluster of very faint stars, like a scattering of pepper grains.

M45 Tau *The Pleiades*
03ʰ 46ᵐ, +23° 56′

A bright and easy naked-eye cluster. The stars lie some 250 light-years away from the Sun, and were probably born only a hundred million years or so ago, which makes it an extremely young cluster. The lowest possible magnification is necessary if all six bright stars are to be included in the field of view, and the best general impression will be obtained with the finder or some other low-power telescope. Note the whiteness of the brightest stars. These are giants, with an absolute magnitude of about − 1, which means that each one is as luminous as 200 Suns. Some hundreds of millions of years hence, these will have evolved into red giants, the cluster will have dispersed, and there will simply be a sprinkle of reddish and yellow stars in this part of the sky.

NGC 1435 Tau *The Merope Nebula*
03ʰ 46ᵐ.1, +23° 56′

This intriguing object, one of the few reflection nebulae bright enough to be seen in amateur instruments, has developed its own folklore as to which type of instrument shows it best. It was discovered in

Nebulosity in the Pleiades Only five of the bright naked-eye Pleiades are shown here; the brightest of all (Alcyone) is beyond the lower left edge of the photograph (*R. W. Arbour, 400-mm Newtonian*)

1859 by an observer in Italy, who was hunting for comets with a 100-mm refractor armed with a low-power eyepiece. He described it as a very faint, extended haze near the star Merope in the Pleiades. Yet Burnham, who made astounding double-star discoveries, was unable to confirm it using a 450-mm refractor; while the well-known comet observer John Bortle has seen it with binoculars! The reason is that some excessively extended but extremely faint objects cannot be seen without the very wide field of view that only a small instrument can give, and that the extra light-grasp of a large aperture is nullified by the impossibility of using a sufficiently low magnification.

A telescope, eye, and night incapable of revealing M33 in Triangulum as a pretty prominent haze has no hope of reaching this object. You will have to put Merope out of the field of view, to avoid dazzle, or else hide it behind a field bar; the presence of other bright stars in the area increases the difficulty of detecting this elusive object. Its apparent size varies with observer and instrument, but a typical estimate is about 20' by 35' arc, extending South from Merope. Photographs show this to be only the brightest patch of a general nebulosity enfolding the Pleiades cluster.

NGC 1501 Cam
04^h 06^m.9, +60° 54'

A "typical" small planetary nebula, about 1' arc across, conspicuous with an aperture of about 250 mm in a good sky. A moderate magnification will be necessary to distinguish it from a faint star, as the telescope sweeps across it. At least one well-known observer has dismissed it as not worth looking at, with a 220-mm aperture telescope; but teasing objects like this are, in the writer's opinion, well worth looking for—it is satisfying to detect them, even if they are not spectacular, and they reinforce one's respect for the more brilliant sights of the night sky.

NGC 1528 Per
04ʰ 15ᵐ.1, +51° 15′

A pretty, irregular group of bright and faint stars.

NGC 1857 Aur
05ʰ 20ᵐ.1, +39° 22′

A fine field of stars, magnitude 7 and fainter, but hardly a cluster. This was included in William Herschel's 7th class of objects, which he defined as "compressed clusters of large and small (i.e., bright and faint) stars"; but the word "compressed" is, of course, relative. This object lies in a beautiful telescopic region of the sky, even though the Milky Way is not particularly conspicuous to the naked eye.

NGC 1907 Aur
05ʰ 28ᵐ.1, +35° 20′

An interesting contrast with the previous object. A 100-mm reflector shows this cluster as a mere nebulous patch containing an 8th-magnitude star. A much larger aperture is required to resolve it into faint stars. Yet this also is included in Herschel's 7th class.

M38 (NGC 1912) Aur
05ʰ 28ᵐ.7, +35° 50′

The most westerly of the three fine Messier open clusters in Auriga, all of which are detectable with binoculars. The brighter stars of this group can be distinguished in the finder. With a 100-mm reflector, it becomes an irregular cluster of bright and faint stars, giving the impression that the lines of bright stars are associated with unresolved matter. The region immediately to the North seems somewhat vacant, as though obscured by dark nebulae. Some observers see a cruciform effect.

M1 (NGC 1952) Tau *The Crab Nebula*
05ʰ 34ᵐ.5, +22° 02′

A planetary nebula, the most dramatic observable object in our galaxy, being the wreckage of a supernova observed in the year 1054. It has, therefore, expanded to its present apparent size of about 6′ by 9′ arc in nine centuries, which, at its distance of 3600 light-years, corresponds to an expansion rate of over 1000 km/sec. Visible in a small telescope as a slightly elongated nebulosity, it rapidly loses distinctness as the magnification increases because the edges are very ill-defined. When looking at this far from spectacular object, remember that at the moment of explosion it was bright enough to be seen in full daylight with the naked eye, and emitted as much light as the entire Galaxy!

M36 (NGC 1960) Aur
05ʰ 36ᵐ.3, +34° 09′

This differs from the previous object in Auriga, M38, in being smaller and containing fewer faint stars. It is well resolved in the finder.

NGC 2022 Ori
05ʰ 42ᵐ.0, +09° 06′

A small planetary nebula, well defined but faint. It will probably be missed with apertures below about 120 mm, but is an interesting challenge for larger apertures.

M78 (NGC 2068) Ori
05ʰ 46ᵐ.8, +00° 04′

A reflection nebula. Two 9th-magnitude stars are involved in it, and in a small telescope it appears irregular, somewhat fan-shaped, with the apex pointing northwest. It appears most distinct with a moderate power, and with adequate aperture it is seen to contain two bright condensations. This object is merely a local brightening in a huge, excessively faint nebulosity involving most of Orion, and of which the Great Nebula is the nucleus.

M37 (NGC 2099) Aur
05ʰ 52ᵐ.3, +32° 33′

In the writer's opinion, this is the finest of the three bright Auriga clusters; with a 100-mm reflector, ×35, it appears as a glittering heap of stars, fading to invisibility; with a higher power, a bright star near the center and lines of other stars are well seen. It is most impressive, after viewing the other, coarser groups; and it appears large and nebulous in binoculars.

M35 (NGC 2168) Gem
06ʰ 08ᵐ.8, +24° 21′

A large, coarse cluster, visible with the naked eye, and partly resolved in the finder. The lowest magnification must be used, to get the best effect; but there is no concentration towards the center to give it a feeling of mass. With apertures of 300 mm or more, the very small, condensed cluster of faint stars catalogued as NGC 2158, that lies about 15′ arc in a southwesterly direction from the center of the cluster, can be made out. If one imagines that the two clusters are similar in real size and brightness, then the awesome remoteness of this dim companion can be appreciated.

NGC 2244 Mon
06ʰ 32ᵐ.3, +04° 53′

Visible in the finder, and recorded with the naked eye by some observers, this is a loose grouping of bright and faint stars, five of which form an elongated X.

NGC 2301 Mon
06ʰ 51ᵐ.8, +00° 28′

A curious group of bright stars and star dust, straggling in a North–South direction, with a third arm winding to the East. Best seen with moderate power, and visible in the finder.

NGC 2392 Gem
07ʰ 29ᵐ.2, +20° 56′

A conspicuous planetary nebula, visible as a disk even with a magnification of ×35 by contrast with a nearby 9th-magnitude star. Higher powers show a distinct central condensation, so that the object resembles a telescopic comet. An excellent bright example of this type of object.

M44 (NGC 2632) Cnc *Praesepe*
08ʰ 40ᵐ, +20°

A very large and conspicuous naked-eye cluster. With a diameter of over 1°, it appears in the telescope as a brilliant field rather than as a cluster, so that the finder may give the best effect. It is interesting to compare this group with its near neighbor in Messier's catalogue, M45 (the Pleiades). Praesepe is twice as far away from the Sun, but, even so, its brighter members appear far fainter than the major Pleiades. This is a clue to the age of Praesepe. Many hundreds of millions of years ago, soon after it was formed, it too would have contained brilliant white giant stars, and must have been a splendid object, but they have long ago burned themselves out.

M67 (NGC 2682) Cnc
08ʰ 51ᵐ.3, + 11° 48′

At an estimated distance of 2700 light-years, M67 is perhaps five times as distant as Praesepe, appearing in a small telescope as a wide scattering of conspicuous stars. It is one of the oldest clusters known, perhaps as old as the Sun, and the stars have remained collected together for so long because the group lies well above the spiral arms of the Galaxy, where the gravitational tides of the massed stars and nebular material have a much reduced disrupting effect.

NGC 2841 UMa
09ʰ 22ᵐ.1, + 50° 59′

A faint galaxy, elliptical in appearance, appearing about 4′ arc long with an aperture of 140 mm. It is not difficult to locate, lying only 17′ arc South and about 10′ arc following a 6th-magnitude star.

M81 (NGC 3031) & M82
(NGC 3034) UMa
09^h 56^m, + 69° 25'

These are two very well-known galaxies. Although not in the Local Group, they are not far beyond it, lying about seven million light-years away. Visible in the finder as two neighboring nebulous patches, even a small telescope shows a clear difference between them when they are included in the same low-power field. M81 is the more conspicuous, elongated, with a clear nucleus; M82 appears almost as a streak of light. It is worth paying considerable attention to these bright galaxies, to see what details of form and brightness can be made out by prolonged gazing. M82 is seen almost edge-on, and evidently is being torn apart by an internal explosion.

M95 (NGC 3351) & M96
(NGC 3368) Leo
10^h 45^m.4, + 11° 45'

Two moderately bright galaxies within 50' arc of each other. There are no convenient guide stars near, but it is helpful to remember that they lie only 15' arc further South in declination than Regulus, which is 9° to the West. In a moderate instrument, both objects are seen as slightly elliptical, about 4' arc across, with a nucleus. They are unlikely to be detected in the finder, or with binoculars

M108 (NGC 3556) UMa
11^h 11^m.6, + 55° 41'

A very elongated nebulosity, being a galaxy seen almost edge-on. Not bright, but worth hunting for the satisfaction of identifying it, and fairly easily located because it lies about two-thirds of the way from β UMa towards the Owl Nebula (see below). M108 is one of the "additional" Messier objects.

M97 (NGC 3587) UMa
The Owl Nebula
11^h 14^m.9, + 55° 02'

A very large, pale, circular nebulosity about 2' arc across, near a 6th-magnitude star. This is one of the largest planetary nebulae in the sky, and in instruments of moderate aperture it is seen to contain two dark openings, but in smaller instruments it is virtually featureless because of its low surface brightness. With 40-mm binoculars, it can just be made out as the faintest hint of nebulosity.

M65 (NGC 3623) & M66 (NGC 3627) Leo
11ʰ 19ᵐ.5, + 13° 03′

These are probably the most conspicuous galaxies in Leo, and are readily seen in a good finder as a large nebulous patch—presumably the two objects merged together. Both appear well elongated, and, being only ½° apart, are easily included in the same low-power field of view. If these are not easily seen, there is little point in looking for other galaxies in this area.

NGC 4026 UMa
11ʰ 59ᵐ.5, + 50° 57′

A small galaxy with a brighter center. Although stated as strongly elongated, a 75-mm telescope shows only the center of the galaxy, so that this elongation of the outer regions is not apparent.

M98 (NGC 4192) Com
12ʰ 13ᵐ.9, + 14° 54′

A large, elongated nebulosity measuring some 8′ arc in length, but it is very faint with apertures below 100 mm. The field is readily located, about ½° preceding the magnitude 5.1 star 6 Comae, that will have to be kept out of the field. To William Herschel, probably using his 450-mm Newtonian, the wings of the nebula extended for more than a quarter of a degree.

M99 (NGC 4254) Com
12ʰ 18ᵐ.9, + 14° 25′

Lying near M98, this galaxy is smaller, less elongated, and perhaps slightly easier to detect. It is, nevertheless, one of the more elusive Messier objects.

NGC 4258 UMa
12ʰ 19ᵐ.0, + 47° 18′

A small galaxy, not very bright with modest apertures, although the central condensation is obvious. With about 300 mm, its length has been extended to 10′ arc.

M61 (NGC 4303) Vir
12ʰ 22ᵐ.0, + 04° 28′

A distant galaxy (about 40 million light-years away), roughly circular in outline. There is a clear central condensation, and it has been described as looking "three-lobed" using a 200-mm aperture telescope.

M85 (NGC 4382) Com
12ʰ 25ᵐ.5, + 18° 12′

A small nebulosity, visible in binoculars. Readily seen in most telescopes as an almost circular hazy patch. Like M84 (below), this object is a so-called SO galaxy, with spiral arms almost absent and its light concentrated into an ellipsoidal mass of reddish stars.

M84 (NGC 4374) & M86 (NGC 4406) Vir
12ʰ 25ᵐ.7, + 12° 56′

Although these two galaxies are readily seen with only a modest telescope and appear close together in the sky, they are not truly related. M84 belongs to the Virgo-Coma group, at a distance of about 40 million light-years, while M86 is a much less luminous and smaller galaxy lying only half the distance away.

M49 (NGC 4472) Vir
12ʰ 29ᵐ.8, + 08° 00′

A relatively bright galaxy, visible with binoculars as a fairly large haze between two 6th-magnitude stars; telescopically it rather resembles an unresolved globular cluster.

M88 (NGC 4501) Com
12ʰ 32ᵐ.0, + 14° 26′

A small but fairly conspicuous spiral galaxy seen from an inclined angle. It lies in a rich field of galaxies; with a moderate aperture, many faint NGC objects will be swept up in the region extending for a couple of degrees southwest of this point.

NGC 4565 Com
12ʰ 36ᵐ.4, + 26° 00′

Faint in small instruments. With moderate apertures, however, this edge-on galaxy may reveal a double or sliced aspect, caused by a band of obscuring interstellar material in its spiral arms, similar to that in NGC 4594 in southern Virgo.

M59 (NGC 4621) & M60
(NGC 4649) Vir
12^h 42^m.9, + 11° 36'

Another pair of galaxies, easily contained together in a low-power field. M60 is the brighter of the two, and may just be detected in the finder. Both galaxies have a central condensation.

M94 (NGC 4736) CVn
12^h 50^m.8, + 41° 07'

A small but bright galaxy, at first sight resembling a hazy 8th-magnitude star or a well-condensed comet. It may benefit from a higher magnification than is usually applied to these objects.

M64 (NGC 4826) Com
12^h 56^m.7, + 21° 40'

This is one of the brightest galaxies in the Virgo-Coma group; in fact, it has been claimed as the second largest and most luminous of all the Messier galaxies, the biggest of all being M77 in Cetus. It is a very easy finder object, revealing a central condensation even with a very low magnification.

M53 (NGC 5024) Com
13^h 13^m.0, + 18° 10'

A bright globular cluster. It is larger than its neighbor M64, and appears considerably brighter and less condensed in the finder. With apertures of 150 mm or more, it begins to be resolved into faint dusty specks of light. At the immense distance of 70,000 light-years it is the second most distant of Messier's globular clusters, and must, indeed, be one of the most remote Galactic objects visible in a very small telescope.

M63 (NGC 5055) CVn
13^h 15^m.8, + 42° 01'

A galaxy of moderate size and brightness. In larger instruments it can be traced to an elliptical outline about 8' arc in length, but small telescopes may show only the brighter center, some 2′ arc across. A star of the 8th magnitude lies very near, to the northeast.

M51 (NGC 5194) &
NGC 5195 CVn
The Whirlpool Galaxy
13ʰ 29ᵐ.8, + 47° 11′

This galaxy has earned its name because of the very open and pronounced face-on spiral structure as seen on photographs. So marked is this structure that it is famous as the first "spiral nebula" to be discovered visually, by Lord Rosse using his 180-cm (72-inch) reflector in 1845. Of course, it is easy to "see" things that are known to exist, and the object is so frequently illustrated in books that it cannot be observed without some preconception of how it should appear. It is, therefore, hard to assess the claims made by some observers that the spiral structure can be detected with an aperture of only 300 mm;

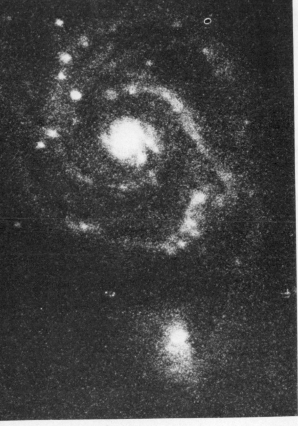

The Whirlpool Galaxy, M51 Lying in Canes Venatici, this is a favorite object with amateur astronomers. Visually, only the two bright nuclei are likely to be detected with small and moderate instruments. (*N. W. Scott, 300-mm Cassegrain*)

the more common appearance is of a bright nucleus surrounded by a ring, with a fainter but more concentrated nebulosity (NGC 5195) to the North. This curious irregular companion galaxy is seen on photographs to be connected to the main spiral by a faint bridge of bright and dark material, but claims to have detected this bridge visually must be treated with some reserve. In the finder, or with binoculars, M51 is easily seen as a fairly even nebulosity, slightly extended to the North.

M3 (NGC 5272) CVn
13^h 42^m.4, + 28° 22′

One of the finest globular clusters, easily detected in the finder as a nebulous disk. However, an aperture of perhaps 200 mm is necessary if the brighter stars in it are to be well seen; with smaller telescopes, some individual stars around the edge may be detected with averted vision. This is a good object on which to test the averted-vision technique; it is remarkable how the cluster appears to sparkle when removed from the direct gaze.

NGC 5322 UMa
13^h 49^m.3, + 60° 11′

A small galaxy with a brighter nucleus. A challenge for telescopes of about 80-mm aperture, with which it is very obscure.

M5 (NGC 5904) Ser
15^h 18^m.6, + 02° 05′

A bright globular cluster, very compressed towards the center. Even a small telescope will reveal individual stars in it, and it can easily be picked up as a hazy disk in binoculars or the finder. Note also the delicate double in the same low-power field, 5 Serpentis (Σ 1930): 5.2, 10.0; 11″ .5; 39°.

M13(NGC 6205) Her
16^h 41^m.7, + 36° 27′

The brightest globular cluster visible from higher northern latitudes, detectable with the naked eye as a very faint, small nebulosity. Individual stars can be made out with apertures of only 75 mm, but the view is transformed when more light is brought to focus; with a 300-mm telescope, in a dark sky, it looks like a true swarm of stars, with a three-dimensional aspect very different from that of an open cluster. Yet we are seeing only at most a few hundred of what is estimated to be a total population of half a million stars! Examine it for the "lanes" mentioned on page 274; the early observers also reported conspicuous branches or lines of stars radiating away onto the dark sky. Undoubtedly these are of physiological rather than astronomical significance (the human eye tends to enjoy chance alignments), but overfamiliarity with photographs of celestial objects such as M13 can inhibit the eye, and the observer, from noticing effects that do not appear on photographs. The stars revealed in a typical long-exposure photograph of M13 are not those seen by the eye, but the crowds of much fainter stars that swamp the bright giants detected visually—the patterns seen by the eye are therefore lost.

NGC 6210 Her
$16^h 44^m.5, + 23° 48'$

Often referred to as Σ5N—the short list of anomalous objects discovered by Struve in his searches are catalogued in this form—this is a remarkable planetary, being very bright and so small that it could easily be mistaken for an ordinary 8th-magnitude star put slightly out of focus. It is, therefore, best sought with a moderate power. Its diameter of about 8″ arc would be missed with a very low magnification, although its intense blue tint might still strike the eye as unusual for a star.

NGC 6229 Her
$16^h 46^m.9, + 47° 32'$

The "third globular" in Hercules, much fainter than the two bright Messier objects, but prettily arranged with a wide pair of magnitude 6 and 7 stars to form a triangle. Rarely observed: it appears, in small instruments, as little more than a faint, nebulous disk, and was in fact classed by its discoverer, William Herschel, as a planetary nebula because he could detect no stars in it.

M92 (NGC 6341) Her
17^h 17^m.1, + 43° 07'

This globular invites comparison with its companion, M13. Although smaller and fainter overall, it appears much more compressed, and is therefore harder to resolve at the center. The outer regions, however, reveal individual stars to a modest aperture, particularly if averted vision is used.

NGC 6543 Dra
17^h 58^m.6, + 66° 38'

A small but conspicuous planetary nebula, somewhat larger than NGC 6210, and therefore easier to detect, despite not being so bright; it has the same characteristic bluish tint. This object is interesting on two counts. First, it lies almost at the North Pole of the ecliptic—in other words, the point on the celestial sphere from which an observer would have a plan view of the Earth's orbit, from the North, as well as an excellent view of the general layout of the solar system. Second, it was the object used by William Huggins, the early astronomical spectroscopist, to discover the composition of the nebulae. His observation, made on August 29, 1864, proved that NGC 6543 and other objects like it were not merely unresolved clusters of stars, but clouds of glowing gas.

NGC 6572 Oph
18^h 12^m.1, + 06° 51'

Another small planetary, and another of Struve's accidental finds. With a small telescope it appears blue-white, bright, and almost starlike with a low power, although its nonstellar character can be confirmed by comparing it with an 8th-magnitude star to the northeast. With a high magnification it acquires a somewhat hazy outline, possibly elongated in a North–South direction.

NGC 6633 Oph
18^h 27^m.5, + 06° 33'

Visible as a cluster with binoculars or a finder; with a low magnification it appears as a bright sprinkle of stars.

M57 (NGC 6720) Lyr
The Ring Nebula
18^h 53^m.6, + 33° 00'

Probably the most famous planetary nebula in the sky. A "celestial smoke-ring" sums up the appearance of this strange object, when viewed with medium power; with a very small aperture and a low power, however, it appears only as a well-defined, elliptical disk about 1' arc across. A faint star close to its outer edge, to the East, is an interesting light-test, and has been claimed with only 75 mm of aperture, the magnitude being about 12. There is a very faint blue star at the center of the nebula, magnitude about 15, that should be a challenge for well-sited telescopes in the 300-mm aperture range.

The Ring Nebula, M57, in Lyra   Although the ring itself is easy to see in a small telescope, the central star is only of the 14th magnitude and is an elusive object (*R. W. Arbour, 400-mm Newtonian*)

M56 (NGC 6779) Lyr
19^h 16^m.6, +30° 11′

One of the fainter examples of Messier globular clusters, readily found as a small, dim haze. With attention, a telescope of 100 mm aperture can reveal some stars near the margin, but it is not a spectacular object. The field, though, is very fine.

NGC 6826 Cyg
19^h 44^m.8, +50° 31′

A fine planetary nebula, looking, as do so many of its class, like a faint star put slightly out of focus. Its diameter, about 20″ arc across, corresponds to that of Mars at a fair opposition. An 11th-magnitude star is situated at the center.

M71 (NGC 6838) Sge
19^h 53^m.7, +18° 46′

A fairly large, faint cluster, that requires a moderate aperture and a high magnification to be properly resolved; with small apertures, or in the finder, it appears nebulous. There is some doubt about the physical nature of this object: is it a true globular or a very compressed Galactic cluster?—but most authorities class it as a globular. About 20′ arc to the southwest, a small telescope reveals a line of 8th- and 9th-magnitude stars set in what appears to be a faint haze. This is catalogued as H20 (the "H" referring to William Herschel) because, for some reason, it did not come to be included in the NGC.

M27 (NGC 6853) Vul
The Dumbbell Nebula
19^h 59^m.7, +22° 42′

The finest planetary nebula in the sky. Even in binoculars, or in the finder, it exhibits a conspicuous, well-defined disk, like a ghostly Full Moon some 8′ arc across. The first telescopic impression is of amazement at its size; with 80 mm of aperture it appears as a circle of luminous haze, partly filled in to resemble a flanged bobbin or reel. It is worth spending considerable time on this object, even if you

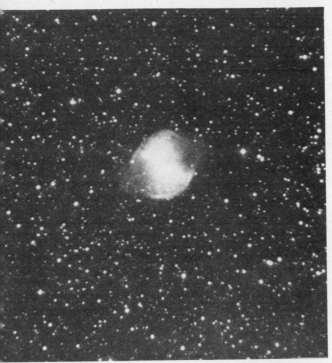

The Dumbbell Nebula, M27
The splendid planetary nebula
in Vulpecula. (*R. W. Arbour,
400-mm Newtonian*)

have only a modest instrument, to try to
delineate the outline and the grades of intensity. Search also for the central star,
magnitude about 13, that is intensely hot,
and no doubt is the seat of this nebular
outburst, calculated to have occurred
around 1500 BC. M27 lies in a splendid
field of stars.

NGC 6866 Cyg
$20^h 03^m.8$, $+43° 59'$

A compact open cluster, representative of
the many groupings to be found in Cygnus. It is surprising, considering its position in the Milky Way, that Cygnus
contains no outstanding clusters—its two
Messier objects, M29 and M39, are among
the least-observed in the list.

M29 (NGC 6913) Cyg
$20^h 23^m.9$, $+38° 32'$

An open cluster, but hardly worth the
search in a vast region so packed with
groups and whorls of stars. This is a collection of six 8th-magnitude stars, in the
form of a square, with fainter members
that are unresolvable with apertures of
about 80 mm.

NGC 6934 Del
20^h 34^m.2, +07° 24′

A small globular cluster, so remote that its brightest individual members are probably not resolved with less than about 200-mm aperture. It lies just to the East of a 9th-magnitude star.

NGC 6960 & NGC 6992 Cyg
The Filamentary Nebula
20^h 45^m.7, +30° 42′
20^h 56^m.4, +31° 42′

These two extremely faint, very extended nebulosities are a real challenge for sky and eye. They are the western and eastern arcs respectively of an immense "bubble" of gaseous material that evidently was expelled in some archaic supernova explosion. The western arc (NGC 6960) lies just to the East of the star 52 Cygni; the eastern arc, NGC 6992, is larger and brighter. Using a low power on a 90-mm refractor, the writer has seen the latter as a nebulous arc—perhaps a third of a circle —fitting comfortably inside the margin of a ×35 eyepiece. With the same instrument, and putting 52 Cygni out of the field, NGC 6960 was only suspected, under excellent conditions. The whole "bubble" measures about 2½° across.

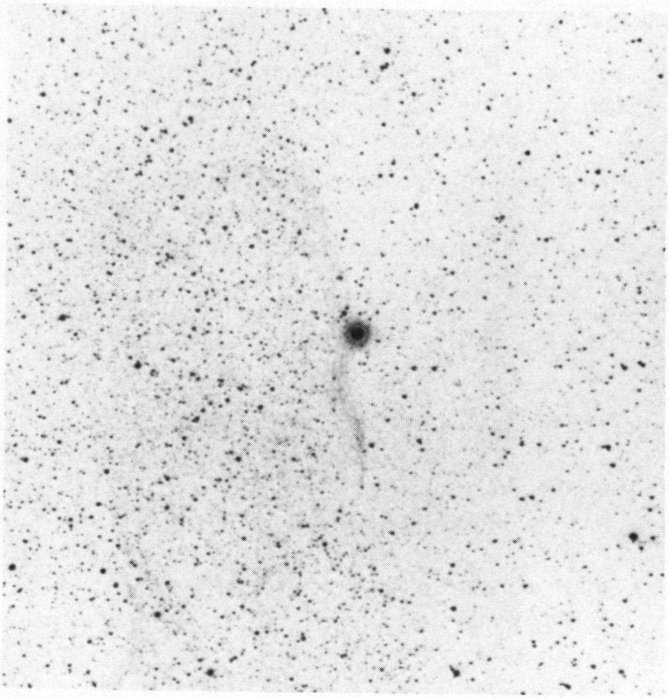

The Filamentary Nebula in Cygnus This is a negative print, which helps to bring out the detail. The star near the brightest part of the nebula is 52 Cygni, magnitude 4.2. (*N. W. Scott, 300-mm Cassegrain*)

The North American Nebula in Cygnus A passing aircraft imprinted the two trails. (*N. W. Scott, 300-mm Cassegrain*)

NGC 7000 Cyg
The North America Nebula
20ʰ 59ᵐ, +44½°

An immense, exceedingly faint nebulous cloud preceding ξ Cygni. It it so large (some 2° across) and of such low surface brightness that it is practically invisible with any magnification exceeding that of binoculars. Sweep steadily across the area on a very dark night, looking for any lightening of the sky background. The region is spectacular on wide-aperture photographs.

NGC 7027 Cyg
21ʰ 07ᵐ.0, +42° 13′

A very small, bright planetary nebula, overlooked by Herschel and Struve. It resembles an 8th-magnitude star with a disk 4″ arc across: about the apparent diameter of Uranus, and the brightness of Neptune.

M15 (NGC 7078) Peg
21ʰ 30ᵐ.0, +12° 09′

A brilliant globular cluster, that should be detectable with the naked eye as a hazy star because its total magnitude is about 6. It invites comparison with M13 in Hercules—although far smaller in apparent size, its massed stars blaze more brightly. Due to its compression, the individual stars are much harder to resolve, and with an aperture of 150 mm the best that can be seen is a "powdery" appearance around the margin. Yet, each of these specks is a star perhaps a hundred times as luminous as the Sun!

M39 (NGC 7092) Cyg
21^h 32^m.2, +48° 25′

A coarse, triangular cluster of 7th and 8th-magnitude stars, perhaps best seen in the finder. A conspicuous double lies at the center.

NGC 7243 Lac
22^h 15^m.2, +49° 52′

An open grouping of moderately bright stars. There are fine fields in the vicinity.

M52 (NGC 7654) Cas
23^h 24^m.1, +61° 36′

A bright cluster, visible in the finder as a hazy patch. The stars are sprinkled in a lozenge shape.

NGC 7662 And
23^h 25^m.9, +42° 32′

An easy planetary nebula, seen as a tiny disk even with low magnification. Using a moderate power, it will be noted as circular, evenly illuminated, and fading suddenly in intensity at the edges, while apertures of 200 mm and more may show it to be annular.

NGC 7789 Cas
23^h 57^m.0, +56° 43′

A large cluster of faint stars, fading away into star-dust. Perhaps it is appropriate to end this circuit of the northern sky with the often-quoted words of W. H. Smyth, who observed the sky for its double stars and other stellar objects during the first half of the 19th century, using a 140-mm refractor, and described his observations in the famous *Bedford Cycle,* named after the English town near which he lived. NGC 7789, he wrote, "is but a mere condensed patch in a vast region of inexpressible splendour, spreading over many fields"; and the writer can offer no better advice to the observer who has brought this cluster into view than to let his telescope respond to Smyth's suggestion and explore the "inexpressible splendour" of this part of the Milky Way for himself, with no guide other than his own inclinations.

SOUTHERN DEEP-SKY OBJECTS

NGC 55 Scl
00ʰ 15ᵐ.2, − 39° 11′

A large, remarkably bright galaxy lying in a barren area of sky. Even a small telescope shows considerable elongation in an East–West direction.

NGC 104 Tuc
00ʰ 24ᵐ.3, − 72° 06′

To the naked eye, this magnificent globular resembles a 4th-magnitude star, the most prominent object near the Small Magellanic Cloud. It was, therefore, originally identified by the number 47 Tucanae. Some 8′ arc across, this is a brilliant object, partly resolved even with a small telescope. By general consensus this is inferior only to ω Centauri among the globular clusters, and some observers would not allow even this comparison.

NGC 253 Scl
00ʰ 47ᵐ.5, − 25° 10′

A galaxy, visible as a nebulosity in the finder. It is considerably elongated in a northeast–southwest direction, and has a star of about magnitude 9½ on its northwestern edge.

NGC 292 Tuc
The Small Magellanic Cloud
00ʰ 52ᵐ, − 73½°

Like an extensive haze to the naked eye; resolved into its brightest stars in the telescope, with a nebulous mass of much fainter ones making it look like what it is: a separate star-system, about 120,000 light-years away. Compared with photographs, however, it often disappoints, because the individual stars are so faint; and its identity as an isolated object is lost if a high magnification is used.

NGC 288 Scl
$00^h 52^m.8$, $-26° 35'$

A cluster of faint stars, appearing large but nebulous with a small telescope, and little condensed at the center. Magnificent with a moderate aperture.

NGC 330 Tuc
$00^h 56^m.3$, $-72° 29'$

A very condensed cluster of stars inside the Small Magellanic Cloud, requiring a large aperture to be resolved.

NGC 362 Tuc
$01^h 02^m.4$, $-70° 41'$

A splendid globular cluster, appearing, like its neighbor 47 Tucanae, near the Small Magellanic Cloud but in no way associated with it. It appears as a 6th-magnitude star with the naked eye, and telescopically is seen as a hazy ball, partly resolved around the margin with moderate apertures.

NGC 584 Cet
$01^h 31^m.4$, $-06° 51'$

A small galaxy, centrally condensed. Not spectacular, but a good sample of the class of object a grade of brightness fainter than those noticed by Messier.

NGC 936 Cet
$02^h 27^m.7$, $-01° 09'$

Classed by Herschel as a planetary, this is a moderately bright galaxy with a much fainter neighbor, NGC 941, about 15' arc further East. This galaxy found itself in Herschel's "very faint" class, which will give you an idea of the limit of his searches, using reflectors of up to 450 mm in aperture.* About 1° East again, halfway towards the star 75 Ceti, is NGC 955, a galaxy classed by Herschel as "moderately bright." This little line of galaxies, readily located by 75 Ceti, therefore constitutes an interesting test sequence.

* Herschel's telescopes used mirrors made of "speculum metal," an alloy of tin and copper. The reflectivity of this material is much lower than that of aluminum, and a 450-mm speculum-metal Newtonian would probably be matched, in light-focusing power, by a modern Newtonian about 300 mm in aperture.

M77 (NGC 1068) Cet
02ʰ 42ᵐ.7, − 00° 01′

The most remote of all the Messier objects, this is a massive galaxy containing about eight times as much material as our own. It lies about 50 million light-years away. Although not particularly conspicuous or spectacular, neither is it the faintest of the Messier objects, and it is easily swept up with a lower power as a small, condensed, nebulous patch. There is a telescopic star involved, to the southeast.

NGC 1097 For
02ʰ 46ᵐ.4, − 30° 16′

A fairly bright galaxy, appearing as an elongated haze with a central condensation.

NGC 1316 For
03ʰ 22ᵐ.7, − 37° 13′

Another galaxy, much less elongated, but also with an obvious central condensation.

NGC 1365 For
03ʰ 33ʰ.7, − 36° 07′

The most westerly of a group of four galaxies lying within about 2° of each other, the others being NGC 1380, 1387, and 1399. This is an interesting group to observe and compare. Notice also the compact cluster NGC 1436, just over the border in Eridanus.

NGC 1535 Eri
04ʰ 14ᵐ.2, − 12° 44′

A small planetary nebula, visible with a small instrument, but not particularly bright; moderate instruments show a central star.

NGC 1763 Dor
04ʰ 56ᵐ.7, − 66° 24′

A bright nebulosity, one of the many fine objects in the Large Magellanic Cloud—the Galaxy's second satellite—that lies about 145,000 light-years away. Supreme amongst this "museum" of deep-sky objects is NGC 2070, the Great Looped

Nebula, also known as 30 Doradus. This is visible without optical aid as a bright condensation in the Cloud.

M79 (NGC 1904) Lep
05ʰ 24ᵐ.2, −24° 31′

A bright but not very large globular cluster, so condensed, and at such a great distance (43,000 light-years) that it is difficult to resolve even the outer stars without a moderate aperture. Prettily situated between two 8th-magnitude stars, forming a North–South line.

M42 (NGC 1976) Ori
The Great Nebula in Orion
05ʰ 35ᵐ.2, −05° 23′

The finest nebula in the sky, and one to which all amateurs turn again and again. Probably the most marvellous aspect of this great object is neither its brightness nor its size, but its range of intensities and the texture of its draperies. Superimposed on the bright nucleus is the top of a dark nebula, at the apex of which shines the Trapezium (θ Orionis, page 248); while extensive wings of material sweep away into the black sky. But this is not merely a low-power object, because with higher magnification you will be able to discern wisps and coruscations that bring to mind John Herschel's description of the nebula as "like the breaking up of a mackerel sky." It is a fact that familiarity breeds contempt, but perhaps this is less true of M42 than of any other celestial object, for its features seem to demand close attention, and be rediscovered, each time a telescope is pointed towards it. Do not fail also to notice M43 (NGC 1982), a much fainter condensation of nebulosity surrounding a 7th-magnitude star about 8′ arc to the northeast, and the complex of nebulosities, of which the brightest component is NGC 1977, about a degree to the North, involving two 5th-magnitude stars.

The Orion Nebula, M42   A photograph of the most famous nebula in the sky, showing the intruding dark nebulae in front of the main luminous mass (*R. W. Arbour, 400-mm Newtonian*)

NGC 2132 Pic
05^h 55^m.2, −59° 54′

An open cluster of bright and faint stars.

NGC 2220 Pup
06^h 21^m.2, −44° 45′

A bright, very open grouping of stars.

NGC 2232 Mon
06^h 27^m.9, −04° 46′

An attractive grouping of stars around the yellowish 6th-magnitude star 10 Monocerotis.

M41 (NGC 2287) CMa
06^h 47^m.0, −20° 44′

A very bright open cluster, visible with the naked eye when high in the sky; it occupies a space about equal to the apparent diameter of the Moon, or ½° across.

M50 (NGC 2323) Mon
07^h 02^m.9, −08° 20′

A bright, scattered cluster, readily visible in finder or binoculars. It contains a prominent red star. This is a fine region of the Milky Way, and many other attractive fields may be noted.

NGC 2360 CMa
07^h 17^m.8, −15° 37′

A cluster of faint stars about 15′ arc across, best seen with a low or moderate power.

M47 (NGC 2422) Pup
07^h 36^m.6, −14° 28′

A coarse group of bright stars, some of the 6th magnitude, visible to the naked eye as a condensation in the Milky Way. The region here is very fine in binoculars, and there are many telescopic groupings to be swept up.

M46 (NGC 2437) Pup
07^h 41^m.8, −14° 48′

Seen with a low power, this open cluster almost fills the field with its sprinkle of stars and star-dust. It is visible with the naked eye, in a good sky. Notice also the small planetary nebula NGC 2438, on the southeastern outskirts, about 10′ arc from the center of the cluster; it appears faint in a small telescope, but with apertures exceeding 200 mm it should be seen annular.

M93 (NGC 2447) Pup
07^h 44^m.5, −23° 52′

A fine bright cluster, containing stars of a wide range of magnitude.

NGC 2451 Pup
07^h 45^m.3, −37° 57′

A brilliant cluster of bright stars surrounding the magnitude 3½ star c Puppis. This grouping is very conspicuous with the naked eye. A much less noticeable object, NGC 2477—a large cloud of faint stars—lies about 2° away to the southeast.

NGC 2516 Car
07^h 58^m.3, −60° 51′

A very bright cluster surrounding a 6th magnitude star.

NGC 2506 Mon
08^h 00^m.0, −10° 36′

The brighter stars in this extensive group can be made out with a small telescope; they are set against a background of stardust. This is a magnificent sight.

NGC 2547 Vel
08^h 10^m.6, −49° 15′

A bright cluster, with a 6th-magnitude star involved, lying on the edge of the great rift in the Milky Way that crosses Vela. This rift is caused by dark nebulae blocking out the stars beyond.

NGC 2539 Pup
08^h 10^m.7, −12° 49′

A bright, scattered cluster near a 4th-magnitude star.

NGC 2548 (M48?) Hya
08^h 13^m.7, −05° 46′

This is one of the "missing" Messier objects because the coordinates given in his original list do not coincide with any known object, but the identity here could be correct if we assume that Messier made a simple error in declination when recording the position. This is an open cluster of faint stars, with a somewhat elongated central grouping apparent when observed with a low power.

NGC 2808 Car
09^h 11^m.9, −64° 52′

A fine globular cluster, just detectable with the naked eye. It has a well-condensed nucleus.

NGC 2792 Vel
09^h 12^m.5, −42° 26'

This small, fairly bright planetary will require a high magnification to be seen well because its apparent disk is only slightly larger than that of Uranus.

NGC 2932 Vel
09^h 35^m.2, −46° 56'

A scattered cluster of bright telescopic stars.

NGC 3115 Sex
10^h 05^h.3, −07° 42'

A small galaxy, very elongated, measuring about 4' arc in length, and readily spotted even with a small aperture.

NGC 3242 Hya
10^h 24^m.7, −18° 38'

A very distinct planetary nebula measuring about 40" arc across, with a well-defined margin. This object is easily located with a low magnification.

NGC 3372 Car
10^h 45^m.0, −59° 41'

An amazing, brilliant nebulosity in a great naked-eye star cloud. At the center is the 6th-magnitude star η Carinae, the famous *Blaze Star,* that rose to zero magnitude in 1842. This is an extraordinary region of the Milky Way, crowded with superb telescopic objects, including the wonderful open cluster NGC 3532 at 11^h 06^m.5, −58° 40', a mass of bright telescopic stars that can be seen with the naked eye.

NGC 3521 Leo
11^h 05^m.8, −00° 02'

This is a typical example of the many galaxies to be found in the Leo-Virgo-Coma region of the sky (most of it lying North of the celestial equator), being the 13th in William Herschel's "bright nebula" class. It is just ½° East of the 6th-magnitude star 62 Leonis, that should be put out of the field of view. The galaxy is seen in very inclined view, and appears elongated, with no definite nucleus.

NGC 3766 Cen
11ʰ 36ᵐ.2, −61° 36′

A large cluster of stars with a great range of brightness. It can be seen with the naked eye as a hazy 4th-magnitude "star," rivaling the nearby star λ.

NGC 3918 Cen
11ʰ 50ᵐ.2, −57° 11′

A very bright, small planetary nebula, with a disk about 12″ arc across. Visible in the finder, although its nonstellar nature will not be obvious.

M68 (NGC 4590) Hya
12ʰ 39ᵐ.5, −26° 45′

A globular cluster about 4′ arc across, and therefore rather smaller and fainter than many others in this list. Its brighter individual stars can be resolved with moderate power, and one of them, on the northern border, has a noticeably reddish tint. This object lies in a star-poor region of the sky.

M104 (NGC 4594) Vir
The Sombrero Galaxy
12ʰ 40ᵐ.0, −11° 37′

A galaxy observed exactly in the plane of the arms, and apparently cut in two by the dark "gust" of its star-forming regions; it has acquired its name from the appearance in the many long-exposure photographs that exist of this object. Even a small telescope will reveal the rounded nucleus and the straight projections of the arms, but the dark cleft will not be evident without a moderate aperture. In some ways, it does look more like a galaxy than do most of its brethren!

NGC 4755 Cru *The Jewel Box*
12ʰ 53ᵐ.6, −60° 21′

Often referred to as κ Crucis, and visible with the naked eye as a prominent "star" near the eastern arm of the Cross. This is a brilliant open cluster of bright stars around the white supergiant star κ, that appears only of the 6th magnitude despite being some 25,000 times as luminous as the Sun. This cluster lies at a distance of about 6,000 light-years, but some reddish stars, much nearer, are superimposed on it, the color effect giving support to its popular name.

NGC 5061 Hya
13ʰ 18ᵐ.2, −26° 52′

A galaxy, appearing small and condensed, with a prominent nucleus.

NGC 5139 Cen ω *Centauri*
13ʰ 26ᵐ.7, −47° 19′

ω Centauri was originally catalogued as a star because of its prominence to the naked eye; it appears as a hazy 4th-magnitude star. In a telescope, it is breathtaking: a ball of faint stars the size of the Full Moon. Although often reproduced in photographs, the original is far more impressive because the fine tracery of the lines of stars can be followed into the center of the mass; photographs always burn the center out into uniform whiteness.

NGC 5662 Cen
14ʰ 35ᵐ.0, −56° 34′

A cluster of bright and faint stars surrounding a reddish one of the 7th magnitude.

NGC 5897 Lib
15ʰ 17ᵐ.4, −21° 02′

A rather inconspicuous globular cluster, appearing in a small instrument as a circular nebulosity.

NGC 6025 TrA
16ʰ 03ᵐ.6, −60° 31′

A very bright cluster, obvious with the naked eye. The brightest stars in this splendid group are of the 7th magnitude.

NGC 6067 Nor
16ʰ 13ᵐ.2, −55° 14′

A fine large cluster near κ, brilliant even to the naked eye, and lying in the superb star cloud that extends across this part of the constellation.

M80 (NGC 6093) Sco
16ʰ 17ᵐ.0, −23° 00′

A fairly conspicuous globular cluster some 5′ arc across. However, the stars are so dim and close-packed that in small telescopes it appears nebulous. This is not an effect of distance alone; in spatial terms the stars really are crowded more closely together than in other examples of its kind.

NGC 6087 Nor
16ʰ 18ᵐ.8, −57° 55′

A very scattered group of bright telescopic stars, containing the Cepheid variable star S Normae, magnitude range 6.1–6.8, period 9¾ days.

M4 (NGC 6121) Sco
16ʰ 23ᵐ.6, −26° 31′

Very easily located only 1½° West of brilliant Antares (α Scorpii), this is a large (20′ arc diameter) but rather dim globular. A moderate telescope will resolve stars right into its center, and it is a fine sight in a dark sky. At a distance of only about 7500 light-years, this is the nearest of all the Messier globulars; it is probably about five times closer than M80.

M12 (NGC 6218) Oph
16ʰ 47ᵐ.3, −01° 58′

A fairly bright globular cluster, that is so loose that it is easily resolved, and looks almost like a small open cluster. This is a fine object.

M10 (NGC 6254) Oph
16ʰ 57ᵐ.2, −04° 07′

Although about the same size and brightness as its neighbor, M12, this globular is more compressed, and the center cannot be resolved with a small telescope.

M62 (NGC 6266) Sco
17ʰ 01ᵐ.2, −30° 08′

A remarkable globular cluster, fairly bright, but very asymmetrical, with the condensed nucleus well to the East of the center. With a very low power, or in the finder, it resembles the head of a telescopic comet, as was noted by Messier in his original list.

M19 (NGC 6273) Oph
17ʰ 02ᵐ.6, −26° 16′

A bright globular cluster. In a small instrument, however, it appears as a nebulosity, with little central condensation.

The sight is magnificent in large telescopes.

M9 (NGC 6333) Oph
17^h 19^m.2, −18° 32′

Another very dense globular, not particularly compressed. A moderate aperture and power are necessary if even the outer stars are to be resolved.

M14 (NGC 6402)
17^h 37^m.5, −03° 16′

Although of about the same size as the two preceding globular clusters, this object appears rather dimmer and less resolvable even with a medium aperture. A coarse double star precedes it in the field, and a coarse triple follows it.

M6 (NGC 6405) Sco
17^h 40^m.0, −32° 13′

Visible with the naked eye. In the finder, this open cluster looks like a butterfly brooch; in a small telescope, lines of magnitude 9 and 10 stars radiate outwards from the center to give it the appearance of a flower with open petals. A line of three small double stars lies to the northwest. Without doubt, this is one of the most beautiful sights in the sky.

M7 (NGC 6475) Sco
17^h 54^m.0, −34° 50′

Easily visible with the naked eye, this is a brilliant, scattered mass of bright stars, filling a low-power field of view. Near the center, the stars appear to be arranged in lines; they include one of yellow tint.

M23 (NGC 6494) Sgr
17^h 57^m.0, −19° 02′

A fine open cluster of 9th-magnitude stars, with a 7th-magnitude star to the northwest. This can partly be resolved in the finder.

M20 (NGC 6514) Sgr
The Trifid Nebula
18^h 02^m.4, −23° 03′

This object is often figured in books, but the visual appearance bears little resemblance to the photographs. Although conspicuous even in a small instrument, it then appears as no more than a rather faint but extensive irregular nebulosity, involving two 8th-magnitude stars. This is an emission-type nebula, in which stars are being formed and which glows because its atoms are excited by very hot white giant stars in its interior. To see the dark lanes that give it the "trifid" appearance will probably require an aperture of about 300 mm.

M8 (NGC 6523) Sgr
The Lagoon Nebula
18ʰ 03ᵐ.7, − 24° 24′

This is an extraordinary object: a coarse, rather spherical cluster of 9th-magnitude stars, with two brighter ones to the West, are wrapped in luminous haze. This haze is the remains of the nebula from which the stars condensed. Patches of nebulosity appear elsewhere in the field, and this object is worth long study on a very dark night. M8 and its neighbor M20 make a distinctive pair in the finder or binoculars.

M21 (NGC 6531) Sgr
18ʰ 04ᵐ.7, − 22° 30′

This open cluster is neither large nor bright, but it lies within a degree of the Trifid Nebula and appears to be linked by two chains of stars, making an attractive sight. The two objects are not truly connected, however, for the cluster is twice as remote as M20, lying about 4000 light-years away.

M24 & NGC 6603 Sgr
18ʰ 18ᵐ.4, − 18° 27′

These two objects, though occupying the same region of the sky, are quite distinct, though often presented as synonymous. The true M24 is a naked-eye star cloud, much larger than the Moon's apparent diameter, that lies north of μ Sagittarii. This is a most superb sight, either in the finder, when its individual size and magnificence can best be appreciated, or in the main telescope, where stars mass and spread in sparkling heaps. NGC 6603 is an insignificant faint cluster, nebulous to low powers, and not really worth the search in all this splendor.

M16 (NGC 6611) Ser
18ʰ 18ᵐ.8, − 13° 48′

With a small telescope, this fine open cluster has a roughly hexagonal shape, the boundary being marked by 9th-magnitude stars and the interior filled with haze that marks the presence of a faint emission nebula.

M18 (NGC 6613) Sgr
18ʰ 19ᵐ.9, − 17° 09′

Visible in the finder as a hazy patch, this is a small cluster of stars of magnitude 9 and fainter.

M17 (NGC 6618) Sgr
The Omega Nebula
18ʰ 20ᵐ.8, − 16° 12′

Also known as the *Horseshoe Nebula,* this is another emission nebula, excited to glow by nearby stars. Very conspicuous even in the finder, this appears as a wide, bright streak, with the western end very obviously overlaid by dark obscuring matter. Some attention, and a moderate aperture, may be required to make out the semicircle of nebulosity to the South that gives it a resemblance to a figure 2, or possibly a swan! There are two 9th-magnitude stars projected upon the nebula.

M28 (NGC 6626) Sgr
18ʰ 24ᵐ.5, − 24° 53′

A globular cluster, visible in the finder. In a small instrument it appears as a bright, nebulous object with a compressed center.

M69 (NGC 6637) Sgr
18ʰ 31ᵐ.4, − 32° 23′

A rather small and faint globular cluster, lying close to the southeast of a magnitude 8½ star.

M25 (IC 4725) Sgr
18ʰ 31ᵐ.9, − 19° 10′

This very loose, bright cluster was omitted from the NGC, but appeared in the *Index Catalogue* of 1908, that for the most part listed new discoveries of faint objects. M25 is not particularly noteworthy in this region of numerous star groupings. An 8th-magnitude star lies near the center.

M22 (NGC 6656) Sgr
18^h 36^m.4, −23° 56′

A superb globular cluster measuring more than ¼° across, appearing large and nebulous in the finder. It is well resolved even with a moderate aperture, and is one of the finest of its class. M22 can be seen with the naked eye as a hazy spot.

M70 (NGC 6681) Sgr
18^h 43^m.2, −32° 18′

A rather small globular cluster with a bright central condensation. There is a curious "tail" of 9th-magnitude stars running off to the North. The outer regions can be resolved with a moderate aperture.

M26 (NGC 6694) Sct
18^h 45^m.3, −09° 25′

A small cluster containing one star of magnitude 9, a few of magnitude 10, and a background haze of much fainter ones.

M11 (NGC 6705) Sct
18^h 51^m.1, −06° 18′

This superb, granular, fan-shaped cluster is the central condensation of a wonderful Milky Way region. In a good sky, the sight is breathtaking even with an aperture of 90 mm; there seems to be no end to the layers of fainter and fainter stars. The cluster has a 9th-magnitude star near the apex, and there is a small double to the North. It is visible with the naked eye as a bright glow.

M54 (NGC 6715) Sgr
18^h 55^m.1, −30° 29′

A small globular cluster of unusual appearance, like a 9th-magnitude star surrounded by haze, the "star" merely being the effect of a sudden concentration at the center.

M55 (NGC 6809) Sgr
19^h 40^m.0, −30° 58′

A very large globular cluster, appearing nebulous in the finder. It is only slightly condensed, and a moderate telescope will resolve it into stars right across its breadth. This is a fine object.

M75 (NGC 6864) Sgr
20ʰ 06ᵐ.1, − 21° 56′

A small globular cluster in a barren field, resembling, as does M54, a hazy star. The individual stars are close and faint, and not easily resolved.

NGC 7009 Aqr
The Saturn Nebula
21ʰ 04ᵐ.2, − 11° 23′

A bright planetary nebula, noticeably nonstellar with a magnification of × 35; a low-power field contains a star of similar brightness for comparison. With such a power, also, it is seen to be noticeably elliptical in a northeast–southwest direction. Higher powers reveal the edge to be somewhat ill-defined. The two very faint rays, that have given it the name, are not visible except on photographs.

M2 (NGC 7089) Aqr
21ʰ 33ᵐ.5, − 00° 50′

A bright globular cluster. There are many faint stars near, and one of these is superimposed upon the outer reaches of the cluster, which itself cannot be resolved even into its peripheral stars without a moderate aperture. It is not particularly condensed at the center.

M30 (NGC 7099) Cap
21ʰ 40ᵐ.3, − 23° 12′

A globular cluster of moderate size, partly resolved with apertures of 200 mm or so. Because this is the last southern object in our cycle, it may be fitting to end, as we did the northern cycle, with another quotation from W. H. Smyth, inspired by his view of this remote and isolated object:

"Here are materials for thinking! What an immensity of space is indicated! Can such an arrangement be intended, as a bungling spouter of the hour insists, for a mere appendage to the speck of a world on which we dwell, to soften the darkness of its petty midnight?"

CHAPTER 11

Windows into Space

The object of this chapter is to present different areas of the sky for detailed inspection. Each of the ten selected regions contains about 300 square degrees, and within each one you will find samples of double stars, clusters, nebulae, and galaxies marked for inspection. These are all described, briefly, in the accompanying lists, unless they have already been referred to in the previous two chapters.

The regions have been chosen to give equal coverage of both celestial hemispheres, and are spread around the sky so that at least one will be above the horizon for part of any clear night. The objects in each one are listed in order of increasing right ascension, from West to East, but their celestial coordinates are not given because they can readily be located on the charts. These charts show stars down to approximately the 7th magnitude.

If no date is given next to a double-star measure, then you can assume that the position will change little before the end of the century. Close binaries may have more than one measure. The positions of some underobserved binaries are uncertain, and these are indicated by a question mark. Wherever possible, a current measure has been extrapolated from the Lick Observatory's *Index Catalogue of Double Stars* (1963), while details of close binary stars have been obtained from Finsen & Worley's *Third Catalogue of Binary Stars* (1970) and Wolfgang Wepner's *291 Doppelstern Ephemeriden fur die Jahre 1975–2000*.

Do not think that *everything* of interest has been included in these lists! There are bound to be other things that catch your attention as you sweep over an area: perhaps richly colored stars, or unusual groups of stars. Take note of these: the eventual aim could be a comprehensive description of each region, based on your own observations, and perhaps supplemented by those of other people. The writer has had to draw on many sources of information, besides his own observations, for the deep-sky notes in this book, because no single observer could possibly cover the whole sky in a comprehensive way unless his nights were always cloudless and he needed no sleep!

Not all of these objects are spectacular. Nor are they all accessible with even moderate apertures. Some of the close binaries and faint nebulae or galaxies may tax the powers of the largest amateur-owned instruments, but they have been included here for two reasons: firstly, large amateur telescopes, particularly on cheap altazimuth mounts, are becoming more popular; secondly, there is a tendency for books to include lists of only the easiest objects in the sky. In struggling to make the most of some of the more elusive objects listed here, you will be exercising the skills that count for so much in astronomical work, and you may derive the satisfaction of observing things that other enthusiasts have never even looked for.

You will almost certainly require several clear nights to examine one region thoroughly, but do not let this deter you, for the project is not meant to be a race. It should, instead, be something to which you can turn your attention from time to time. Come back to the objects again in a year or two, perhaps with a different telescope, and certainly with more experience: compare your observations, and reflect upon the differences.

R.A. 5^h–6^h, Dec. $+10°$–$-10°$

REGION 1

This equatorial region is confined almost entirely to Orion, although parts of Eridanus and Monoceros are included. Here you will find many nebulous masses and double stars, but few clusters and no galaxies.

Double stars

Σ630	6.5	7.7	14″.2	50°		
Σ642	5.3	8.5	52.8	9	66 Eri	
OΣ98	5.9	6.7	0.68	5	(1985)	14 Ori
			0.79	321	(2000)	

Σ649	5.7	9.7	21.6	81	
Σ652	6.3	7.8	1.7	183	
Σ654	4.6	8.4	7.0	64	ρ Ori
OΣ517	6.9	7.1	0.48	239	(1985)
			0.53	244	(2000)
Σ668	0.1	8?	9.5	202	*Rigel,* β Ori (See page 248)
h 2259	3.7	10.9	36.0	60	
β189	6.3	11.0	4.4	285	
h697	5.6	11.3	32.7	60	AB
	5.6	10.8	37.6	110	AC
Σ696	5.0	7.1	32.1	28	
Σ701	6.0	7.8	5.9	142	
WNC 2	6.1	7.3	2.8	158	
DA 5	3.8	4.8	1.6	78	η Ori
KNT 3	4.7	10.3	2.7	323	ψ Ori
h 2268	7.0	9.6	26.0	300	
Σ 725	5.0	10.2	12.7	87	
Σ 728	4.5	6.0	0.96	43	(1985) 32 Ori
			1.05	40	(2000)
Σ 729	5.8	7.1	1.8	27	33 Ori AB
	5.8	13.4	95	52	AC
Σ14	2.5	6.6	52.6	359	δ Ori
Σ 734	6.7	8.4	1.8	354	AB
	6.7	8.4	29.4	243	AC
h 118	6.2	9.8	27.5	264	
Σ 738	3.4	5.6	4.4	43	λ Ori AB (See page 227)
	3.4	11.2	28.6	184	AC
Σ 747	3.7	5.6	35.7	223	AB
	3.7	11.2	78	271	AC
DA 4	4.7	7.9	1.6	213	
Σ 748					θ Ori (See page 248)
Σ 752	2.9	7.0	11.3	141	ι Ori (See page 250)
Σ 750	6.5	8.5	4.2	60	
Σ 754	5.7	8.9	5.2	287	
Σ 762					σ Ori (See page 250)
Σ 774	2.0	4.2	2.5	164	ζ Ori (See page 251)
β 1052	6.7	7.7	0.7	160?	
Σ 789	6.2	8.7	14.0	150	
Σ 790	6.5	8.8	6.9	89	
J 251	5.9	11.9	18.0	295	
Σ 795	6.1	6.1	1.6	212	52 Ori (See page 251)
56 Ori	5.0	13.5	43.4	121	
Σ 816	6.8	9.3	4.3	289	

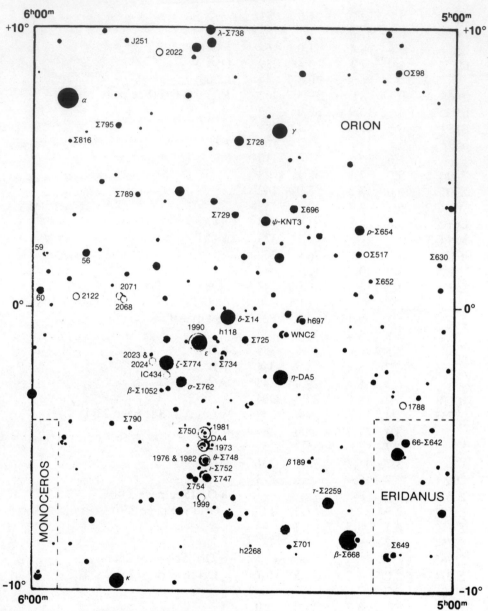

Region 1   Orion (Eridanus, Monoceros)

α Ori	0.5v	13.5	77.0	348	*Betelgeuse*
59 Ori	6.1	9.7	36.7	204	
60 Ori	5.2	11.8	19.1	30	

Clusters

1981 A fine bright cluster, visible in the finder.

2112 A rather small cluster of faint stars, attractive with apertures of 150 mm and above.

Nebulae

1788 A small nebulosity, faint with an aperture of 150 mm.

1976 M42, the Great Nebula in Orion, described on page 302.

1982 M43, a separate cloud of the Great Nebula, described on page 302.

1973 A complex area of nebulosity to the North of M42, containing altogether three separate NGC entries.

1990 An excessively faint, extensive nebulosity involving ε Orionis. It will probably elude apertures smaller than 300 mm.

1999 A small nebulosity containing a 10th magnitude star, originally listed as a planetary because of its appearance. A testing object for apertures below 200 mm.

2022 A small planetary nebula with no central star, not difficult with a moderate aperture.

IC 434 A very faint nebulosity extending South of ζ Orionis, and containing on its eastern edge the minute dark irregular projection known as the *Horsehead Nebula,* prominent on photographs but very minute. It is normally considered a difficult object with a 200- to 300-mm telescope, although the comet-hunter Leslie Peltier recorded it with his 150-mm refractor.

2023, 2024 Two further patches of faint nebulosity to the East of ζ Orionis. A field bar may be necessary, to reduce the glare of the bright star.

2068 M78, a nebula described on page 282. There are two very faint nebulous patches (2064, 2067) to the southwest.

2071 A 10th-magnitude star involved in nebulosity. Faint with an aperture of 150 mm.

REGION 2

This region includes part of the Milky Way where it crosses Taurus and Gemini. Parts of Orion and Auriga are also to be found here. It is a good hunting-ground for the double-star observer.

Double stars

Σ 623	6.8	8.3	20″.5	205°	
103 Tau	5.6	8.6	35.3	197	
Σ 645	6.1	8.6	11.8	27	
Σ 680	6.2	10.1	9.0	204	
h 365	4.8	10.9	37.8	348	114 Tau AB
	4.8	10.4	58.8	194	AC
	4.8	11.6	74	280	AD
Σ 716	5.8	6.6	4.8	205	118 Tau
Σ 749	6.4	6.5	1.1	325	
Σ 764	6.8	7.4	26.0	14	
OΣ 118	6.1	7.6	0.6	318	
β 1054	5.6	11.6	15.0	232	
OΣ 129	6.3	11.0	10.0	209	
β 1241	5.8	14.4	18.4	63	
β 1059	3.2	11	73	77°	μ Gem
15 Gem	6.6	8.0	26.5	204	
β 1192	4.1	13.9	53.9	13	ν Gem AQ
	4.1	12.6	53.5	253	AR
Ho 514	6.8	12.5	20.3	118	AB
	6.8	12.8	39?	35?	AC
OΣ 149	7.1	8.7	0.62	316	(1985)
			0.72	295	(2000)
Pou 1700	6.8	14.3	19.5	48	
Ho 625	6.7	11.9	13.4	194	AB
	6.7	9.3	47.4	251	AC
OΣ 152	6.0	7.8	0.9	36	54 Aur
h 397	6.5	11.7	30.7	46	
OΣ 160	6.7	9.7	1.6	190	
OΣ 161	6.8	11.1	20.1	171	
39 Gem	6.1	12.1	30?	40?	

Clusters

1746   A large, scattered group of faint stars.
2129   A small cluster of bright and faint telescopic stars, including one of the 7th magnitude.

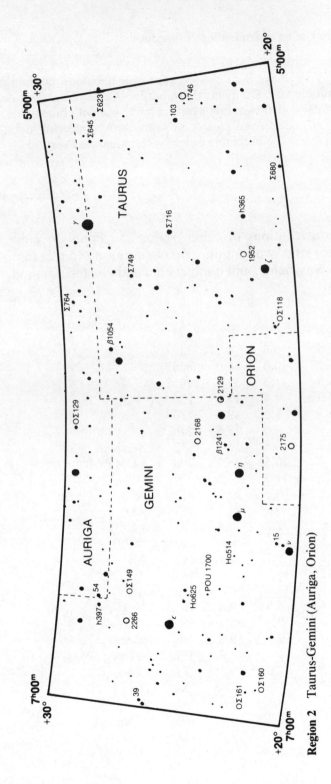

Region 2 Taurus-Gemini (Auriga, Orion)

2168   M35 described on page 283.
2175   A very small and faint cluster, appearing nebulous, containing an 8th-magnitude star. On its northern border is a large but faint nebulosity, NGC 2174, that can, however, be made out with binoculars.
2266   A small, condensed cluster of faint stars, appearing nebulous in a small telescope but a fine object with moderate apertures.

R.A. 7ʰ–8ʰ, Dec. − 10°− − 30°

REGION 3

The fine constellations of Canis Major and Puppis occupy most of this area, together with a thin strip of southern Monoceros. Here are fine Milky Way fields, and double stars and clusters abound.

Double stars

β 572	6.6	10.6	5″.2	143°	
β 328	5.6	6.8	0.7?	110?	
I 772	6.9	12.0	16.0	326	
β 329	6.0	11.3	38?	105?	
h 3934	7.0	9.4	13.6	236	
h 3938	6.5	8.3	19.7	250	
Σ 1064	6.9	8.9	15.6	240	
h 3945	4.8	6.8	26.5	52	See page 253.
h 3948	4.4	10.5	8.2	90	τ CMa AB
	4.4	11.2	14.5	79	AC
Σ 1097	6.1	8.5	20.0	313	AC
	6.1	9.5	23.2	157	AD
	6.1	12.2	32.3	43	AE
h 2391	6.5	11.0	16.7	290	
Σ 1104	6.1	7.7	2.1	17	AB
	6.1	10.7	20.4	188	AC
	6.1	11.2	50?	30?	AD
n Pup	5.9	6.0	9.6	114	
p Pup	4.6	9.3	38.4	156	
Σ 1120	5.6	9.5	19.6	36	In M47.
κ Pup	4.5	4.7	9.9	318	See page 253.
Hu 710	7.1	7.6	0.36	75	(1985)
			0.38	54	(2000)
Σ 1138	6.1	6.8	16.8	339	2 Pup
Σ 1146	5.6	7.7	2.0?	360?	5 Pup
See 86	4.6	12.6	27.0	198	o Pup

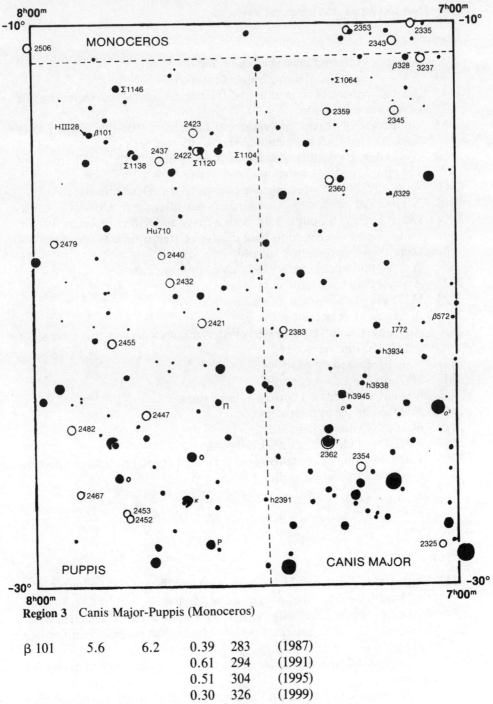

Region 3 Canis Major-Puppis (Monoceros)

β 101	5.6	6.2	0.39	283	(1987)
			0.61	294	(1991)
			0.51	304	(1995)
			0.30	326	(1999)
H III 28	6.9	10.5	24.0	195	

Clusters

2335 A rather scattered cluster of stars, mostly faint.

2343 A loose group, containing only a few stars.

2345 A small cluster of moderate and faint telescopic stars, including one of the 8th magnitude.

2353 An attractive cluster of bright and faint stars that fills a low-power field. There is a 6th-magnitude star on its southern edge.

2354 A cluster consisting mainly of faint stars. With an aperture of 150 mm, it appears nebulous.

2360 A small cluster containing some fairly bright telescopic stars.

2362 A group of faint stars surrounding the 4th-magnitude star τ CMa, which itself is a double star, with a magnitude 10 companion 10″ arc away. This is believed to be a cluster of stellar infants, possibly not more than one million years old.

2383 A small and very compressed cluster of faint stars.

2421 A small cluster of faint stars.

2422 M47; see page 304.

2423 A cluster of very faint stars.

2432 An elongated and compressed cluster of faint stars, an attractive sight with a 150-mm telescope.

2437 M46, described on page 304.

2447 M93, described on page 305.

2453 A small, compressed cluster of faint stars.

2455 A loose cluster of faint stars.

2479 A small cluster of faint stars.

2482 A scattered cluster of rather faint stars.

2506 This small cluster contains some very faint stars, and a large aperture is necessary to see it well.

Nebulae

2327 A small and faint nebulosity containing an 8th-magnitude double star, but barely visible with an aperture of 250 mm.

2359 An easy, bright nebulosity, visible with a 60-mm aperture telescope.

2440 A small planetary nebula, very ill-defined, but readily found with a low power.

2452 A 300-mm telescope may be needed to reveal this very faint, round, planetary.

2467 A dim planetary nebula on the edge of a small cluster, containing an 8th-magnitude star.

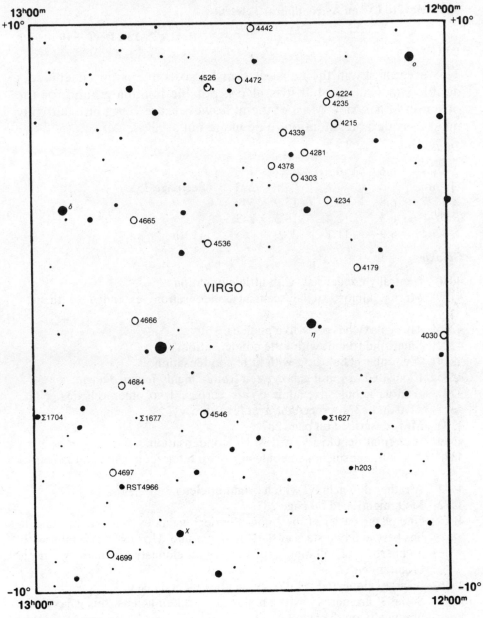

Region 4 Virgo

Galaxy

2325 An object for large apertures only, this galaxy appears of the 12th
      magnitude. Slightly condensed, it appears so faint partly because of
      dimming by interstellar dust: this is why so few galaxies can be seen
      near the great circle of the Milky Way.

REGION 4

This area of sky in the constellation Virgo contains the magnificent double star Gamma, but it is also a notable hunting-ground for the observer of galaxies. Most of them, however, are very faint, and may need considerable attention to be made out at all.

Double stars

Σ 1627	6.6	6.9	20″.1	196°	
Gamma	3.5	3.5	3.5	293	See page 255.
Σ 1677	6.8	8.3	15.9	349	
RST 4966	6.3	12.4	16.3	243	
Σ 1704	5.9	11.1	20.9	55	44 Vir

Galaxies

4030   Centrally condensed, with little elongation.

4179   Strong, almost stellar, central condensation, extended southeast–northwest.

4215   This galaxy has a starlike nucleus.

4224   Small and round, with little condensation.

4234   A circular nebulosity, with little condensation.

4235   Elongated: a spiral galaxy seen from a highly inclined angle.

4281   A large nebulosity, some 5′ arc across. Two other galaxies, NGC 4270 and 4273, precede it by a few ′ arc.

4303   M61, described on page 287.

4339   A circular nebulosity, with slight condensation.

4378   A small, conspicuous nebulosity, with a telescopic star to the southeast.

4442   A rather dim galaxy, with a bright nucleus.

4472   M49, mentioned on page 287.

4526   One of the easier of the ''non-Messier'' galaxies to locate because it lies between two stars of the 7th magnitude. However, it is faint with an aperture of 150 mm, appearing as an elongated nebulosity with a brighter center.

4536   A large, elongated nebulosity, centrally condensed.

4546   Some 3′ arc across, with an almost starlike nucleus, this galaxy lies practically on the ecliptic.

4665   A fairly condensed galaxy with a magnitude 10 star to the southwest.

4666   A very elongated nebulosity about 5′ arc across, with a strong central condensation.

4684   A faint nebulosity, little condensed.

4697   One of the more conspicuous galaxies in this region. It has a bright nucleus, and appears slightly elongated.

4699   A fairly bright, well-condensed object.

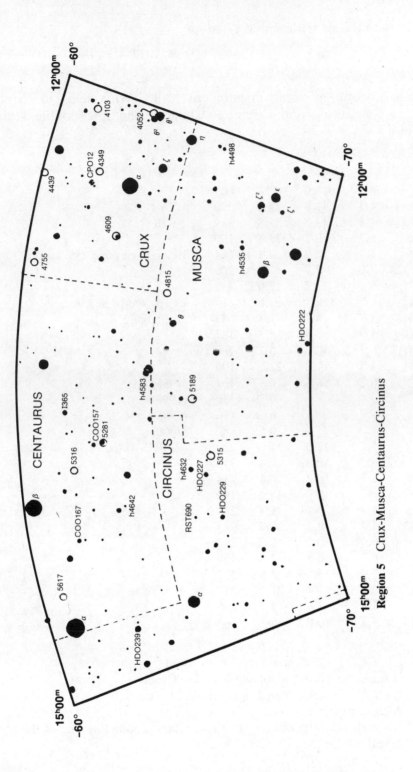

Region 5 Crux-Musca-Centaurus-Circinus

REGION 5

This area contains some magnificent Milky Way fields in Crux, Musca, Centaurus, and Circinus. Clusters and fine double stars abound.

Double stars

h 4498	6.2	8.0	8″.7	60°	
h 4501	4.3	11.8	44.0	299	η Cru
h 4512	4.3	13.3	33.8	339	ζ Cru
ζ² Mus	5.3	10.7	32.4	130	
α Cru	1.4	2.9	4.1	114	*Acrux;* see page 255.
CPO 12	7.0	7.7	1.9?	200?	
h 4535	6.5	12.2	17.0	339	
h 4547	4.7	8.5	26.8	10	ι Cru; see page 258.
β Mus	3.7	4.0	1.3	33	See page 258.
θ Mus	5.9	7.5	5.3	187	
HDO 222	6.1	12.0	30.7	280	
h 4583	5.5	11.2	22?	190?	
I 365	6.2	6.5	0.37	151	(1985) AB
			0.39	130	(1990)
			0.38	110	(1995)
			0.25	82	(2000)
	5.6	12.0	45.0	229	(AB)C
COO 157	6.2	10.2	9.4	318	
h 4632	6.2	10.2	6.4	14	
HDO 227	6.1	12.0	46.2	197	
β Cen	0.9	4.1	1.3	240	
h 4642	6.7	10.2	26.4	335	
RST 690	6.9	10.2	1.8	282	
COO 167	6.7	8.5	2.8	159	
HDO 229	5.9	13.0	23.8	307	
α Cen	0.0	1.4	21.2	212	(1985) See page 259.
α Cir	3.4	8.8	15.7	230	
HDO 239	5.3	10.8	36.3	64	

Clusters

4052 A scattered cluster involving θ¹ and θ² Crucis.
4103 A cluster of moderately bright telescopic stars.
4349 A cluster of faint stars.
4439 A small, rather poor cluster of faint stars, containing one of the 8th magnitude.

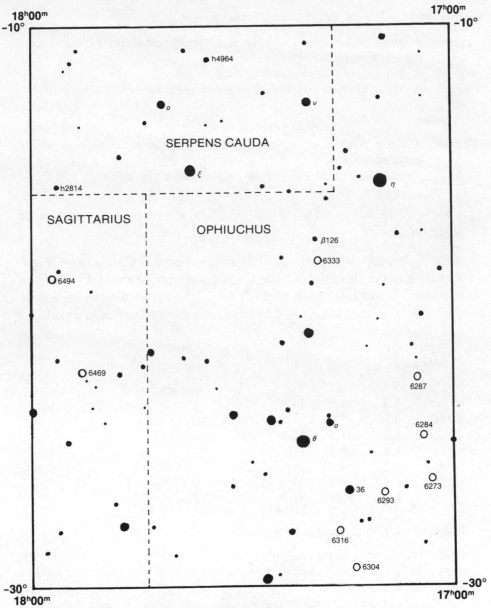

Region 6 Ophiuchus (Serpens Cauda, Sagittarius)

4609 A loose cluster of telescopic stars, with one of the 6th magnitude. This is interesting because it lies in the *Coalsack,* a large nebula blotting out part of this Milky Way region.

4755 The famous *Jewel Box,* containing some naked-eye stars, mentioned on page 307.

4815 A cluster on the edge of the Coalsack, its members ranging from the 9th magnitude to extreme faintness.

5281   A small, rather sparse cluster of moderate telescopic stars, containing a 6th-magnitude star.

5316   A grouping of 10th-magnitude stars.

5617   A coarse cluster, containing some binocular stars.

Nebulae

5189   A fairly prominent curved nebulosity, in the same field as a 7th-magnitude star.

5315   A faint (magnitude 10½) almost stellar planetary nebula.

R.A. 17ʰ–18ʰ, Dec. − 10°– − 30°

REGION 6

This region contains parts of Ophiuchus, Serpens Cauda, and Sagittarius, and the direction of the Galactic center is about 1° South of the star 3 Sagittarii. Much of this part of the sky is heavily veiled by dark interstellar clouds, but elsewhere the Milky Way is brilliant.

Double stars

36 Oph	5.1	5.1	4″.7	152°	See page 264.
o Oph	5.4	6.9	10.3	355	
β 126	6.3	7.4	1.9	262	AB
	6.3	11.3	11.4	140	AC
ν Ser	4.4	8.4	45.5	27	
h 4964	5.7	8.7	55.0	225	
h 2814	6.1	8.6	20.8	157	AB
	6.1	11.5	33.7	349	AC

Clusters (All globular, unless stated otherwise.)

6273   M19, described on page 309.

6284   Very condensed, appearing like a nebulous star in small and moderate instruments.

6287   A faint object, weakly condensed, about 5′ arc across. (Note that this, and NGC 6304, were classed by William Herschel as nebulae—in other words, he failed to resolve them into stars.)

6293   Bright, some 5′ arc across, and well consensed.

6304   A cluster showing little central condensation: individual stars very faint.

6316   Distinguished by a very intense central condensation, and almost starlike with low powers.

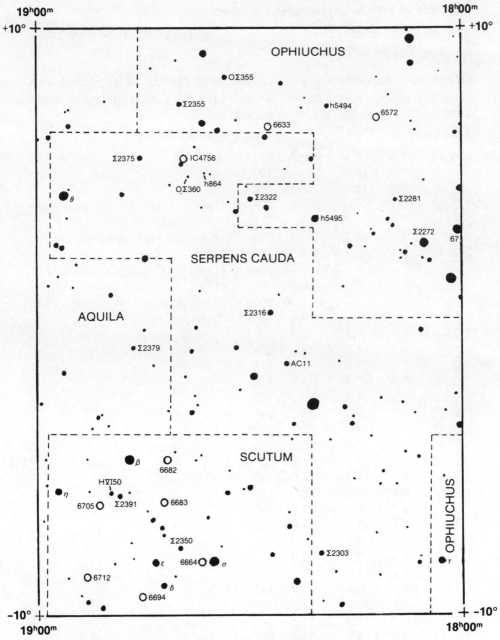

Region 7 Ophiuchus-Serpens Cauda-Scutum (Aquila)

6333 M9, described on page 310.
6469 A scattered open cluster of bright and faint stars.
6494 M23, described on page 310.

REGION 7

Ophiuchus, Serpens Cauda, Scutum, and Aquila all appear in this area. It contains a fine region of the Milky Way, and there are many double stars and clusters to be found.

Double stars

67 Oph	3.9	8.5	54″.5	143°	AC	
	3.9	10.9	45.8	179	AE	
	8.5	12.5	8.4	128	CD	
Σ 2262	5.3	6.0	1.8	278	(1985)	τ Oph AB; see page 264.
	5.3	9.3	100	127		AC
Σ 2272	4.3	6.3	2.21	293	(1984)	70 Oph; see page 238.
Σ 2281	5.9	7.4	0.38	314	(1985)	73 Oph
			0.44	304	(1990)	
			0.50	296	(1995)	
			0.56	290	(2000)	
h 5494	5.6	11.4	39.6	70		
Σ 2303	6.6	9.1	1.9?	240?		
h 5495	4.9	11.5	28.1	286		
AC 11	6.8	7.0	0.79	355		
Σ 2316	5.4	7.7	3.8	318		
Σ 2322	6.5	11.8	20.1	168		
OΣ 355	5.4	9.7	38.7	248		
Σ 2342	6.7	9.3	32?	360?		
h 864	6.8	11.0	18.5	330		
OΣ 360	6.6	10.1	1.6	282		
Σ 2350	6.1	11.1	20?	190?		
Σ 2355	6.4	11.4	22?	130?		
δ Sct	4.7	9.2	52.6	130		
ε Sct	5.1	13.7	37.6	195		
Σ 2375	6.2	6.6	2.4	118		
Σ 2379	6.1	7.9	13.0	121	AB	
	6.1	11.3	23?	146?	AC	
Σ 2391	6.5	9.8	37.9	332	AB	
	9.8	14.3	12.6	107	BC	
H VI 50	6.2	12.3	23.3	360		
Σ 2417	4.5	4.5	22.4	104	θ Ser	

Clusters

6633 A coarse grouping of bright and moderate stars, visible with the naked eye.

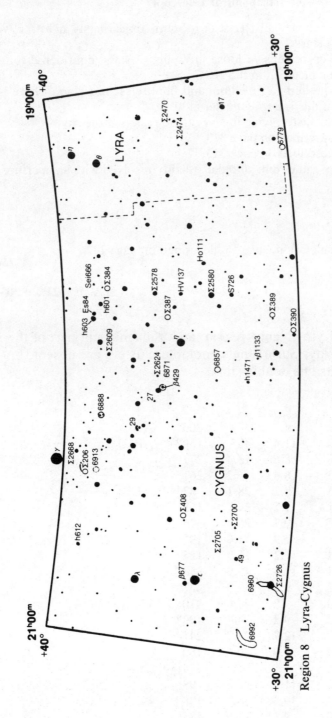

Region 8 Lyra-Cygnus

6664   A loose cluster, just East of α Scuti, appearing as an attractive swirl of faint stars.

IC 4756   A very extensive loose grouping of stars, considered too coarse to be included in the NGC.

6682   A fine grouping of medium and faint telescopic stars. There is some doubt if this is a true cluster at all.

6683   A small cluster, lying in a very fine region of the sky.

6694   M26, described on page 313.

6705   M11, described on page 313.

6712   A large, rather dim globular, readily resolved with an aperture of 150 mm.

Nebula

6572   The planetary nebula Σ6, described on page 292.

R.A. 19ʰ–21ʰ, +30°–+40°

REGION 8

Part of the Lyra-Cygnus region, and a magnificent part of the northern Milky Way. Surprisingly, notable clusters are absent, but there are numerous fine double stars.

Double stars

17 Lyr	5.0	9.1	3″.4	300°	AB	
	5.0	11.4	48?	50?	AP	
Σ 2470	6.6	8.6	13.4	271		
Σ 2474	6.5	8.6	16.2	261		
Σ 2487	4.5	9.2	28.1	82	η Lyr	
Ho 111	6.1	11.7	24.7	168		
Sei 666	6.7	9.7	25.4	326		
OΣ 384	7.6	7.9	1.0	195		
Σ 2578	6.4	7.2	15.0	125	AB	
	6.4	11.4	45.9	358	AC	
H V 137	6.2	9.2	39.0	27		
Σ 2580	5.0	9.2	26.0	70	17 Cyg	
S 726	6.2	9.2	28?	190?		
h 601	5.7	12.7	24.7	241		
OΣ 387	6.9	7.9	0.61	160	(1985)	
			0.53	134	(2000)	
Es 84	6.2	11.2	22.8	98		
h 603	5.4	10.4	56?	108?	19 Cyg	
OΣ 389	7.0	9.3	12.5	183		

OΣ 390	6.5	8.8	9.7	22
h 1455	4.0	10.5	46.0	327	η Cyg
Σ 2609	6.6	7.7	2.0	22
β 1133	6.8	9.5	0.9	338
Σ 2624	7.2	7.8	1.7	173
h 1471	5.7	11.3	34?	9?
β 429	7.0	9.6	11.4	300	AD
	7.0	8.0	36.1	28	AF
	7.0	13.8	30.1	57	AH
27 Cyg	5.5	11.6	30?	130?
29 Cyg	5.0	12.2	36.0	325
Σ 2668	6.3	9.2	3.3	280
OΣ 206	6.6	9.1	43.0	256	AB
	6.6	10.8	23.4	262	AP
OΣ 408	6.7	9.7	1.6	190
Σ 2700	7.0	8.8	23.9	285
Σ 2705	7.4	8.4	3.1	261	AB
	7.4	12.9	53.8	3	AD
49 Cyg	5.9	8.0	2.7	47
h 612	6.5	9.8	48.3	9
Σ 2726	4.3	9.5	5.9	69	52 Cyg
β 677	5v	10.2	9.9	121	T Cyg
OΣ 413	4.8	6.1	0.87	10	λ Cyg

Clusters

6779   M56, described on page 294.
6871   A coarse cluster, containing a double star.
6913   M29, described on page 295.

Nebulae

6857   A planetary nebula measuring some 45″ arc across, but extremely faint.
6888   An extended, very faint nebulosity involving two 7th-magnitude stars, but barely visible even with an aperture of 200 mm.
6960, etc.   The *Filamentary Nebula,* described on page 296.

R.A. 21ʰ–24ʰ, Dec. +50°–+60°

REGION 9

This far northern part of the Milky Way includes part of five constellations: Cygnus, Cepheus, Lacerta, Andromeda, and Cassiopeia. There are no galaxies here, but plenty of clusters and doubles, and magnificent sweeping with a low power.

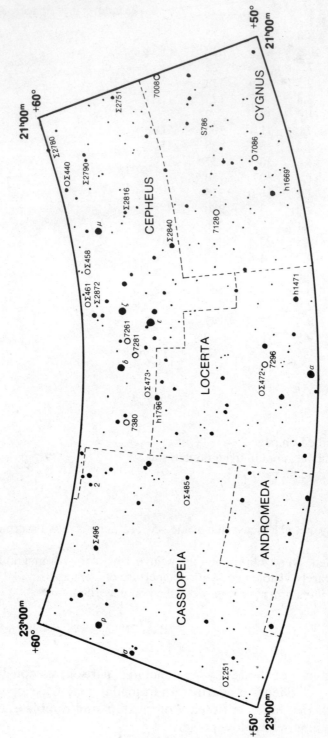

Region 9 Cygnus-Cepheus-Lacerta-Cassiopeia (Andromeda)

Double stars

Σ 2751	6.1	7.1	1″.5	353°		
Σ 2780	6.0	7.0	1.0	216		
S 786	6.8	8.9	48.5	301		
Σ 2790	5.8	10.1	4.5	45		
OΣ 440	6.4	10.7	11.2	180		
h 1669	7.0	11.2	15.6	240		
Σ 2816	5.8	7.7	11.7	121	AB	
	5.8	7.8	19.9	339	AC	
Σ 2840	5.5	7.3	18.0	196	AB	
	5.5	13.2	55.2	348	AC	
OΣ 458	7.0	8.5	0.8	349		
OΣ 461	6.7	11.4	11.1	298		
Σ 2872	6.6	7.3	21.6	316		
h 1471	5.4	10.4	30?	290?		
OΣ 472	6.6	11.5	13.9	312		
OΣ 473	6.8	10.1	14.8	357		
β 703	3.8	11.8	42?	290?	α Lac	
h 1796	5.5	10.8	32.0	10		
OΣ 485	6.6	9.8	19.5	50	AB	
	6.6	10.0	56.6	260	AC	
2 Cas	5.6	13.0	19.9	322		
OΣ 496	5.0	8.9	43.4	114		
OΣ 251	7.0	9.4	46?	205?		
σ Cas	5.0	7.1	3.0	326		

Clusters

7086   A small assemblage of stars, fading out to star-dust.

7128   A small and very condensed cluster of faint stars, appearing nebulous and unresolved with small or moderate instruments.

7261   A tiny group of faint stars involved with a wide 8th-magnitude double star.

7281   A very scattered group of moderate telescopic stars.

7296   A scattered cluster containing only a few faint stars, rather linear in distribution.

7380   A loose group.

Nebula

7008   A fairly large but rather dim planetary nebula of irregular shape, visible with a 75-mm telescope.

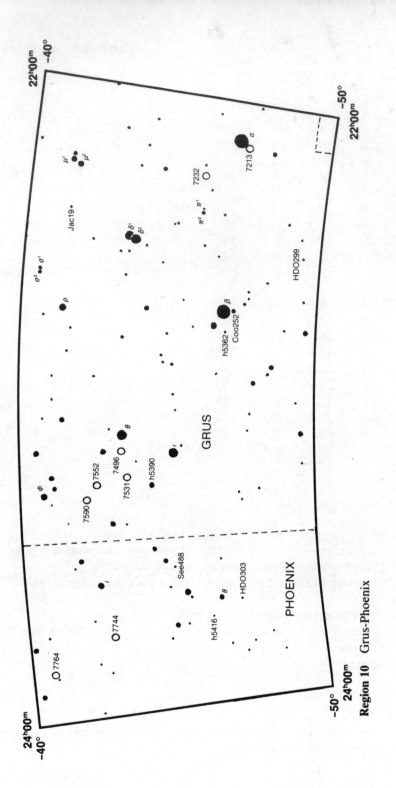

Region 10   Grus–Phoenix

R.A. 22^h–24^h, Dec. −40°−−50°

REGION 10

The constellations of Grus and Phoenix make up this area. Being far from the Milky Way, it contains a number of galaxies, but clusters and nebulae are absent.

Double stars

I 135	6.5	10.7	2″.7	201°	π^1 Gru
I 382	5.8	11.3	4.6	218	π^2 Gru
Jac 19	6.7	8.4	20?	70?	
HDO 299	6.8	12.0	19.8	284	
Coo 252	6.3	10.3	7.8	128	
h 5362	6.9	9.9	10.5	141	
h 5390	5.9	10.7	22.5	44	
See 488	7.0	12	18.9	337	
θ Phe	6.6	7.2	4.0	278	
HDO 303	7.0	10.0	2.0	65	
h 5416	6.8	10.4	45.2	215	

Galaxies

7213 A small round nebulosity, with slight condensation, in the same field as α Gru.

7232 A faint, elongated nebulosity, with a much fainter galaxy, NGC 7233, about 2′ arc to the East.

7496 A faint, almost circular nebulosity, with feeble condensation; there is a 12th-magnitude star near the center.

7531 A small nebulosity.

7552 Brighter than most in this region, well elongated in a North–South direction.

7590 A rather faint nebulosity, with condensation; two others, NGC 7582 and the much fainter 7599, are in the same low-power field.

7744 A small nebulosity with an intense nucleus.

7764 Fairly conspicuous, with slight central condensation: one of the brighter objects in this part of the sky.

Some Useful Publications

YEARBOOKS & ALMANACS

The Astronomical Almanac A standard annual publication that contains ephemerides for the Sun and Moon, the planets and their satellites, eclipse information, minor planet data, and other useful information. Some of the contents are abstracted in *Astronomical Phenomena for the Year* ————, that describes the visibility of the planets, rising and setting of the Sun and Moon, eclipses, etc. These publications are obtainable from the U.S. Government Printing Office, Washington, DC 20402.

Observer's Handbook This contains a calendar of sky events, lists of bright stars and deep-sky objects, finder charts for the planets and brighter asteroids, information on some planetary satellites, etc. Obtainable from Sky Publishing Corp., 49 Bay State Road, Cambridge, MA 02238.

Handbook of the British Astronomical Association This, probably, is the most comprehensive annual publication available to the amateur. It includes tables or charts showing the positions of all the major planets, and the satellites of Jupiter and Saturn; ephemerides for a number of asteroids and all expected returning comets; occultation predictions, and much other useful information. It can be purchased directly from the Association at Burlington House, Piccadilly, London W1V 0NL.

ATLASES AND CATALOGUES

A. Becvar, *Atlas Borealis, Atlas Eclipticalis, Atlas Australis*. Three separate atlases covering the northern ($+90°-+30°$), equatorial ($+30°--30°$), and southern ($-30°--90°$) regions of the celestial sphere. Stars down to magnitude 9½ are shown, and color-coded by spectral type. Deep-sky objects are not included. (Sky Publishing Corp.; see above.)

A. Hirshfield & R. W. Sinnott, *Sky Catalogue 2000.0*. A listing of 45,269 stars over the whole sky, down to visual magnitude 8.0. Photoelectric magnitudes are given where possible, together with estimated distances, spectral type, and proper motion. (Sky Publishing Corp.; see above.)

R. H. Lampkin, *Naked Eye Stars*. A very useful catalogue of 3,047 stars of magnitude 5.5 and brighter. (Gall & Inglis, Edinburgh, Scotland.)

W. Tirion, *Sky Atlas 2000.0*. A chart of the whole sky, showing about 43,000 stars down to magnitude 8.0 and about 2,500 deep-sky objects. There are twenty-six sections, with generous overlap. (Cambridge University Press.)

A. P. Norton, *Norton's Star Atlas and Reference Handbook*. A very useful atlas, showing stars to magnitude 6 or fainter, as well as the brighter deep-sky objects. The best chart for the beginner because the stars are clearly set out and labeled, and it is convenient for use outside. (Gall & Inglis; see above.)

Making Magnitude Estimates

Variable stars—stars that change in brightness—can be counted by the thousand. Some are visible with the naked eye; others, even when at their brightest, are but dim telescopic objects. Some enthusiasts observe nothing else, and may amass thousands of observations or *estimates* in the course of a year.

It is possible that you, too, will join the ranks of the variable-star observers. On the other hand, variable-star work is a specialized field, and perhaps is less appropriate for inclusion in an introductory book such as this one than the observational work described in the preceding chapters. Considerations of space make a detailed treatment impossible, and a sketchy one would mislead by omission. Therefore, this brief appendix will simply outline one way in which the brightness of a celestial object may be measured. It is applicable to a star, or to the satellite of a planet, or to an asteroid, or even, as mentioned on page 184, to a comet.

You first need a *sequence*. This is a selection of stars, assumed to be of constant brightness, lying in or near the telescopic field of the object being observed (which for convenience we shall call the "variable." The stars in the sequence must be of known magnitude. Proper variable-star charts include a sequence for each star. Without such a chart you will need to consult a catalogue, the most suitable being *Sky Catalogue 2000.0*. This gives magnitude values for stars down to 8.0 over the whole sky.

Assuming that you have some idea of the brightness range of the variable, locate a sequence of stars that will embrace this range. When selecting magnitudes, try, if possible, to use what are known as "V" or visual magnitudes. Don't use "B" or blue-light magnitudes because a red or yellow star will be graded too faint on the B system, compared with its visual appearance—these errors can exceed one whole magnitude. If photoelectric V magnitudes can be obtained, so much the better. These are usually given in catalogues to two decimal places, although the hundredths can be rounded off to the nearest tenth, this being the limit of normal visual accuracy.

Now compare the variable with the stars of the sequence. Find the two stars that are the closest in brightness to the variable, one being too bright, and the other too faint. Then try to judge where the variable fits in between them as regards its own brightness.

Suppose that the brighter and fainter comparison stars are A and B respectively, and that the variable is V. If V were judged to be as much fainter than A as it is brighter than B, then the estimate would be written as A_1V_1B (estimate 1). If it were judged to be one part fainter than A and two parts brighter than B, the record would be A_1V_2B (estimate 2). If it were still nearer to A in brightness, the observation might be A_1V_3B (estimate 3): in other words, 3 times as far from B as from A in brightness.

It now remains to calculate the magnitude of the variable. Let the magnitude of A be 6.8 and of B be 7.4, so that there is an interval of 0.6 of a magnitude between them. Estimate 1, clearly, would be reduced as V = 7.1. In estimate 2, the interval of 0.6 has been divided into 3 parts of 0.2 each, and V would work out at magnitude 7.0. Estimate 3 divides this interval into 4 parts of 0.15 each, so that V would theoretically be 6.95, but normal practice would round this off to 7.0 (i.e., to the nearest even number).

This is known as the *fractional method*. If you can find another useful pair of comparison stars, then make another estimate as a check on the first. If the two sets of observations disagree badly, try to find out why. If they correspond to within a couple of tenths of a magnitude, take the mean and feel pleased with your work.

Note the two following important precautions:

1. Try not to use strongly colored stars for comparison purposes, unless of course the variable itself is tinted yellow or red. The eye responds differently to different tints; in particular, a red star appears to brighten if stared at.
2. Do not look ''in between'' two stars when comparing them because the image of each will fall on a different part of the retina, and the retina is not equally sensitive all over. Bring one star to the center of the eye, memorize its appearance, and then bring the other to the same position. You may have to repeat this process several times before you are satisfied that you have got the brightness interval right.

Even if you should happen to find a comparison star that appears exactly equal in brightness to the variable, always make at least one fractional estimate as well, as a check.

Using this method, you will be able to estimate the brightness of Uranus and Neptune (pages 170 and 172), of asteroids, of suitable comets (page 184), and of any other objects that may interest you.

Where to Find the Planets

The purpose of this Appendix is to aid the ready location of all the major planets except Pluto during the period 1988–2000. The information takes different forms.

Mercury and Venus   The tables give their positions in the sky relative to the sun, in terms of right ascension and declination. These positions are at 5-day intervals for Mercury and at 10-day intervals for Venus. Their magnitude is also indicated.

Mars, Jupiter, and Saturn   Their positions on the celestial sphere at 20-day intervals are given, together with their magnitude.

Uranus and Neptune   Their positions on the celestial sphere at 40-day intervals are given. Their paths are also shown on special star maps.

Mercury

Books about observational astronomy usually mention only the dates of Mercury's greatest elongation from the Sun. This is the time when it is most likely to be seen with the unaided eye before dawn (western elongation) or after sunset (eastern elongation). However, it is also interesting and challenging to try to locate the planet in the daylight sky, using a telescope of perhaps 200mm or larger.

The aim of this section is, therefore, to list the position of Mercury *relative to the Sun* at 5-day intervals, omitting the unobservable periods near inferior and superior conjunction. For example, on June 23, 1988, Mercury is 0 hours 57 minutes West of the Sun in right ascension, and 5° South in declination. Having brought the Sun to the center of the field of view (taking appropriate precautions), note the readings on the right ascension and declination circles and shift the telescope through the appropriate angles until it is pointing at the position of Mercury.

The planet appears much fainter at the beginning of western elongation and the end of morning elongation, since at these times it is near superior conjunction, on the far side of the Sun from the Earth. Even a moderate telescope is unlikely to reveal it, unless the sky is exceptionally transparent. However, the data will allow the challenge to be accepted! (G.E. = greatest elongation.)

Date	in R.A.	in Dec.		Mag.	
		Angular Distance from Sun			
1988					
Western (morning) elongation					
Jun 23	0ʰ 57ᵐ	5°.0	S		
28	1 17	4.6	S	+1.5	
Jul 3	1 28	3.4	S		
8	1 31	1.9	S	+0.5	G.E. 21°, Jul 6
13	1 26	0.6	S		
18	1 12	1.6	N	−0.6	
23	0 52	2.8	N		
Eastern (evening) elongation					
Aug 22	1 06	5.7	S		
27	1 16	7.7	S	−0.1	
Sep 1	1 24	9.4	S		
6	1 29	11.0	S	+0.2	
11	1 33	12.1	S		
16	1 34	12.9	S	+0.4	G.E. 27°, Sep 15
21	1 31	12.5	S		
26	1 22	12.4	S	+0.7	
Oct 1	1 04	10.6	S		
Western (morning) elongation					
Oct 21	0 59	7.1	N		
26	1 06	8.5	N	−0.2	G.E. 18°, Oct 26
31	1 04	8.1	N		
Nov 5	0 57	6.9	N	−0.7	
1988/89					
Eastern (evening) elongation					
Dec 25	0 58	1.2	S	−0.7	
30	1 10	0.1	S		
Jan 4	1 18	1.4	N	−0.6	
9	1 21	2.9	N		G.E. 19°, Jan 9
14	1 13	4.4	N	+0.2	
Western (morning) elongation					
Feb 3	1 15	1.3	S	+1.1	
8	1 35	3.7	S		
13	1 44	5.8	S	+0.4	
18	1 45	7.6	S		G.E. 26°, Feb 18
23	1 41	8.7	S	+0.2	
28	1 34	9.5	S		

Date	Angular Distance from Sun in R.A.		in Dec.		Mag.
Mar 5	1	25	9.8	S	0.0
10	1	14	9.6	S	
15	1	02	8.8	S	−0.3

Eastern (evening) elongation

Apr 19	0	57	6.5	N	
24	1	11	7.8	N	−0.3
29	1	20	8.3	N	G.E. 21°, May 1
May 4	1	21	7.6	N	+0.8
9	1	13	6.3	N	
14	0	56	4.1	N	+2.0

Western (morning) elongation

Jun 3	0	56	6.1	S	+2.1
8	1	16	6.9	S	
13	1	29	6.8	S	+1.1
18	1	35	5.8	S	G.E. 23°, Jun 18
23	1	34	4.3	S	+0.3
28	1	27	2.4	S	
Jul 3	1	12	0.6	S	−0.6
8	0	52	1.6	N	

Eastern (evening) elongation

Aug 2	1	01	3.2	S	−0.5
7	1	15	5.2	S	
12	1	25	7.4	S	0.0
17	1	32	9.2	S	
22	1	36	10.9	S	+0.3
27	1	37	12.1	S	G.E. 27°, Aug 29
Sep 1	1	35	13.0	S	+0.5
6	1	29	13.1	S	
11	1	16	12.4	S	+1.0
16	0	54	10.3	S	

Western (morning) elongation

Oct 6	1	00	7.2	N	
11	1	04	8.6	N	−0.3 G.E. 18°, Oct 10
16	0	58	8.1	N	

Eastern (evening) elongation

Dec 5	0	57	3.2	S	
10	1	09	2.8	S	−0.5
15	1	19	2.0	S	
20	1	26	0.8	S	−0.4
25	1	25	0.6	N	G.E. 20°, Dec 23
30	1	15	2.1	N	+0.3

1990
Western (morning) elongation

Jan 14	0	48	1.8	N	
19	1	21	0.4	N	+0.8
24	1	39	1.5	S	

Date	Angular Distance from Sun in R.A.		in Dec.		Mag.	
Jan 29	1	45	3.3	S	+0.2	G.E. 25°, Feb 1
Feb 3	1	44	4.9	S		
8	1	39	5.6	S	+0.1	
13	1	30	6.2	S		
18	1	20	7.7	S	−0.1	
23	1	08	7.9	S		

Eastern (evening) elongation

Date						
Apr 4	0	56	7.0	N		
9	1	08	8.7	N	−0.3	
14	1	11	9.2	N		G.E. 20°, Apr 13
19	1	06	8.6	N	+2.6	
24	0	52	6.9	N		

Western (morning) elongation

Date						
May 14	0	56	6.3	S		
19	1	15	8.4	S	+1.5	
24	1	28	9.2	S		
29	1	35	9.2	S	+0.9	G.E. 25°, May 31
Jun 3	1	37	8.2	S		
8	1	33	6.7	S	+0.2	
13	1	25	4.7	S		
18	1	10	2.6	S	−0.6	

Eastern (evening) elongation

Date						
Jul 18	1	09	2.2	S	−0.5	
23	1	23	4.1	S		
28	1	33	6.3	S	+0.1	
Aug 2	1	39	8.2	S		
7	1	42	10.0	S	+0.4	
12	1	41	11.3	S		G.E. 27°, Aug 11
17	1	36	12.3	S	+0.7	
22	1	25	12.3	S		
27	1	08	11.5	S	+1.3	

Western (morning) elongation

Date						
Sep 16	0	51	3.8	N	+1.3	
21	1	03	6.9	N		
26	1	03	8.1	N	−0.4	G.E. 18°, Sep 24
Oct 1	0	54	7.7	N		

Eastern (evening) elongation

Date						
Nov 20	1	06	5.1	S		
25	1	17	4.8	S	−0.3	
30	1	26	4.4	S		
Dec 5	1	31	3.3	S	−0.2	G.E. 21°, Dec 6
10	1	30	2.1	S		
15	1	15	0.5	S	+0.5	

1990/91
Western (morning) elongation

Date						
Dec 30	0	55	3.0	N		
Jan 4	1	24	2.5	N	+0.5	

| Date | Angular Distance from Sun | | | | Mag. | |
	in R.A.		in Dec.			
Jan 9	1	39	1.1	N		
14	1	43	0.4	S	0.0	G.E. 24°, Jan 14
19	1	41	2.1	S		
24	1	33	3.5	S	−0.1	
29	1	25	4.7	S		
Feb 3	1	13	5.4	S	−0.2	
8	1	02	5.9	S		

Eastern (evening) elongation

Date						
Mar 20	0	56	7.1	N		
25	1	05	7.9	N	−0.2	G.E. 19°, Mar 27
30	1	03	9.5	N		
Apr 4	0	52	8.7	N	+1.3	

Western (morning) elongation

Date						
Apr 24	0	53	5.7	S	+2.0	
29	1	15	8.8	S		
May 4	1	28	10.5	S	+1.2	
9	1	36	11.4	S		
14	1	37	11.2	S	+0.7	G.E. 26°, May 12
19	1	37	10.3	S		
24	1	30	8.7	S	+0.1	
29	1	21	6.7	S		
Jun 3	1	05	4.3	S	−0.7	

Eastern (evening) elongation

Date						
Jun 28	0	54	0.7	N		
Jul 3	1	08	0.6	S	−0.5	
8	1	29	1.4	S		
13	1	40	4.4	S	+0.2	
18	1	44	6.3	S		
23	1	47	8.2	S	+0.6	G.E. 27°, Jul 5
28	1	42	9.6	S		
Aug 2	1	34	10.5	S	+1.0	
7	1	19	10.6	S		
12	0	57	9.8	S	+1.7	

Western (morning) elongation

Date						
Sep 1	1	00	3.6	N	+1.1	
6	1	09	6.2	N		G.E. 18°, Sep 7
11	1	04	7.3	N	−0.5	
16	0	53	7.2	N		

Eastern (evening) elongation

Date						
Oct 26	0	53	5.6	S		
31	1	04	6.5	S	−0.2	
Nov 5	1	13	6.8	S		
10	1	23	7.0	S	−0.2	
15	1	29	6.7	S		
20	1	33	6.0	S	−0.1	G.E. 22°, Nov 19
25	1	28	4.7	S		
30	1	12	3.0	S	+0.7	

		Angular Distance from Sun			
Date		**in R.A.**	**in Dec.**		**Mag.**

1991/92
Western (morning) elongation

Dec	15	0	58	4.1	N		
	20	1	22	4.5	N	+0.3	
	25	1	34	3.7	N		G.E. 22°, Dec 27
	30	1	35	2.2	N	−0.2	
Jan	4	1	32	0.6	N		
	9	1	23	0.9	S	−0.2	
	14	1	16	2.2	S		
	19	1	04	3.2	S	−0.3	
	24	0	53	3.9	S		

Eastern (evening) elongation

Mar	4	1	01	7.2	N		
	9	1	04	8.8	N	−0.2	G.E. 18°, Mar 9
	14	0	59	9.4	N		

Western (morning) elongation

Apr	8	1	13	7.6	S	+1.5	
	13	1	26	10.2	S		
	18	1	35	11.9	S	+0.9	
	23	1	37	12.4	S		G.E. 27°, Apr 23
	28	1	37	12.3	S	+0.5	
May	3	1	31	11.3	S		
	8	1	25	9.9	S	0.0	
	13	1	13	8.0	S		
	18	1	00	5.7	S	−0.7	

Eastern (evening) elongation

Jun	12	0	59	2.2	N		
	17	1	19	1.3	N	−0.5	
	22	1	35	0.0			
	27	1	44	1.8	S	+0.3	
Jul	2	1	50	3.7	S		
	7	1	49	5.5	S	+0.8	G.E. 26°, Jul 6
	12	1	43	7.0	S		
	17	1	29	8.0	S	+1.3	
	22	1	09	8.2	S		

Western (morning) elongation

Aug	11	0	52	0.1	S		
	16	1	10	2.8	N	+0.9	
	21	1	13	5.8	N		G.E. 18°, Aug 21
	26	1	07	6.1	N	−0.5	
	31	0	51	6.0	N		

Eastern (evening) elongation

Oct	5	0	53	5.8	S	−0.3	
	10	1	04	7.2	S		
	15	1	11	8.1	S	−0.1	
	20	1	21	8.9	S		
	25	1	28	9.3	S	0.0	

Date	Angular Distance from Sun				Mag.	
	in R.A.		in Dec.			
Oct 30	1	33	9.3	S		G.E. 24°, Oct 31
Nov 4	1	33	8.6	S	+0.1	
9	1	26	7.4	S		
14	1	05	5.1	S	+0.9	

1992/93
Western (morning) elongation

Nov 29	0	56	5.3	N		
Dec 4	1	18	6.4	N	+0.1	
9	1	25	5.7	N		G.E. 21°, Dec 9
14	1	25	4.4	N	−0.3	
19	1	19	2.8	N		
24	1	13	1.3	N	−0.4	
29	1	03	0.1	S		
Jan 3	0	54	1.3	S	−0.4	

Eastern (evening) elongation

Feb 12	0	56	4.3	N	−1.0	
17	1	05	6.6	N		
22	1	05	8.2	N	−0.1	G.E. 18°, Feb 21
27	0	55	8.6	N		

Western (morning) elongation

Mar 19	1	06	5.1	S		
24	1	26	8.6	S	+1.1	
29	1	35	10.7	S		
Apr 3	1	40	12.2	S	+0.6	G.E. 28°, Apr 5
8	1	38	12.7	S		
13	1	34	12.6	S	+0.3	
18	1	26	11.8	S		
23	1	19	10.5	S	−0.1	
28	1	06	8.6	S		
May 3	0	53	6.4	S	−0.7	

Eastern (evening) elongation

May 28	0	59	3.6	N		
Jun 2	1	19	3.4	N	−0.4	
7	1	36	2.6	N		
12	1	44	1.2	N	+0.4	
17	1	49	0.5	S		G.E. 25°, Jun 17
22	1	44	2.2	S	+1.0	
27	1	35	3.7	S		
Jul 2	1	16	4.9	S	+1.7	

Western (morning) elongation

Jul 27	1	07	0.8	S		
Aug 1	1	19	1.5	N	+0.7	
6	1	19	3.2	N		G.E. 19°, Aug 4
11	1	10	4.6	N	−0.5	
16	0	51	4.8	N		

		Angular Distance from Sun			
Date		in R.A.	in Dec.		Mag.

Eastern (evening) elongation

Sep	15	0 52	5.2	S	
	20	0 59	6.8	S	−0.2
	25	1 11	8.5	S	
	30	1 19	9.6	S	0.0
Oct	5	1 26	10.6	S	
	10	1 31	11.1	S	+0.1
	15	1 34	11.2	S	G.E. 25°, Oct 14
	20	1 30	10.6	S	+0.3
	25	1 20	9.3	S	
	30	0 56	6.7	S	+1.2

Western (morning) elongation

Nov	14	0 56	6.2	N	
	19	1 13	7.6	N	0.0
	24	1 16	7.2	N	G.E. 20°, Nov 22
	29	1 14	6.0	N	−0.5
Dec	4	1 07	4.4	N	
	9	0 59	2.8	N	−0.5

1994
Eastern (evening) elongation

Jan	23	0 54	1.7	N	
	28	1 04	3.6	N	−0.8
Feb	2	1 11	5.6	N	G.E. 18°, Feb 4
	7	1 07	7.1	N	0.0
	12	0 53	7.4	N	

Western (morning) elongation

Feb	27	0 55	1.8	S	+1.7
Mar	4	1 21	5.3	S	
	9	1 35	8.3	S	+0.8
	14	1 41	10.1	S	
	19	1 41	11.5	S	+0.4 G.E. 28°, Mar 19
	24	1 38	12.0	S	
	29	1 31	12.1	S	+0.2
Apr	3	1 23	11.4	S	
	8	1 13	10.4	S	−0.1
	13	1 01	8.7	S	

Eastern (evening) elongation

May	13	0 57	4.9	N	
	18	1 17	5.3	N	−0.4
	23	1 31	5.1	N	
	28	1 39	4.1	N	+0.5 G.E. 23°, May 30
Jun	2	1 40	2.7	N	
	7	1 32	1.0	N	+1.3
	12	1 16	0.7	S	
	17	0 52	2.3	S	+2.3

	Angular Distance from Sun				
Date	in R.A.		in Dec.		Mag.

Western (morning) elongation

Date		in R.A.		in Dec.		Mag.	
Jul	7	1	07	3.6	S	+ 1.7	
	12	1	22	2.4	S		
	17	1	28	0.6	S	+ 0.6	G.E. 21°, Jul 17
	22	1	25	1.2	N		
	27	1	13	2.7	N	− 0.6	
Aug	1	0	53	3.4	N		

Eastern (evening) elongation

Date		in R.A.		in Dec.		Mag.	
Aug	31	1	00	5.7	S		
Sep	5	1	10	7.5	S	− 0.2	
	10	1	18	9.3	S		
	15	1	25	10.6	S	+ 0.1	
	20	1	30	11.7	S		
	25	1	32	12.4	S	+ 0.3	G.E. 26°, Sep 26
	30	1	32	12.6	S		
Oct	5	1	26	12.0	S	+ 0.5	
	10	1	11	10.5	S		

Western (morning) elongation

Date		in R.A.		in Dec.		Mag.	
Oct	30	0	58	6.8	N		
Nov	4	1	08	8.4	N	− 0.2	G.E. 19°, Nov 6
	9	1	08	8.0	N		
	14	1	05	6.8	N	− 0.6	
	19	0	55	5.1	N		

1995
Eastern (evening) elongation

Date		in R.A.		in Dec.		Mag.	
Jan	3	0	52	0.6	S	− 0.7	
	8	1	03	0.7	N		
	13	1	13	2.3	N	− 0.7	
	18	1	17	4.1	N		G.E. 19°, Jan 19
	23	1	11	5.5	N	+ 0.1	

Western (morning) elongation

Date		in R.A.		in Dec.		Mag.	
Feb	12	1	07	1.6	S	+ 1.3	
	17	1	07	4.6	S		
	22	1	41	6.9	S	+ 0.5	
	27	1	45	8.8	S		G.E. 27°, Mar 1
Mar	4	1	43	9.9	S	+ 0.3	
	9	1	37	10.8	S		
	14	1	28	10.8	S	+ 0.1	
	19	1	18	10.6	S		
	24	1	06	9.6	S	− 0.2	
	29	0	54	8.3	S		

Eastern (evening) elongation

Date		in R.A.		in Dec.		Mag.	
Apr	28	0	55	5.8	N		
May	3	1	13	7.0	N	− 0.4	
	8	1	23	7.1	N		
	13	1	28	6.6	N	+ 0.7	G.E. 22°, May 12

Date		Angular Distance from Sun			Mag.
		in R.A.	in Dec.		
May	18	1 23	5.2	N	
	23	1 11	3.4	N	+1.7

Western (morning) elongation

Jun	17	1 05	5.8	S	
	22	1 22	5.5	S	+1.3
	27	1 32	4.5	S	G.E. 22°, Jun 29
Jul	2	1 34	3.0	S	+0.4
	7	1 28	1.1	S	
	12	1 14	0.6	N	−0.6
	17	1 05	1.9	N	

Eastern (evening) elongation

Aug	11	0 56	3.4	S	−0.6
	16	1 09	5.4	S	
	21	1 19	7.5	S	−0.1
	26	1 26	9.2	S	
	31	1 32	11.0	S	+0.2
Sep	5	1 35	12.2	S	
	10	1 34	13.0	S	+0.4 G.E. 27°, Sep 9
	15	1 30	13.2	S	
	20	1 20	12.6	S	+0.8
	25	1 01	10.6	S	

Western (morning) elongation

Oct	15	0 58	7.2	N	
	20	1 05	8.6	N	−0.3 G.E. 18°, Oct 20
	25	1 02	8.4	N	
	30	0 53	7.1	N	−0.8

1995/6
Eastern (evening) elongation

Dec	14	0 50	2.4	S	
	19	1 02	1.8	S	−0.6
	24	1 13	0.0		
	29	1 21	0.5	N	−0.5
Jan	3	1 24	2.0	N	G.E. 19°, Jan 2
	8	1 15	3.5	N	+0.2

Western (morning) elongation

Jan	28	1 16	0.5	S	+1.0
Feb	2	1 36	2.7	S	
	7	1 44	4.6	S	+0.3
	12	1 46	6.5	S	G.E. 26°, Feb 11
	17	1 41	7.7	S	+0.1
	22	1 34	8.7	S	
	27	1 24	8.9	S	0.0
Mar	3	1 13	9.0	S	
	8	1 00	8.4	S	−0.3

Eastern (evening) elongation

Apr	12	0 56	7.5	N	
	17	1 09	8.1	N	−0.3

	Date	Angular Distance from Sun in R.A.		in Dec.		Mag.	
Apr	22	1	16	8.6	N		G.E. 20°, Apr 23
	27	1	15	8.2	N	+0.9	
May	2	1	05	6.7	N		

Western (morning) elongation

	Date					Mag.	
May	27	1	04	6.9	S	+1.9	
Jun	1	1	21	8.0	S		
	6	1	32	7.9	S	+1.0	
	11	1	36	7.7	S		G.E. 24°, Jun 10
	16	1	35	5.6	S	+0.3	
	21	1	26	3.5	S		
	26	1	13	1.6	S	−0.6	
Jul	1	0	54	0.2	N		

Eastern (evening) elongation

	Date					Mag.	
Jul	26	0	56	2.7	S	−0.6	
	31	1	17	4.8	S		
Aug	5	1	28	6.8	S	0.0	
	10	1	35	8.9	S		
	15	1	38	11.5	S	+0.4	
	20	1	40	12.0	S		G.E. 27°, Aug 21
	25	1	37	12.8	S	+0.6	
	30	1	28	13.1	S		
Sep	4	1	14	12.2	S	+1.1	
	9	0	51	10.2	S		

Western (morning) elongation

	Date					Mag.	
Sep	29	1	01	6.9	N		
Oct	4	1	03	8.4	N	−0.3	G.E. 18°, Oct 3
	9	0	57	8.0	N		

Eastern (evening) elongation

	Date					Mag.	
Nov	28	1	00	3.9	S		
Dec	3	1	12	3.7	S	−0.4	
	8	1	21	3.0	S		
	13	1	28	1.8	S	−0.3	G.E. 20°, Dec 15
	18	1	28	0.5	S		
	23	1	17	0.9	N	+0.3	

1997
Western (morning) elongation

	Date					Mag.	
Jan	12	1	22	1.4	N	+0.7	
	17	1	39	0.2	S		
	22	1	44	2.0	S	+0.1	G.E. 24°, Jan 24
	27	1	44	3.5	S		
Feb	1	1	37	5.1	S	0.0	
	6	1	28	6.0	S		
	11	1	18	6.7	S	−0.1	
	16	1	06	6.8	S		
	21	0	53	6.7	S	−0.4	

	Angular Distance from Sun			
Date		**in R.A.**	**in Dec.**	**Mag.**

Eastern (evening) elongation

Date	h	m	Dec	N/S	Mag.	
Mar 28	0	56	7.1	N		
Apr 2	1	05	8.7	N	− 0.3	
7	1	09	9.5	N		G.E. 19°, Apr 6
12	1	01	8.9	N	+ 1.1	

Western (morning) elongation

Date	h	m	Dec	N/S	Mag.	
May 7	1	04	7.1	S		
12	1	20	9.3	S	+ 1.4	
17	1	31	10.1	S		
22	1	36	10.2	S	+ 0.8	G.E. 25°, May 22
27	1	37	9.2	S		
Jun 1	1	32	7.7	S	+ 0.2	
6	1	23	5.7	S		
11	1	10	3.6	S	− 0.6	
16	0	50	1.3	S		

Eastern (evening) elongation

Date	h	m	Dec	N/S	Mag.	
Jul 6	0	51	0.1	N		
11	1	10	1.4	S	− 0.5	
16	1	25	3.4	S		
21	1	35	5.4	S	+ 0.1	
26	1	43	7.5	S		
31	1	45	9.2	S	+ 0.5	
Aug 5	1	43	10.8	S		G.E. 27°, Aug 4
10	1	36	11.6	S	+ 0.8	
15	1	24	11.9	S		
20	1	05	10.9	S	+ 1.4	

Western (morning) elongation

Date	h	m	Dec	N/S	Mag.	
Sep 9	0	53	3.6	N	+ 1.3	
14	1	07	6.4	N		G.E. 18°, Sep 16
19	1	04	7.8	N	− 0.4	
24	0	53	7.5	N		

Eastern (evening) elongation

Date	h	m	Dec	N/S	Mag.	
Nov 8	0	59	5.4	S	− 0.3	
13	1	10	5.8	S		
18	1	19	5.8	S	− 0.3	
23	1	28	5.4	S		
28	1	32	4.5	S	− 0.2	G.E. 22°, Nov 28
Dec 3	1	31	3.4	S		
8	1	28	1.7	S	+ 0.5	

1997/8

Western (morning) elongation

Date	h	m	Dec	N/S	Mag.	
Dec 23	0	53	3.4	N		
28	1	33	3.4	N	+ 0.4	
Jan 2	1	36	2.4	N		
7	1	39	0.8	N	− 0.1	G.E. 23°, Jan 6

Date	Angular Distance from Sun in R.A.		in Dec.		Mag.	
Jan 12	1	37	0.8	S		
17	1	30	2.3	S	−0.1	
22	1	21	3.5	S		
27	1	10	4.4	S	−0.2	
Feb 1	0	58	4.8	S		

Eastern (evening) elongation

Mar 13	0	58	7.2	N		
18	1	04	8.9	N	−0.3	G.E. 18°, Mar 20
23	1	02	9.6	N		

Western (morning) elongation

Apr 17	1	02	6.3	S	+1.8	
22	1	20	9.3	S		
27	1	31	11.1	S	+1.0	
May 2	1	34	11.8	S		
7	1	37	11.8	S	+0.6	G.E. 27°, May 4
12	1	39	10.9	S		
17	1	28	9.4	S	+0.1	
22	1	17	7.4	S		
27	1	04	5.1	S	−0.6	

Eastern (evening) elongation

Jun 21	0	55	1.4	N		
26	1	16	0.2	N	−0.5	
Jul 1	1	31	1.4	S		
6	1	41	3.2	S	+0.2	
11	1	47	5.2	S		
16	1	48	7.0	S	+0.6	G.E. 26°, Jul 17
21	1	44	8.6	S		
26	1	34	9.5	S	+1.1	
31	1	17	9.7	S		
Aug 5	0	52	8.9	S	+1.9	

Western (morning) elongation

Aug 25	1	03	3.1	N	+1.1	
30	1	10	5.5	N		G.E. 18°, Aug 31
Sep 4	1	05	6.9	N	−0.4	
9	0	53	6.6	N		

Eastern (evening) elongation

Oct 19	0	57	6.3	S		
24	1	07	7.1	S	−0.2	
29	1	16	7.9	S		
Nov 3	1	25	8.0	S	−0.1	
8	1	31	7.9	S		
13	1	34	7.1	S	0.0	G.E. 23°, Nov 11
18	1	29	6.0	S		
23	1	11	4.1	S	+0.8	

Date		Angular Distance from Sun in R.A.		in Dec.		Mag.	

1998/9
Western (morning) elongation

Date		in R.A.		in Dec.		Mag.	
Dec	8	0	55	4.5	N		
	13	1	20	5.3	N	+0.3	
	18	1	30	4.7	N		G.E. 22°, Dec 20
	23	1	07	3.3	N	−0.2	
	28	1	26	1.7	N		
Jan	2	1	20	0.2	N	−0.3	
	7	1	11	1.1	S		
	12	1	00	2.3	S	−0.4	

Eastern (evening) elongation

Feb	21	0	50	4.4	N	−1.0	
	26	1	01	6.9	N		
Mar	3	1	04	8.5	N	−0.2	G.E. 18°, Mar 3
	8	0	57	9.1	N		

Western (morning) elongation

Mar	28	0	56	4.1	S		
Apr	2	1	18	8.0	S	+1.3	
	7	1	31	10.5	S		
	12	1	37	12.0	S	+0.8	
	17	1	38	12.7	S		G.E. 28°, Apr 16
	22	1	36	12.6	S	+0.4	
	27	1	31	11.7	S		
May	2	1	22	10.4	S	0.0	
	7	1	12	8.5	S		
	12	0	58	6.3	S	−0.7	

Eastern (evening) elongation

Jun	6	0	57	2.7	N		
	11	1	18	2.3	N	−0.5	
	16	1	34	1.1	N		
	21	1	44	0.5	S	+0.3	
	26	1	49	2.3	S		G.E. 26°, Jun 28
Jul	1	1	47	4.1	S	+0.8	
	6	1	40	5.6	S		
	11	1	25	6.6	S	+1.4	
	16	1	04	7.1	S		

Western (morning) elongation

Aug	5	0	59	0.3	S		
	10	1	13	2.2	S	+0.9	
	15	1	16	4.3	S		G.E. 19°, Aug 14
	20	1	08	5.4	S	−0.5	
	25	0	53	5.8	S		

Eastern (evening) elongation

Sept	29	0	56	6.1	S	−0.3	
Oct	4	1	06	7.5	S		
	9	1	14	8.8	S	−0.1	
	14	1	22	9.5	S		

Date	Angular Distance from Sun in R.A.		in Dec.		Mag.	
Oct 19	1	29	10.1	S	0.0	
24	1	32	10.0	S		G.E. 24°, Oct 24
29	1	32	9.6	S	+0.2	
Nov 3	1	23	8.3	S		
8	1	04	6.0	S	+1.0	

Western (morning) elongation

Date	in R.A.		in Dec.		Mag.	
Nov 23	0	56	5.7	N		
28	1	15	6.9	N	+0.1	
Dec 3	1	22	6.5	N		G.E. 20°, Dec 2
8	1	21	5.2	N	−0.4	
13	1	16	3.7	N		
18	1	07	2.1	N	−0.4	
23	0	59	0.6	N		

2000
Eastern (evening) elongation

Date	in R.A.		in Dec.		Mag.	
Feb 6	0	58	3.9	N	−0.9	
11	1	07	5.9	N		
16	1	07	7.7	N	−0.1	G.E. 18°, Feb 15
21	0	55	8.0	N		

Western (morning) elongation

Date	in R.A.		in Dec.		Mag.	
Mar 12	1	11	5.3	S		
17	1	29	8.4	S	+1.0	
22	1	39	10.8	S		
27	1	41	11.9	S	+0.6	G.E. 28°, Mar 28
Apr 1	1	38	12.7	S		
6	1	33	12.5	S	+0.3	
11	1	26	11.9	S		
16	1	17	9.7	S	−0.1	
21	1	04	9.0	S		
26	0	51	6.7	S	−0.7	

Eastern (evening) elongation

Date	in R.A.		in Dec.		Mag.	
May 21	0	57	4.0	N		
26	1	17	4.3	N	−0.4	
31	1	33	3.6	N		
Jun 5	1	42	2.5	N	+0.4	
10	1	45	1.0	N		G.E. 24°, Jun 9
15	1	41	0.7	S	+1.1	
20	1	23	2.3	S		
25	1	08	3.6	S	+1.9	

Western (morning) elongation

Date	in R.A.		in Dec.		Mag.	
Jul 15	0	53	3.1	S	+2.0	
20	1	14	1.4	S		
25	1	23	0.5	N	+0.7	G.E. 20°, Jul 27
30	1	23	2.5	N		
Aug 4	1	12	3.8	N	−0.5	
9	1	05	4.4	N		

	Angular Distance from Sun			
Date	in R.A.		in Dec.	Mag.

Eastern (evening) elongation

Date		in R.A.		in Dec.		Mag.	
Sep	8	0	53	5.2	S		
	13	1	05	7.1	S	−0.2	
	18	1	13	8.5	S		
	23	1	21	10.0	S	0.0	
	28	1	26	10.9	S		
Oct	3	1	31	11.7	S	+0.2	
	8	1	31	11.8	S		G.E. 26°, Oct 6
	13	1	28	11.4	S	+0.4	
	18	1	16	9.9	S		
	23	0	53	7.3	S	+1.3	

Western (morning) elongation

Date		in R.A.		in Dec.		Mag.	
Nov	7	0	56	6.5	N		
	12	1	10	7.9	N	−0.1	
	17	1	14	7.8	N		G.E. 19°, Nov 15
	22	1	09	6.4	N	−0.5	
	27	1	03	5.0	N		
Dec	2	0	54	3.3	N	−0.6	

Venus

Venus is sufficiently brilliant to be detected within a few degrees of the Sun's limb, if the sky is a reasonably deep blue. This section, therefore, gives continuous 10-day positions, right through inferior and superior conjunction. (G.E. = greatest elongation, S.C. = superior conjunction, I.F. = inferior conjunction.)

	Angular Distance from Sun			
Date	in R.A.		in Dec.	Mag.

1988
Western (morning) elongation

Date		in R.A.		in Dec.		Mag.	
Jun	18	0ʰ	33ᵐ	2°.4	S	−3.1	
	28	1	33	4.4	S	−3.8	
Jul	8	2	16	4.6	S	−4.1	
	18	2	44	3.0	S	−4.2	
	28	3	01	0.4	S	−4.2	
Aug	7	3	09	2.9	N	−4.1	
	17	3	10	6.3	N	−4.0	G.E. Aug 22
Aug	27	3	06	9.5	N	−3.9	
Sep	6	2	59	12.2	N	−3.8	
	16	2	50	14.3	N	−3.8	
	26	2	41	15.7	N	−3.7	
Oct	6	2	32	16.3	N	−3.6	
	16	2	24	16.3	N	−3.6	

| | | Angular Distance from Sun | | | | |
Date		in R.A.		in Dec.		Mag.
Oct	26	2	18	15.6	N	−3.5
Nov	5	2	12	14.3	N	−3.5
	15	2	07	12.5	N	−3.5
	25	2	03	10.3	N	−3.4
Dec	5	1	58	7.9	N	−3.4
	15	1	52	5.2	N	−3.4
	25	1	45	2.6	N	−3.4

1989

Jan	4	1	36	0.2	N	−3.4
	14	1	25	1.8	S	−3.3
	24	1	13	3.3	S	−3.3
Feb	3	1	01	4.3	S	−3.4
	13	0	49	4.8	S	−3.4
	23	0	38	4.7	S	−3.4
Mar	5	0	27	4.2	S	−3.4
	15	0	17	3.4	S	−3.4
	25	0	08	2.4	S	−3.5
Apr	4	0	01	1.3	S	−3.5	S.C. Apr 5

Eastern (evening) elongation

Apr	14	0	10	0.1	S	−3.5
	24	0	20	0.9	S	−3.4
May	4	0	31	1.8	N	−3.4
	14	0	42	2.3	N	−3.4
	24	0	54	2.4	N	−3.4
Jun	3	1	07	1.9	N	−3.3
	13	1	19	1.0	N	−3.3
	23	1	31	0.5	S	−3.3
	3	1	41	2.3	S	−3.3
	13	1	49	4.5	S	−3.3
	23	1	56	6.8	S	−3.4
Aug	2	2	03	9.1	S	−3.4
	12	2	08	11.2	S	−3.4
	22	2	14	13.2	S	−3.4
Sep	1	2	20	14.8	S	−3.5
	11	2	27	16.0	S	−3.5
	21	2	36	16.7	S	−3.6
Oct	1	2	45	16.8	S	−3.7
	11	2	55	16.1	S	−3.8
	21	3	05	14.8	S	−3.9
	31	3	14	12.6	S	−4.0
Nov	10	3	20	9.8	S	−4.1	G.E. Nov 8
	20	3	22	6.4	S	−4.2
	30	3	17	2.8	S	−4.3
Dec	10	3	03	0.7	N	−4.4
	20	2	37	3.7	N	−4.4
	30	1	58	5.8	N	−4.2

1990

Jan	9	1	02	6.7	N	−3.7

Date	Angular Distance from Sun in R.A.	in Dec.		Mag.	

Western (morning) elongation

Date	in R.A.	in Dec.		Mag.	
Jan 19	0 06	6.0	N	−3.2	I.C. Jan 19
29	1 10	3.8	N	−3.8	
Feb 8	2 00	0.5	N	−4.2	
18	2 31	3.5	S	−4.3	
28	2 49	7.4	S	−4.3	
Mar 10	2 56	11.0	S	−4.2	
20	2 58	14.0	S	−4.1	
30	2 55	16.1	S	−4.0	G.E. Mar 30
Apr 9	2 52	17.5	S	−3.9	
19	2 48	17.9	S	−3.8	
29	2 44	17.6	S	−3.7	
May 9	2 40	16.5	S	−3.6	
19	2 37	14.8	S	−3.6	
29	2 34	12.5	S	−3.5	
Jun 8	2 30	9.9	S	−3.5	
18	2 26	7.0	S	−3.4	
28	2 19	4.0	S	−3.4	
Jul 8	2 10	1.1	S	−3.4	
18	1 59	1.5	N	−3.3	
28	1 47	3.6	N	−3.3	
Aug 7	1 34	5.2	N	−3.3	
17	1 20	6.2	N	−3.3	
27	1 07	6.6	N	−3.3	
Sep 6	0 54	6.5	N	−3.4	
16	0 43	6.0	N	−3.4	
26	0 33	5.1	N	−3.4	
Oct 6	0 23	4.0	N	−3.4	
16	0 14	2.8	N	−3.5	
26	0 05	1.6	N	−3.5	

Eastern (evening) elongation

Date	in R.A.	in Dec.		Mag.	
Nov 5	0 04	0.5	N	−3.5	S.C. Nov 1
15	0 14	0.4	S	−3.5	
25	0 25	1.0	S	−3.4	
Dec 5	0 36	1.2	S	−3.4	
15	0 46	0.9	S	−3.4	
25	0 57	0.2	S	−3.4	

1991

Date	in R.A.	in Dec.		Mag.	
Jan 4	1 07	1.0	N	−3.3	
14	1 15	2.5	N	−3.3	
24	1 23	4.2	N	−3.3	
Feb 3	1 29	6.0	N	−3.3	
13	1 35	7.8	N	−3.4	
23	1 41	9.4	N	−3.4	
Mar 5	1 49	10.9	N	−3.4	
15	1 56	12.1	N	−3.4	
25	2 05	12.7	N	−3.4	
Apr 4	2 15	13.0	N	−3.5	
14	2 28	12.6	N	−3.5	
24	2 40	11.5	N	−3.6	
May 4	2 53	9.8	N	−3.6	

		Angular Distance from Sun			
Date	in R.A.		in Dec.		Mag.
May 14	3	04	7.4	N	−3.7
24	3	11	4.4	N	−3.8
Jun 3	3	16	1.1	N	−3.8
13	3	16	2.5	S	−3.9	G.E. Jun 13
23	3	11	5.9	S	−4.0
Jul 3	3	01	9.0	S	−4.1
13	2	44	11.5	S	−4.2
23	2	20	13.1	S	−4.2
Aug 2	1	43	13.3	S	−4.0
12	0	53	11.8	S	−3.6

Western (morning) elongation

Aug 22	0	07	8.1	S	−3.1	I.C. Aug 22
Sep 1	1	05	2.8	S	−3.6
11	1	50	2.8	N	−4.0
21	2	20	8.0	N	−4.2
Oct 1	2	38	12.1	N	−4.3
11	2	47	15.1	N	−4.2
21	2	51	17.0	N	−4.1
31	2	53	17.8	N	−4.1	G.E. Nov 2
Nov 10	2	54	17.6	N	−4.0
20	2	55	16.6	N	−3.9
30	2	55	14.6	N	−3.8
Dec 10	2	54	10.1	N	−3.7
20	2	52	9.0	N	−3.7
30	2	49	5.6	N	−3.6

1992

Jan 9	2	43	2.1	N	−3.5
19	2	34	1.3	S	−3.5
29	2	23	4.2	S	−3.4
Feb 8	2	11	6.6	S	−3.4
18	1	58	8.4	S	−3.4
28	1	44	9.5	S	−3.4
Mar 9	1	32	10.0	S	−3.3
19	1	21	9.8	S	−3.3
29	1	11	9.2	S	−3.3
Apr 8	1	02	8.3	S	−3.3
18	0	54	6.0	S	−3.3
28	0	46	5.6	S	−3.3
May 8	0	37	4.0	S	−3.4
18	0	29	2.6	S	−3.4
28	0	19	1.3	S	−3.4
Jun 7	0	08	0.3	S	−3.5

Eastern (evening) elongation

June 17	0	04	0.3	N	−3.5	S.C. Jun 13
27	0	16	0.5	N	−3.5
Jul 7	0	28	0.1	N	−3.4
17	0	40	0.7	S	−3.4
27	0	50	1.9	S	−3.4
Aug 6	0	59	3.3	S	−3.3

Date		Angular Distance from Sun				Mag.	
		in R.A.		in Dec.			
Aug	16	1	08	4.9	S	−3.3	
	26	1	16	6.5	S	−3.3	
Sep	5	1	25	8.0	S	−3.3	
	15	1	34	9.4	S	−3.3	
	25	1	43	10.4	S	−3.3	
Oct	5	1	54	10.9	S	−3.4	
	15	2	05	11.0	S	−3.4	
	25	2	19	10.5	S	−3.4	
Nov	4	2	32	9.2	S	−3.5	
	14	2	45	7.2	S	−3.5	
	24	2	56	4.5	S	−3.6	
Dec	4	3	05	1.2	S	−3.6	
	14	3	10	1.3	N	−3.7	
	24	3	13	6.0	N	−3.8	
1993							
Jan	3	3	11	9.6	N	−3.9	
	13	3	08	13.0	N	−4.0	
	23	3	02	15.8	N	−4.1	G.E. Jan 19
Feb	2	2	53	18.0	N	−4.2	
	12	2	43	21.5	N	−4.3	
	22	2	26	20.0	N	−4.3	
Mar	4	2	02	19.5	N	−4.3	
	14	1	26	17.3	N	−4.1	
	24	0	37	12.8	N	−3.6	

Western (morning) elongation

Date						Mag.	
Apr	3	0	21	6.1	N	−3.2	I.C. Apr 1
	13	1	14	1.3	S	−3.7	
	23	1	55	7.7	S	−4.1	
May	3	2	22	11.9	S	−4.2	
	13	2	46	14.3	S	−4.2	
	23	2	52	14.9	S	−4.1	
Jun	2	3	01	14.1	S	−4.0	
	12	3	06	12.3	S	−3.9	G.E. Jun 10
	22	3	08	9.7	S	−3.8	
Jul	2	3	08	6.5	S	−3.7	
	12	3	05	3.1	S	−3.7	
	22	2	59	0.4	N	−3.6	
Aug	1	2	49	3.7	N	−3.5	
	11	2	38	6.5	N	−3.5	
	21	2	25	8.7	N	−3.5	
	31	2	13	10.4	N	−3.4	
Sep	10	1	59	11.4	N	−3.4	
	20	1	47	11.8	N	−3.4	
	30	1	36	11.6	N	−3.4	
Oct	10	1	26	10.8	N	−3.4	
	20	1	18	9.7	N	−3.4	
	30	1	11	8.3	N	−3.4	
Nov	9	1	04	6.7	N	−3.4	
	19	0	56	5.0	N	−3.4	
	29	0	49	3.3	N	−3.4	
Dec	9	0	40	1.8	N	−3.4	

Date	Angular Distance from Sun in R.A.		in Dec.		Mag.	
Dec 19	0	31	0.5	N	−3.4	
29	0	20	0.5	S	−3.5	
1994						
Jan 8	0	09	0.9	S	−3.5	

Eastern (evening) elongation

Jan 18	0	02	0.9	S	−3.5	S.C. Jan 17
28	0	12	0.5	S	−3.5	
Feb 7	0	22	0.3	N	−3.4	
17	0	31	1.3	N	−3.4	
27	0	39	2.5	N	−3.4	
Mar 9	0	47	3.7	N	−3.4	
19	0	56	4.9	N	−3.4	
29	1	05	6.0	N	−3.3	
Apr 8	1	15	6.7	N	−3.3	
18	1	27	7.2	N	−3.3	
28	1	39	7.2	N	−3.3	
May 8	1	54	6.5	N	−3.4	
18	2	06	5.3	N	−3.4	
28	2	19	3.5	N	−3.4	
Jun 7	2	29	1.1	N	−3.4	
17	2	38	1.8	S	−3.5	
27	2	44	4.8	S	−3.5	
Jul 7	2	48	7.9	S	−3.6	
17	2	50	11.1	S	−3.6	
27	2	51	13.8	S	−3.7	
Aug 6	2	50	16.3	S	−3.8	
16	2	49	18.3	S	−3.9	
26	2	47	19.7	S	−4.0	G.E. Aug 25
Sep 5	2	44	20.6	S	−4.1	
15	2	38	20.8	S	−4.2	
25	2	26	20.1	S	−4.3	
Oct 5	2	07	18.4	S	−4.3	
15	1	34	15.5	S	−4.1	
25	0	46	10.8	S	−3.6	

Western (morning) elongation

Nov 4	0	14	4.3	S	−3.1	I.C. Nov 2
14	1	11	2.5	N	−3.8	
24	1	56	7.8	N	−4.2	
Dec 4	2	29	10.6	N	−4.4	
14	2	52	11.1	N	−4.4	
24	3	07	9.7	N	−4.3	
1995						
Jan 3	3	16	7.2	N	−4.2	
13	3	19	3.9	N	−4.1	G.E. Jan 13
23	3	19	0.1	N	−4.0	
Feb 2	3	14	3.6	S	−3.9	
12	3	06	7.1	S	−3.8	
22	2	56	10.1	S	−3.7	
Mar 4	2	44	12.3	S	−3.6	

Date	Angular Distance from Sun in R.A.		in Dec.		Mag.	
Mar 14	2	32	13.9	S	−3.6	
24	2	21	14.8	S	−3.5	
Apr 3	2	11	14.9	S	−3.4	
13	2	02	14.3	S	−3.4	
23	1	55	12.4	S	−3.4	
May 3	1	48	11.8	S	−3.3	
13	1	42	10.0	S	−3.3	
23	1	35	8.0	S	−3.3	
Jun 2	1	28	5.8	S	−3.3	
12	1	20	3.6	S	−3.3	
22	1	10	1.6	S	−3.3	
Jul 2	1	00	0.0		−3.3	
12	0	47	1.2	N	−3.4	
22	0	34	2.0	N	−3.4	
Aug 1	0	22	2.2	N	−3.4	
11	0	10	1.9	N	−3.5	

Eastern (evening) elongation

Date	Angular Distance from Sun in R.A.		in Dec.		Mag.	
Aug 21	0	02	1.3	N	−3.5	S.C. Aug 20
31	0	13	0.2	N	−3.5	
Sep 10	0	22	0.9	S	−3.4	
20	0	32	2.1	S	−3.4	
30	0	41	3.2	S	−3.4	
Oct 10	0	52	4.3	S	−3.4	
20	1	02	5.1	S	−3.3	
30	1	11	5.4	S	−3.3	
Nov 9	1	26	5.3	S	−3.3	
19	1	38	4.6	S	−3.3	
29	1	50	3.3	S	−3.3	
Dec 9	2	01	1.5	S	−3.4	
19	2	11	0.8	N	−3.4	
29	2	18	3.5	N	−3.4	

1996

Date	Angular Distance from Sun in R.A.		in Dec.		Mag.	
Jan 8	2	23	5.7	N	−3.4	
18	2	27	7.0	N	−3.5	
28	2	30	11.6	N	−3.5	
Feb 7	2	32	14.0	N	−3.6	
17	2	35	15.9	N	−3.6	
27	2	38	17.5	N	−3.7	
Mar 8	2	43	18.3	N	−3.8	
18	2	48	18.7	N	−3.9	
28	2	53	18.5	N	−4.0	G.E. Apr 1
Apr 7	2	57	17.5	N	−4.0	
17	2	58	15.9	N	−4.1	
27	2	54	13.6	N	−4.2	
May 7	2	41	11.0	N	−4.2	
17	2	15	7.9	N	−4.1	
27	1	32	4.7	N	−3.8	
Jun 6	0	32	1.1	N	−3.0	I.C. Jun 10

Western (morning) elongation

Date	Angular Distance from Sun in R.A.		in Dec.		Mag.	
Jun 16	0	35	2.3	S	−3.1	
26	1	28	4.6	S	−3.8	

Date	Angular Distance from Sun in R.A.		in Dec.		Mag.	
Jul 6	2	17	5.0	S	−4.1	
16	2	45	3.5	S	−4.2	
26	3	02	0.9	S	−4.2	
Aug 5	3	09	2.3	N	−4.1	
15	3	10	5.7	N	−4.0	G.E. Aug 19
25	3	07	9.0	N	−3.9	
Sep 4	2	59	11.8	N	−3.8	
14	2	51	13.9	N	−3.8	
24	2	41	15.4	N	−3.7	
Oct 4	2	32	16.2	N	−3.6	
14	2	24	16.2	N	−3.6	
24	2	17	15.6	N	−3.5	
Nov 3	2	10	14.4	N	−3.5	
13	2	06	12.7	N	−3.5	
23	2	02	10.6	N	−3.4	
Dec 3	1	56	8.2	N	−3.4	
13	1	49	5.6	N	−3.4	
23	1	44	3.0	N	−3.4	

1997

Jan 2	1	35	0.6	N	−3.4	
12	1	25	1.4	S	−3.3	
22	1	13	3.0	S	−3.3	
Feb 1	1	01	4.1	S	−3.4	
11	0	49	4.5	S	−3.4	
21	0	37	4.6	S	−3.4	
Mar 3	0	28	4.1	S	−3.4	
13	0	17	3.3	S	−3.4	
23	0	08	2.4	S	−3.5	

Eastern (evening) elongation

Apr 2	0	02	1.3	S	−3.5	S.C. Apr 2
12	0	11	0.1	S	−3.5	
22	0	20	0.9	N	−3.4	
May 2	0	31	1.9	N	−3.4	
12	0	43	2.4	N	−3.4	
22	0	55	2.5	N	−3.4	
Jun 1	1	07	2.2	N	−3.3	
11	1	19	1.3	N	−3.3	
21	1	31	0.1	S	−3.3	
Jul 1	1	41	2.0	S	−3.3	
11	1	50	4.2	S	−3.3	
21	1	57	6.5	S	−3.4	
31	2	03	8.8	S	−3.4	
Aug 10	2	09	11.0	S	−3.4	
20	2	14	13.1	S	−3.4	
30	2	21	14.7	S	−3.5	
Sep 9	2	27	16.1	S	−3.5	
19	2	35	16.8	S	−3.6	
29	2	44	16.9	S	−3.7	
Oct 9	2	54	16.5	S	−3.8	
19	3	04	15.3	S	−3.9	
29	3	13	13.2	S	−4.0	
Nov 8	3	19	10.5	S	−4.1	G.E. Nov 6

Date	Angular Distance from Sun				Mag.	
	in R.A.		in Dec.			
Nov 18	3	21	7.2	S	−4.2	
28	3	16	3.7	S	−4.3	
Dec 8	3	02	0.2	S	−4.4	
18	2	37	2.9	N	−4.4	
28	1	57	5.1	N	−4.2	

1998

Date						
Jan 7	1	00	6.1	N	−3.7	

Western (morning) elongation

Date						
Jan 17	0	08	5.6	N	−3.2	I.C. Jan 16
27	1	13	3.8·	N	−3.8	
Feb 6	2	02	0.6	N	−4.2	
16	2	33	3.1	N	−4.3	
26	2	50	7.0	S	−4.3	
Mar 8	2	58	10.7	S	−4.2	
18	2	59	13.7	S	−4.1	
28	2	56	15.9	S	−4.0	G.E. Mar 27
Apr 7	2	54	17.3	S	−3.9	
17	2	47	17.8	S	−3.8	
27	2	43	17.6	S	−3.7	
May 7	2	40	16.6	S	−3.6	
17	2	36	15.0	S	−3.6	
27	2	33	12.8	S	−3.5	
Jun 6	2	29	10.3	S	−3.4	
16	2	25	7.4	S	−3.4	
26	2	19	4.5	S	−3.4	
Jul 6	2	10	1.6	S	−3.3	
16	1	59	1.1	N	−3.3	
26	1	47	3.3	N	−3.3	
Aug 5	1	34	4.9	N	−3.3	
15	1	20	5.9	N	−3.3	
25	1	07	6.4	N	−3.3	
Sep 4	0	54	6.4	N	−3.4	
14	0	43	5.8	N	−3.4	
24	0	32	5.0	N	−3.4	
Oct 4	0	23	4.0	N	−3.4	
14	0	13	3.0	N	−3.5	
24	0	05	1.6	N	−3.5	

Eastern (evening) elongation

Date						
Nov 3	0	05	0.4	N	−3.5	S.C. Oct 30
13	0	14	0.4	S	−3.5	
23	0	25	1.0	S	−3.4	
Dec 3	0	36	1.3	S	−3.4	
13	0	47	1.0	S	−3.4	
23	0	58	0.4	S	−3.4	

1999

Date						
Jan 2	1	07	0.8	N	−3.3	
12	1	16	2.3	N	−3.3	
22	1	23	4.0	N	−3.3	
Feb 1	1	30	5.9	N	−3.3	

Date	Angular Distance from Sun in R.A.		in Dec.		Mag.	
Feb 11	1	36	7.7	N	−3.4	
21	1	42	9.4	N	−3.4	
Mar 3	1	49	10.9	N	−3.4	
13	1	56	12.1	N	−3.4	
23	2	05	12.9	N	−3.4	
Apr 2	2	16	13.2	N	−3.5	
12	2	27	13.0	N	−3.5	
22	2	40	12.0	N	−3.6	
May 2	2	54	10.3	N	−3.6	
12	3	02	8.0	N	−3.7	
22	3	11	5.0	N	−3.8	
Jun 1	3	17	1.7	N	−3.9	
11	3	17	0.8	S	−3.9	G.E. Jun 11
21	3	13	5.2	S	−4.0	
Jul 1	3	02	8.5	S	−4.1	
11	2	45	10.9	S	−4.2	
21	2	19	12.6	S	−4.2	
31	1	71	12.9	S	−4.0	
Aug 10	0	54	11.4	S	−3.6	

Western (morning) elongation

Date	Angular Distance from Sun in R.A.		in Dec.		Mag.	
Aug 20	0	08	7.9	N	−3.1	I.C. Aug 20
30	1	07	2.7	S	−3.6	
Sep 9	1	52	2.7	N	−4.1	
19	2	22	7.8	N	−4.2	
29	2	38	11.9	N	−4.3	
Oct 9	2	48	14.9	N	−4.2	
19	2	51	16.8	N	−4.1	
29	2	53	17.8	N	−4.0	G.E. Oct 30
Nov 8	2	53	17.8	N	−4.0	
18	2	53	16.8	N	−3.9	
28	2	53	15.0	N	−3.8	
Dec 8	2	53	12.4	N	−3.7	
18	2	51	9.5	N	−3.7	
28	2	48	6.1	N	−3.6	

2000

Date	Angular Distance from Sun in R.A.		in Dec.		Mag.	
Jan 7	2	43	2.7	N	−3.5	
17	2	34	0.7	S	−3.5	
27	2	23	3.7	S	−3.4	
Feb 6	2	10	6.2	S	−3.4	
16	1	58	8.1	S	−3.4	
26	1	45	9.2	S	−3.4	
Mar 7	1	33	9.8	S	−3.3	
17	1	22	9.7	S	−3.3	
27	1	11	9.1	S	−3.3	
Apr 6	1	02	8.2	S	−3.3	
16	0	53	7.1	S	−3.3	
26	0	45	5.6	S	−3.3	
May 6	0	37	4.2	S	−3.4	
16	0	28	2.7	S	−3.4	
26	0	19	1.3	S	−3.4	
Jun 5	0	07	0.4	S	−3.5	

Date	Angular Distance from Sun				
	in R.A.		in Dec.		Mag.

Eastern (evening) elongation

Date		in R.A.		in Dec.		Mag.	
Jun	15	0	04	0.3	N	− 3.5	S.C. Jun 11
	25	0	17	0.4	N	− 3.5	
Jul	5	0	29	0.1	N	− 3.4	
	15	0	44	0.6	S	− 3.4	
	25	0	50	1.7	S	− 3.4	
Aug	4	1	00	3.2	S	− 3.3	
	14	1	09	4.7	S	− 3.3	
	24	1	17	6.4	S	− 3.3	
Sep	3	1	25	7.9	S	− 3.3	
	13	1	34	9.4	S	− 3.3	
	23	1	43	10.4	S	− 3.3	
Oct	3	1	53	11.1	S	− 3.4	
	13	2	05	11.3	S	− 3.4	
	23	2	19	10.8	S	− 3.4	
Nov	2	2	31	9.6	S	− 3.5	
	12	2	44	7.8	S	− 3.5	
	22	2	56	5.2	S	− 3.6	
Dec	2	3	05	1.9	S	− 3.7	
	12	3	11	1.7	N	− 3.7	
	22	3	13	5.4	N	− 3.8	
Jan	1	3	13	9.1	N	− 3.9	

Mars

The right ascension and declination of Mars, for epoch 2000, are given for every 20th day throughout the year, even though for several months around the time of conjunction it is too faint and near the Sun to be observed.

Date		R.A.		Dec.	Mag.	
1988						
Jun	18	23^h	15^m	−8°.2	− 0.4	
Jul	8	23	56	− 4.5	− 0.8	
	28	00	29	− 1.8	− 1.2	
Aug	17	00	49	− 0.3	− 1.7	Stationary, Aug 26
Sep	6	00	48	− 0.6	− 2.2	
	26	00	30	− 1.9	− 2.5	Opposition, Sep 28
Oct	16	00	09	− 2.6	− 2.1	Stationary, Oct 30
Nov	5	00	04	− 1.5	− 1.5	
	25	00	18	+ 1.2	− 0.8	
Dec	15	00	45	+ 4.9	− 0.3	
1989						
Jan	4	01	40	+ 9.1	+ 0.2	
	24	02	01	+ 13.3	+ 0.6	
Feb	13	02	46	+ 17.2	+ 0.9	
Mar	5	03	35	+ 20.6	+ 1.2	

Date	R.A.		Dec.	Mag.	
Mar 25	04	26	+23.0	+1.5	
Apr 14	05	19	+24.5	+1.6	
May 4	06	14	+24.8	+1.8	
24	07	08	+23.9	+1.9	
Jun 13	08	01	+21.8	+2.0	
Jul 3	08	53	+18.8	+2.0	
23	09	43	+15.0	+2.0	
Aug 12	10	31	+10.5	+2.0	
Sep 1	11	18	+5.6	+2.0	
21	12	05	+0.4	+1.9	Conjunction, Sep 29
Oct 11	12	53	−4.9	+1.9	
31	13	42	−10.0	+1.9	
Nov 20	14	34	−14.7	+1.9	
Dec 10	15	29	−18.7	+1.8	
30	16	26	−21.7	+1.8	

1990

Date	R.A.		Dec.	Mag.	
Jan 19	17	28	−23.5	−+1.7	
Feb 8	18	31	−23.7	+1.6	
28	19	34	−22.5	+1.4	
Mar 20	20	36	−19.7	+1.3	
Apr 9	21	36	−15.6	+1.1	
29	22	34	−10.7	+1.0	
May 19	23	30	−5.2	+0.8	
Jun 8	00	24	+0.5	+0.7	
28	01	17	+6.0	+0.5	
Jul 18	02	09	+10.8	+0.3	
Aug 7	02	58	+14.9	+0.1	
27	03	44	+17.9	−0.2	
Sep 16	04	22	+20.0	−0.5	
Oct 6	04	47	+21.4	−0.9	
26	04	53	+22.4	−1.3	Stationary, Oct 20
Nov 15	04	34	+22.8	−1.7	
Dec 5	04	02	+22.5	−1.6	Opposition, Nov 27
25	03	42	+22.0	−1.0	

1991

Date	R.A.		Dec.	Mag.	
Jan 4	03	40	+22.0	−0.7	Stationary, Jan 1
24	03	52	+22.6	−0.1	
Feb 13	04	19	+23.8	+0.4	
Mar 5	04	56	+24.8	+0.8	
25	05	40	+25.3	+1.1	
Apr 14	06	26	+25.1	+1.4	
May 4	07	15	+24.0	+1.6	
24	08	04	+21.9	+1.8	
Jun 13	08	53	+18.9	+1.9	
Jul 3	09	41	+15.2	+1.9	
23	10	28	+10.7	+2.0	
Aug 12	11	14	+5.9	+2.0	
Sep 1	12	01	+0.7	+2.0	
21	12	49	−4.6	+1.9	
Oct 11	13	38	−9.8	+1.8	
31	14	30	−14.6	+1.7	Conjunction, Nov 8

Date		R.A.		Dec.	Mag.	
Nov	20	15	25	− 18.7	+ 1.7	
Dec	10	16	25	− 21.9	+ 1.7	
	30	17	27	− 23.7	+ 1.6	
1992						
Jan	19	18	32	− 23.9	+ 1.6	
Feb	8	19	37	− 22.4	+ 1.5	
	28	20	41	− 19.3	+ 1.5	
Mar	19	21	43	− 14.9	+ 1.4	
Apr	8	22	43	− 9.5	+ 1.3	
	28	23	40	− 3.6	+ 1.3	
May	18	0	37	+ 2.5	+ 1.2	
Jun	7	1	32	+ 8.3	+ 1.1	
	27	2	28	+ 13.4	+ 1.0	
Jul	17	3	25	+ 17.7	+ 1.0	
Aug	6	4	21	+ 20.8	+ 0.9	
	26	5	16	+ 22.6	+ 0.8	
Sep	15	6	07	+ 23.4	+ 0.6	
Oct	5	6	53	+ 23.3	+ 0.4	
	25	7	30	+ 22.9	+ 0.1	
Nov	14	7	55	+ 22.7	− 0.2	
Dec	4	8	01	+ 23.3	− 0.7	Stationary, Nov 29
	24	7	44	+ 25.0	− 1.1	
1993						
Jan	13	7	13	+ 26.6	− 1.2	Opposition, Jan 7
Feb	2	6	44	+ 27.0	− 0.7	
	22	6	40	+ 26.6	− 0.1	Stationary, Feb 15
Mar	14	6	56	+ 25.8	+ 0.4	
Apr	3	7	25	+ 24.5	+ 0.8	
	23	8	02	+ 22.7	+ 1.1	
May	13	8	41	+ 20.1	+ 1.4	
Jun	2	9	24	+ 16.8	+ 1.6	
	22	10	07	+ 12.8	+ 1.7	
Jul	12	10	52	+ 8.3	+ 1.8	
Aug	1	11	36	+ 3.3	+ 1.8	
	21	12	22	− 1.9	+ 1.8	
Sep	10	13	10	− 7.1	+ 1.8	
	30	14	00	− 12.2	+ 1.8	
Oct	20	14	53	− 16.7	+ 1.7	
Nov	9	15	50	− 20.5	+ 1.7	
	29	16	54	− 23.1	+ 1.6	
Dec	19	17	57	− 24.2	+ 1.5	Conjunction, Dec 27
1994						
Jan	8	19	04	− 23.5	+ 1.4	
	28	20	10	− 21.1	+ 1.4	
Feb	17	21	14	− 17.2	+ 1.4	
Mar	9	22	15	− 12.1	+ 1.4	
	29	23	14	− 6.1	+ 1.4	
Apr	18	0	11	+ 0.1	+ 1.4	
May	8	1	08	+ 6.2	+ 1.4	
	28	2	05	+ 11.8	+ 1.4	
Jun	17	3	02	+ 16.6	+ 1.4	

Date	R.A.		Dec.	Mag.	
Jul 7	4	01	+20.3	+1.4	
27	5	00	+22.6	+1.4	
Aug 16	5	58	+23.6	+1.4	
Sep 5	6	54	+23.3	+1.3	
25	7	46	+22.0	+1.2	
Oct 15	8	34	+20.0	+1.1	
Nov 4	9	16	+17.6	+0.9	
24	9	50	+15.4	+0.6	
Dec 14	10	14	+13.9	+0.2	
1995					
Jan 3	10	24	+13.8	−0.2	Stationary, Jan 4
23	10	15	+15.4	−0.7	
Feb 12	9	47	+18.1	−1.0	Opposition, Feb 12
Mar 4	9	19	+20.0	−0.7	
24	9	07	+20.1	−0.2	Stationary, Mar 25
Apr 13	9	14	+18.8	+0.3	
May 3	9	36	+16.5	+0.7	
23	10	06	+13.3	+1.0	
Jun 12	10	42	+9.5	+1.2	
Jul 2	11	20	+5.0	+1.4	
22	12	02	+0.2	+1.5	
Aug 11	12	46	−4.8	+1.6	
31	13	34	−9.9	+1.6	
Sep 20	14	24	−14.7	+1.6	
Oct 10	15	19	−18.9	+1.6	
30	16	19	−22.1	+1.6	
Nov 19	17	22	−24.0	+1.5	
Dec 9	18	28	−24.3	+1.5	
29	19	35	−22.7	+1.4	
1996					
Jan 18	20	41	−19.4	+1.4	
Feb 7	21	44	−14.8	+1.3	
27	22	44	−9.1	+1.3	Conjunction, Mar 4
Mar 18	23	43	−2.9	+1.3	
Apr 7	0	40	+3.4	+1.4	
27	1	36	+9.4	+1.5	
May 17	2	34	+14.7	+1.5	
Jun 6	3	32	+18.9	+1.6	
26	4	32	+21.9	+1.6	
Jul 16	5	31	+23.6	+1.7	
Aug 5	6	30	+23.8	+1.7	
25	7	27	+22.7	+1.7	
Sep 14	8	21	+20.5	+1.7	
Oct 4	9	13	+17.6	+1.6	
24	9	57	+14.1	+1.5	
Nov 13	10	40	+10.4	+1.3	
Dec 3	11	17	+6.9	+1.1	
23	11	49	+3.8	+0.8	
1997					
Jan 12	12	14	+1.6	+0.4	
Feb 1	12	26	+0.8	0.0	Stationary, Feb 6
21	12	23	+1.6	−0.6	

Date		R.A.		Dec.	Mag.	
Mar	13	12	01	+4.0	−1.0	Opposition, Mar 17
Apr	2	11	32	+6.5	−0.9	
	22	11	16	+7.3	−0.5	Stationary, Apr 29
May	12	11	18	+6.2	0.0	
Jun	1	11	35	+3.6	+0.4	
	21	12	04	0.0	+0.6	
Jul	11	12	38	−4.2	+0.9	
	31	13	19	−8.7	+1.0	
Aug	20	14	04	−13.3	+1.1	
Sep	9	14	55	−17.6	+1.2	
	29	15	50	−21.2	+1.3	
Oct	19	16	52	−23.7	+1.3	
Nov	8	17	56	−24.7	+1.3	
	28	19	03	−23.9	+1.3	
Dec	18	20	09	−21.4	+1.4	

1998

Jan	7	21	13	−17.3	+1.4	
	27	22	15	−12.0	+1.4	
Feb	16	23	14	−5.9	+1.4	
Mar	8	0	13	+0.4	+1.4	
	28	1	07	+6.6	+1.5	
Apr	17	2	04	+12.3	+1.5	
May	7	3	01	+17.1	+1.5	Conjunction, May 12
	27	4	00	+20.8	+1.6	
Jun	16	4	59	+23.1	+1.7	
Jul	6	5	59	+24.0	+1.8	
	26	6	58	+23.6	+1.8	
Aug	15	7	54	+21.8	+1.9	
Sep	4	8	47	+19.0	+1.9	
	24	9	38	+15.4	+1.9	
Oct	14	10	26	+11.3	+1.8	
Nov	3	11	11	+6.9	+1.8	
	23	11	53	+2.5	+1.6	
Dec	13	12	34	−1.8	+1.4	

1999

Jan	2	13	13	−5.6	+1.2	
	22	13	47	−8.9	+0.9	
Feb	11	14	17	−11.4	+0.5	
Mar	3	14	37	−13.0	0.0	
	23	14	41	−13.5	−0.6	Stationary, Mar 18
Apr	12	14	27	−12.7	−1.2	
May	2	13	58	−11.0	−1.4	Opposition, Apr 24
	22	13	35	−9.7	−1.1	
Jun	11	13	31	−10.1	−0.7	Stationary, Jun 5
Jul	1	13	45	−12.0	−0.3	
	21	14	13	−14.9	0.0	
Aug	10	14	54	−18.2	+0.3	
	30	15	40	−21.4	+0.5	
Sep	19	16	34	−23.8	+0.6	
Oct	9	17	35	−25.1	+0.8	
	29	18	38	−24.9	+0.9	
Nov	18	19	43	−22.9	+1.0	

Date		R.A.		Dec.	Mag.	
Dec	8	20	47	− 19.3	+ 1.1	
	28	21	49	− 14.4	+ 1.2	
2000						
Jan	17	22	47	− 8.6	+ 1.3	
Feb	6	23	44	− 2.4	+ 1.4	
	26	0	39	+ 3.9	+ 1.5	
Mar	17	1	35	+ 9.8	+ 1.6	
Apr	6	2	31	+ 15.0	+ 1.6	
	26	3	28	+ 19.2	+ 1.7	
May	16	4	27	+ 22.2	+ 1.8	
Jun	5	5	26	+ 23.9	+ 1.8	
	25	6	25	+ 24.1	+ 1.8	Conjunction, Jul 1
Jul	15	7	22	+ 23.0	+ 1.9	
Aug	4	8	17	+ 20.8	+ 1.9	
	24	9	10	+ 17.5	+ 2.0	
Sep	13	10	01	+ 13.5	+ 2.0	
Oct	3	10	48	+ 9.0	+ 2.0	
	23	11	34	+ 4.1	+ 2.0	
Nov	12	12	30	− 0.7	+ 1.9	
Dec	2	13	05	− 5.5	+ 1.8	
	22	13	50	− 9.9	+ 1.6	

Jupiter

The right ascension and declination of Jupiter, for epoch 2000, are given for every 20th day throughout the year. Unlike Mars, Jupiter is lost in the bright sky near the Sun for only a few weeks around the time of conjunction.

Date		R.A.		Dec.	Mag.	
1988						
Jun	18	3^h	25^m	+ 17°.8	− 1.6	
Jul	8	3	42	+ 18.7	− 1.7	
	28	3	57	+ 19.5	− 1.8	
Aug	17	4	08	+ 20.0	− 1.9	
Sep	6	4	16	+ 20.3	− 2.0	
	26	4	19	+ 20.3	− 2.2	Stationary, Sep 24
Oct	16	4	15	+ 20.2	− 2.3	
Nov	5	4	07	+ 19.8	− 2.4	
	25	3	56	+ 19.3	− 2.4	Opposition, Nov 23
Dec	15	3	46	+ 18.9	− 2.3	
1989						
Jan	4	3	38	+ 18.6	− 2.2	
	24	3	37	+ 18.5	− 2.1	Stationary, Jan 20
Feb	13	3	40	+ 18.8	− 2.0	
Mar	5	3	49	+ 19.4	− 1.8	
	25	4	02	+ 20.1	− 1.7	

Date		R.A.		Dec.	Mag.	
Apr	14	4	17	+20.8	−1.6	
May	4	4	35	+21.5	−1.5	
	24	4	55	+22.2	−1.5	
Jun	13	5	14	+22.6	−1.5	Conjunction, Jun 9
Jul	3	5	34	+22.9	−1.5	
	23	5	53	+23.1	−1.5	
Aug	12	6	11	+23.1	−1.6	
Sep	1	6	26	+23.0	−1.7	
	21	6	38	+22.9	−1.8	
Oct	11	6	46	+22.8	−1.9	
	31	6	48	+22.7	−2.1	Stationary, Oct 29
Nov	20	6	44	+22.8	−2.2	
Dec	10	6	36	+23.0	−2.3	
	30	6	25	+23.2	−2.3	Opposition, Dec 27

1990

Date		R.A.		Dec.	Mag.	
Jan	19	6	14	+23.3	−2.2	
Feb	8	6	06	+23.4	−2.1	
	28	6	04	+23.5	−2.0	Stationary, Feb 24
Mar	20	6	08	+23.5	−1.8	
Apr	9	6	17	+23.5	−1.7	
	29	6	29	+23.4	−1.6	
May	19	6	45	+23.2	−1.5	
Jun	8	7	03	+22.8	−1.4	
	28	7	22	+22.3	−1.4	
Jul	18	7	41	+21.7	−1.4	Conjunction, Jul 15
Aug	7	7	59	+20.9	−1.4	
	27	8	17	+20.0	−1.4	
Sep	16	8	33	+19.1	−1.5	
Oct	6	8	47	+18.3	−1.6	
	26	8	58	+17.7	−1.7	
Nov	15	9	04	+17.3	−1.8	Stationary, Nov 30
Dec	5	9	05	+17.3	−1.9	
	25	8	02	+17.6	−2.1	

1991

Date		R.A.		Dec.	Mag.	
Jan	14	8	53	+18.3	−2.1	
Feb	3	8	43	+19.0	−2.1	Opposition, Jan 29
	23	8	34	+19.6	−2.1	
Mar	15	8	27	+20.0	−2.0	
Apr	4	8	25	+20.1	−1.9	Stationary, Mar 30
	24	8	29	+19.9	−1.7	
May	14	8	37	+19.4	−1.6	
Jun	3	8	48	+18.6	−1.5	
	23	9	02	+17.6	−1.4	
Jul	13	9	18	+16.5	−1.3	
Aug	2	9	35	+15.2	−1.3	
	22	9	52	+13.7	−1.3	Conjunction, Aug 17
Sep	11	10	09	+12.3	−1.3	
Oct	1	10	25	+10.8	−1.3	
	21	10	39	+9.5	−1.4	
Nov	10	10	51	+8.4	−1.5	
	30	11	00	+7.5	−1.6	
Dec	20	11	05	+7.1	−1.7	Stationary, Dec 31

Date		R.A.		Dec.	Mag.	
1992						
Jan	9	11	05	+7.2	−1.9	
	29	11	01	+7.7	−2.0	
Feb	18	10	53	+8.6	−2.0	
Mar	9	10	43	+9.6	−2.0	Opposition, Feb 29
	29	10	34	+10.5	−2.0	
Apr	18	10	29	+10.9	−1.9	
May	8	10	49	+10.9	−1.8	Stationary, May 1
	28	10	32	+10.5	−1.6	
Jun	17	10	40	+9.7	−1.5	
Jul	7	10	51	+8.6	−1.4	
	27	11	04	+7.2	−1.3	
Aug	16	11	18	+5.6	−1.2	
Sep	5	11	34	+4.0	−1.2	
	25	11	50	+2.3	−1.2	Conjunction, Sep 17
Oct	15	12	09	+0.6	−1.2	
Nov	4	12	20	−1.0	−1.3	
	24	12	34	−2.4	−1.4	
Dec	14	12	45	−3.5	−1.5	
1993						
Jan	3	12	53	−4.2	−1.6	
	23	12	56	−4.5	−1.7	Stationary, Jan 29
Feb	12	12	56	−4.3	−1.9	
Mar	4	12	50	−3.7	−1.9	
	24	12	42	−2.8	−2.0	Opposition, Mar 30
Apr	13	12	32	−1.8	−2.0	
May	3	12	25	−1.0	−1.9	
	23	12	20	−0.6	−1.8	Stationary, Jun 1
Jun	12	12	20	−0.7	−1.7	
Jul	2	12	25	−1.3	−1.6	
	22	12	33	−2.2	−1.5	
Aug	11	12	44	−3.4	−1.4	
	31	12	57	−4.9	−1.3	
Sep	20	13	12	−6.5	−1.2	
Oct	10	13	28	−8.1	−1.2	Conjunction, Oct 18
	30	13	44	−9.6	−1.2	
Nov	19	14	01	−11.1	−1.2	
Dec	9	14	16	−12.4	−1.3	
	29	14	29	−13.5	−1.4	
1994						
Jan	18	14	40	−14.4	−1.5	
Feb	7	14	48	−14.9	−1.6	
	27	14	50	−15.0	−1.8	Stationary, Feb 28
Mar	19	14	49	−14.8	−1.9	
Apr	8	14	42	−14.3	−2.0	
	28	14	33	−13.5	−2.0	Opposition, Apr 30
May	18	14	23	−12.8	−2.0	
Jun	7	14	16	−12.2	−1.9	
	27	14	12	−12.0	−1.8	Stationary, Jul 2
Jul	17	14	13	−12.2	−1.7	
Aug	6	14	18	−12.7	−1.6	
	26	14	28	−13.6	−1.5	

Date	R.A.		Dec.	Mag.	
Sep 15	14	40	− 14.6	− 1.4	
Oct 5	14	55	− 15.8	− 1.3	
25	15	11	− 17.0	− 1.3	
Nov 14	15	29	− 18.1	− 1.2	Conjunction, Nov 17
Dec 4	15	47	− 19.2	− 1.3	
24	16	05	− 20.0	− 1.3	
1995					
Jan 13	16	22	− 20.8	− 1.4	
Feb 2	16	36	− 21.6	− 1.5	
22	16	47	− 21.6	− 1.6	
Mar 14	16	55	− 21.8	− 1.7	
Apr 3	16	57	− 21.8	− 1.9	Stationary, Apr 1
23	16	54	− 21.7	− 2.0	
May 13	16	46	− 21.5	− 2.1	
Jun 2	16	36	− 21.2	− 2.1	Opposition, Jun 1
22	16	26	− 20.9	− 2.1	
Jul 12	16	18	− 20.7	− 2.0	
Aug 1	16	15	− 20.6	− 1.9	Stationary, Aug 2
21	16	17	− 20.8	− 1.8	
Sep 10	16	24	− 21.1	− 1.7	
30	16	35	− 21.6	− 1.5	
Oct 20	16	50	− 22.1	− 1.5	
Nov 9	17	07	− 22.5	− 1.4	
29	17	25	− 22.9	− 1.3	
Dec 19	17	45	− 23.1	− 1.3	Conjunction, Dec 18
1996					
Jan 8	18	05	− 23.2	− 1.4	
28	18	24	− 23.1	− 1.4	
Feb 17	18	41	− 22.9	− 1.5	
Mar 8	18	56	− 22.7	− 1.6	
28	19	07	− 22.4	− 1.7	
Apr 17	19	14	− 22.2	− 1.9	
May 7	19	17	− 22.2	− 2.0	Stationary, May 4
27	19	13	− 22.3	− 2.1	
Jun 16	19	05	− 22.6	− 2.2	
Jul 6	18	55	− 22.9	− 2.2	Opposition, Jul 4
26	18	44	− 23.2	− 2.2	
Aug 15	18	37	− 23.3	− 2.1	
Sep 4	18	34	− 23.4	− 2.0	Stationary, Sep 3
24	18	37	− 23.4	− 1.9	
Oct 14	18	45	− 23.3	− 1.8	
Nov 3	18	58	− 23.0	− 1.6	
23	19	13	− 22.7	− 1.6	
Dec 13	19	31	− 22.1	− 1.5	
1997					
Jan 2	19	50	− 21.4	− 1.5	
22	20	10	− 20.5	− 1.4	Conjunction, Jan 19
Feb 11	20	29	− 19.4	− 1.5	
Mar 3	20	47	− 18.4	− 1.5	
23	21	04	− 17.3	− 1.6	

Date		R.A.		Dec.	Mag.	
Apr	12	21	28	− 16.3	− 1.7	
May	2	21	29	− 15.5	− 1.8	
	22	21	36	− 15.0	− 2.0	
Jun	11	21	38	− 14.9	− 2.1	Stationary, Jun 10
Jul	1	21	36	− 15.2	− 2.2	
	21	21	29	− 15.9	− 2.3	
Aug	10	21	19	− 16.7	− 2.4	Opposition, Aug 9
	30	21	09	− 17.4	− 2.3	
Sep	19	21	02	− 17.9	− 2.3	
Oct	9	20	59	− 18.1	− 2.1	Stationary, Oct 8
	29	21	02	− 17.9	− 2.0	
Nov	18	21	10	− 17.3	− 1.9	
Dec	8	21	22	− 16.4	− 1.8	
	28	21	37	− 15.2	− 1.7	

1998

Date		R.A.		Dec.	Mag.	
Jan	17	21	53	− 13.7	− 1.6	
Feb	6	22	11	− 12.1	− 1.6	
	26	22	29	− 10.4	− 1.5	Conjunction, Feb 23
Mar	18	22	47	− 8.7	− 1.6	
Apr	7	23	05	− 7.0	− 1.6	
	27	23	20	− 5.4	− 1.7	
May	17	23	34	− 4.0	− 1.8	
Jun	6	23	45	− 2.9	− 1.9	
	26	23	54	− 2.2	− 2.1	
Jul	16	23	55	− 2.0	− 2.2	Stationary, Jul 18
Aug	5	23	53	− 2.3	− 2.3	
	25	23	47	− 3.0	− 2.4	
Sep	14	23	38	− 4.1	− 2.5	Opposition, Sep 16
Oct	4	23	29	− 5.1	− 2.4	
	24	23	21	− 5.8	− 2.3	
Nov	13	23	19	− 6.0	− 2.2	Stationary, Nov 14
Dec	3	23	21	− 5.7	− 2.1	
	23	23	28	− 4.8	− 1.9	

1999

Date		R.A.		Dec.	Mag.	
Jan	12	23	39	− 3.6	− 1.8	
Feb	1	23	53	− 2.1	− 1.7	
	21	0	08	− 0.3	− 1.6	
Mar	13	0	25	+ 1.6	− 1.6	
Apr	2	0	43	+ 3.5	− 1.6	Conjunction, Apr 1
	22	1	01	+ 5.3	− 1.6	
May	12	1	18	+ 7.1	− 1.6	
Jun	1	1	34	+ 8.6	− 1.7	
	21	1	49	+ 9.9	− 1.8	
Jul	11	2	00	+ 10.9	− 1.9	
	31	2	09	+ 11.6	− 2.1	
Aug	20	2	13	+ 11.9	− 2.2	Stationary, Aug 25
Sep	9	2	11	+ 11.7	− 2.3	
	29	2	05	+ 11.1	− 2.4	
Oct	19	1	56	+ 10.2	− 2.5	Opposition, Oct 23
Nov	8	1	41	+ 9.3	− 2.4	
	28	1	38	+ 8.7	− 2.4	
Dec	18	1	35	+ 8.4	− 2.2	Stationary, Dec 21

Date		R.A.		Dec.	Mag.	
2000						
Jan	7	1	36	+8.7	−2.0	
	27	1	43	+9.5	−1.9	
Feb	16	1	54	+10.6	−1.8	
Mar	7	2	08	+11.9	−1.7	
	27	2	24	+13.3	−1.6	
Apr	16	2	42	+14.8	−1.6	
May	6	3	01	+16.2	−1.6	Conjunction, May 8
	26	3	20	+17.5	−1.6	
Jun	15	3	38	+18.6	−1.6	
Jul	5	3	55	+19.5	−1.7	
	25	4	12	+20.3	−1.7	
Aug	14	4	24	+20.8	−1.9	
Sep	3	4	34	+21.1	−2.0	
	23	4	39	+21.2	−2.1	Stationary, Sep 29
Oct	13	4	38	+21.2	−2.2	
Nov	2	4	32	+20.9	−2.3	
	22	4	21	+20.6	−2.4	Opposition, Nov 28
Dec	12	4	10	+20.1	−2.4	
Jan	1	4	01	+19.8	−2.3	

Saturn

The right ascension and declination of Saturn, for epoch 2000, are given for every 20th day throughout the year. The brightness of Saturn is greatly affected by the tilt of the rings toward the Earth; they appear most open in 1988, and edge-on in 1995–96.

Date		R.A.		Dec.	Mag.	
1988						
Jun	18	17^h	58^m	−22°.3	+0.2	Opposition, Jun 20
Jul	8	17	52	−22.3	+0.3	
	28	17	47	−22.3	+0.4	
Aug	17	17	44	−22.4	+0.5	
Sep	6	17	43	−22.4	+0.6	Stationary, Aug 30
	26	17	46	−22.5	+0.7	
Oct	16	17	50	−22.6	+0.8	
Nov	5	17	58	−22.6	+0.8	
	25	18	07	−22.7	+0.7	
Dec	15	18	16	−22.6	+0.7	Conjunction, Dec 26
1989						
Jan	4	18	26	−22.6	+0.7	
	24	18	37	−22.5	+0.7	
Feb	13	18	45	−22.3	+0.8	
Mar	5	18	53	−22.2	+0.8	
	25	18	57	−22.1	+0.8	
Apr	14	19	01	−22.0	+0.7	Stationary, Apr 23

Date	R.A.	Dec.	Mag.	
May 4	19 00	−22.0	+0.6	
24	18 58	−22.1	+0.5	
Jun 13	18 53	−22.2	+0.3	
Jul 3	18 46	−22.4	+0.2	Opposition, Jul 2
23	18 40	−22.5	+0.3	
Aug 12	18 35	−22.6	+0.4	
Sep 1	18 32	−22.7	+0.6	Stationary, Sep 11
21	18 32	−22.8	+0.6	
Oct 11	18 35	−22.8	+0.7	
31	18 41	−22.7	+0.8	
Nov 20	18 48	−22.6	+0.8	
Dec 10	18 57	−22.5	+0.7	
30	19 07	−22.2	+0.7	
1990				
Jan 19	19 17	−22.0	+0.7	Conjunction, Jan 6
Feb 8	19 27	−21.7	+0.8	
28	19 35	−21.4	+0.8	
Mar 20	19 43	−21.1	+0.8	
Apr 9	19 47	−21.0	+0.8	
29	19 49	−20.9	+0.7	Stationary, May 5
May 19	19 49	−20.9	+0.6	
Jun 8	19 46	−21.1	+0.5	
28	19 41	−21.3	+0.4	
Jul 18	19 35	−21.6	+0.3	Opposition, Jul 14
Aug 7	19 29	−21.8	+0.4	
27	19 24	−22.0	+0.5	
Sep 16	19 22	−22.1	+0.6	Stationary, Sep 23
Oct 6	19 22	−22.1	+0.7	
26	19 25	−22.1	+0.8	
Nov 15	19 31	−21.9	+0.8	
Dec 5	19 39	−21.6	+0.8	
25	19 48	−21.3	+0.8	
1991				
Jan 14	19 58	−20.9	+0.7	Conjunction, Jan 18
Feb 3	20 08	−20.4	+0.8	
23	20 17	−19.9	+0.9	
Mar 15	20 25	−19.5	+0.9	
Apr 4	20 32	−19.2	+0.9	
24	20 36	−18.9	+0.9	
May 14	20 38	−18.9	+0.8	Stationary, May 17
Jun 3	20 37	−19.0	+0.7	
23	20 34	−19.2	+0.5	
Jul 13	20 28	−19.5	+0.4	
Aug 2	20 22	−19.9	+0.3	Opposition, Jul 27
22	20 17	−20.3	+0.4	
Sep 11	20 13	−20.5	+0.5	
Oct 1	20 11	−20.6	+0.7	Stationary, Oct 5
21	20 11	−20.6	+0.7	
Nov 10	20 15	−20.4	+0.8	
30	20 21	−20.1	+0.9	
Dec 20	20 29	−19.7	+0.9	

Date		R.A.		Dec.	Mag.	
1992						
Jan	9	20	38	− 19.1	+ 0.8	
	29	20	47	− 18.5	+ 0.8	Conjunction, Jan 29
Feb	18	20	57	− 17.9	+ 0.9	
Mar	9	21	06	− 17.3	+ 1.0	
	29	21	14	− 16.8	+ 1.0	
Apr	18	21	20	− 16.4	+ 1.0	
May	8	21	24	− 16.1	+ 0.9	
	28	21	25	− 16.1	+ 0.9	Stationary, May 29
Jun	17	21	24	− 16.2	+ 0.7	
Jul	7	21	20	− 16.5	+ 0.6	
	27	21	16	− 17.0	+ 0.5	
Aug	16	21	10	− 17.4	+ 0.4	Conjunction, Aug 7
Sep	5	21	04	− 17.8	+ 0.5	
	25	21	00	− 18.1	+ 0.6	
Oct	15	20	59	− 18.2	+ 0.7	Stationary, Oct 16
Nov	4	21	00	− 18.1	+ 0.8	
	24	21	04	− 17.8	+ 0.9	
Dec	14	21	10	− 17.4	+ 0.9	
1993						
Jan	3	21	18	− 16.8	+ 1.0	
	23	21	26	− 16.1	+ 0.9	
Feb	12	21	36	− 15.4	+ 0.9	Conjunction, Feb 9
Mar	4	21	46	− 14.6	+ 1.0	
	24	21	54	− 13.9	+ 1.1	
Apr	13	22	02	− 13.3	+ 1.1	
May	3	22	07	− 12.9	+ 1.1	
	23	22	11	− 12.6	+ 1.0	
Jun	12	22	12	− 12.6	+ 1.0	Stationary, Jun 11
Jul	2	22	11	− 12.8	+ 0.8	
	22	22	07	− 13.2	+ 0.7	
Aug	11	22	02	− 13.7	+ 0.6	Opposition, Aug 19
	31	21	56	− 14.2	+ 0.6	
Sep	20	21	51	− 14.7	+ 0.6	
Oct	10	21	47	− 15.0	+ 0.7	
	30	21	46	− 15.0	+ 0.8	Stationary, Oct 28
Nov	19	21	48	− 14.9	+ 0.9	
Dec	9	21	52	− 14.5	+ 1.0	
	29	21	58	− 13.9	+ 1.1	
1994						
Jan	18	22	06	− 13.2	+ 1.1	
Feb	7	22	15	− 12.4	+ 1.1	
	27	22	24	− 11.5	+ 1.1	Conjunction, Feb 21
Mar	19	22	33	− 10.7	+ 1.2	
Apr	8	22	41	− 9.9	+ 1.2	
	28	22	49	− 9.3	+ 1.3	
May	18	22	54	− 8.8	+ 1.3	
Jun	7	22	57	− 8.5	+ 1.2	
	27	22	57	− 8.5	+ 1.1	Stationary, Jun 24
Jul	17	22	56	− 8.8	+ 1.0	
Aug	6	22	53	− 9.2	+ 0.9	
	26	22	47	− 9.8	+ 0.7	Opposition, Sep 1

Date		R.A.		Dec.	Mag.	
Sep	15	22	42	− 10.4	+ 0.7	
Oct	5	22	37	− 10.9	+ 0.8	
	25	22	34	− 11.1	+ 0.9	
Nov	14	22	33	− 11.2	+ 1.0	Stationary, Nov 9
Dec	4	22	35	− 10.9	+ 1.1	
	24	22	39	− 10.5	+ 1.2	
1995						
Jan	13	22	46	− 9.8	+ 1.2	
Feb	2	22	53	− 9.0	+ 1.2	
	22	23	02	− 8.1	+ 1.2	
Mar	14	23	11	− 7.1	+ 1.3	Conjunction, Mar 6
Apr	3	23	20	− 6.2	+ 1.4	
	23	23	28	− 5.4	+ 1.4	
May	13	23	35	− 4.7	+ 1.5	
Jun	2	23	41	− 4.3	+ 1.4	
	22	23	44	− 4.0	+ 1.3	
Jul	12	23	44	− 4.1	+ 1.2	Stationary, Jul 7
Aug	1	23	43	− 4.3	+ 1.1	
	21	23	39	− 4.8	+ 1.0	
Sep	10	23	34	− 5.4	+ 0.9	Opposition, Sep 14
	30	23	28	− 6.0	+ 0.9	
Oct	20	23	23	− 6.5	+ 1.0	
Nov	9	23	20	− 6.8	+ 1.1	
	29	23	20	− 6.7	+ 1.2	Stationary, Nov 22
Dec	19	23	22	− 6.4	+ 1.3	
1996						
Jan	8	23	26	− 5.9	+ 1.4	
	28	23	33	− 5.2	+ 1.4	
Feb	17	23	41	− 4.3	+ 1.4	
Mar	8	23	50	− 3.3	+ 1.3	Conjunction, Mar 17
	28	23	59	− 2.3	+ 1.3	
Apr	17	0	08	− 1.4	+ 1.3	
May	7	0	16	− 0.6	+ 1.2	
	27	0	23	+ 0.1	+ 1.2	
Jun	16	0	28	+ 0.5	+ 1.1	
Jul	6	0	31	+ 0.8	+ 1.0	
	26	0	31	+ 0.7	+ 0.9	Stationary, Jul 20
Aug	15	0	29	+ 0.4	+ 0.8	
Sep	4	0	25	− 0.1	+ 0.8	
	24	0	20	− 0.7	+ 0.7	Opposition, Sep 26
Oct	14	0	14	− 1.3	+ 0.8	
Nov	3	0	10	− 1.8	+ 0.9	
	23	0	07	− 2.0	+ 1.0	
Dec	13	0	07	− 1.9	+ 1.1	Stationary, Dec 4
1997						
Jan	2	0	09	− 1.6	+ 1.2	
	22	0	14	− 1.0	+ 1.2	
Feb	11	0	20	− 0.2	+ 1.2	
Mar	3	0	29	+ 0.7	+ 1.1	
	23	0	38	+ 1.7	+ 1.0	Conjunction, Mar 30
Apr	12	0	47	+ 2.6	+ 1.0	

Date		R.A.		Dec.	Mag.	
May	2	0	56	+3.6	+1.0	
	22	1	04	+4.4	+1.0	
Jun	11	1	11	+5.0	+0.9	
Jul	1	1	16	+5.4	+1.0	
	21	1	19	+5.6	+0.9	
Aug	10	1	19	+5.5	+0.7	Stationary, Aug 2
	30	1	17	+5.2	+0.6	
Sep	19	1	13	+4.7	+0.5	
Oct	9	1	07	+4.1	+0.4	Opposition, Oct 10
	29	1	01	+3.6	+0.5	
Nov	18	0	57	+3.1	+0.6	
Dec	8	0	54	+3.0	+0.8	Stationary, Dec 17
	28	0	55	+3.1	+0.9	

1998

Jan	17	0	57	+3.4	+0.9	
Feb	6	1	02	+4.1	+0.9	
	26	1	09	+4.9	+0.9	
Mar	18	1	17	+5.8	+0.9	
Apr	7	1	27	+6.7	+0.8	Conjunction, Apr 13
	27	1	36	+7.6	+0.7	
May	17	1	41	+8.5	+0.8	
Jun	6	1	54	+9.2	+0.7	
	26	2	01	+9.8	+0.7	
Jul	16	2	06	+10.2	+0.7	
Aug	5	2	09	+10.3	+0.6	
	25	2	09	+10.2	+0.5	Stationary, Aug 16
Sep	14	2	07	+9.9	+0.4	
Oct	4	2	02	+9.5	+0.3	
	24	1	56	+8.9	+0.2	Opposition, Oct 23
Nov	13	1	50	+8.4	+0.3	
Dec	3	1	46	+8.1	+0.4	
	23	1	43	+7.9	+0.5	Stationary, Dec 30

1999

Jan	12	1	44	+8.1	+0.6	
Feb	1	1	47	+8.4	+0.7	
	21	1	52	+9.0	+0.7	
Mar	13	1	59	+9.8	+0.7	
Apr	2	2	08	+10.6	+0.6	
	22	2	18	+11.5	+0.5	Conjunction, Apr 27
May	12	2	28	+12.3	+0.5	
Jun	1	2	37	+13.1	+0.6	
	21	2	46	+13.7	+0.6	
Jul	11	2	53	+14.2	+0.6	
	31	2	59	+14.5	+0.5	
Aug	20	3	01	+14.6	+0.4	Stationary, Aug 30
Sep	9	3	01	+14.5	+0.3	
	29	2	59	+14.3	+0.2	
Oct	19	2	54	+13.9	+0.1	
Nov	8	2	48	+13.4	0.0	Opposition, Nov 6
	28	2	41	+13.0	+0.1	
Dec	18	2	37	+12.7	+0.2	

Date		R.A.		Dec.	Mag.	
2000						
Jan	7	2	35	+12.6	+0.4	Stationary, Jan 13
	27	2	35	+12.8	+0.5	
Feb	16	2	38	+13.1	+0.5	
Mar	7	2	44	+13.7	+0.6	
	27	2	52	+14.3	+0.5	
Apr	16	2	02	+15.0	+0.5	
May	6	3	12	+15.8	+0.4	Conjunction, May 10
	26	3	22	+16.4	+0.4	
Jun	15	3	32	+17.0	+0.4	
Jul	5	3	41	+17.5	+0.5	
	25	3	49	+17.9	+0.4	
Aug	14	3	54	+18.1	+0.4	
Sep	3	3	57	+18.2	+0.3	Stationary, Sep 12
	23	3	57	+18.1	+0.2	
Oct	13	3	54	+17.9	+0.1	
Nov	2	3	49	+17.6	−0.1	
	22	3	42	+17.3	−0.2	Opposition, Nov 19
Dec	12	3	36	+17.0	0.0	
Jan	1	3	31	+16.8	+0.1	

Uranus

The right ascension and declination of Uranus, for epoch 2000, are given for every 40th day throughout the year. During the period 1988–2000, this planet will be passing slowly through Sagittarius and Capricornus, shining at about magnitude 5.7.

The planet follows its path around the Zodiac in a series of very flat "loops," effectively backtracking in a westerly direction along its previous course for some five months in the year, and then continuing in an easterly direction for seven. The chart shows its approximate position in the sky at the time of each opposition; but for positive identification, its accurate location should be plotted on a star map using the 40-day positions given here.

Date		R.A.		Dec.		
1988						
Jun	8	17^h	58^m	−23°.6		Opposition, Jun 20
Jul	18	17	52	−23.6		
Aug	27	17	48	−23.6		Stationary, Sep 5
Oct	6	17	50	−23.6		
Nov	15	17	56	−23.7		
Dec	25	18	07	−23.6		Conjunction, Dec 22
1989						
Feb	3	18	16	−23.6		
Mar	15	18	23	−23.6		

THE POSITION OF URANUS AT OPPOSITION, 1988–2000

Date	R.A.		Dec.	
Apr 24	18	23	−23.6	Stationary, Apr 9
Jun 3	18	19	−23.6	
Jul 13	18	01	−23.7	Opposition, Jun 24
Aug 22	18	07	−23.7	Stationary, Sep 10
Oct 1	18	07	−23.7	
Nov 10	18	13	−23.7	
Dec 20	18	23	−23.6	Conjunction, Dec 27

1990

Date	R.A.		Dec.	
Jan 29	18	33	−23.5	
Mar 10	18	40	−23.4	
Apr 19	18	43	−23.4	Stationary, Apr 13
Mar 29	18	39	−23.5	
Jul 8	18	32	−23.6	Opposition, Jun 29
Aug 17	18	26	−23.6	
Sep 26	18	25	−23.6	Stationary, Sep 14
Nov 5	18	30	−23.6	
Dec 15	18	38	−23.5	Conjunction, Dec 31

1991

Date	R.A.		Dec.	
Jan 24	18	49	−23.3	
Mar 5	18	57	−23.1	
Apr 14	19	01	−23.1	Stationary, Apr 18
May 24	18	58	−23.1	
Jul 3	18	52	−23.3	Opposition, Jul 4
Aug 12	18	46	−23.4	
Sep 21	18	43	−23.4	Stationary, Sep 19
Oct 31	18	47	−23.4	
Dec 10	18	55	−23.2	

1992

Jan	19	19	05	−23.0	Conjunction, Jan 5
Feb	28	19	14	−22.7	
Apr	8	19	18	−22.6	Stationary, Apr 22
May	18	19	17	−22.7	
Jun	27	19	12	−22.8	Opposition, Jul 7
Aug	6	19	05	−23.0	
Sep	15	19	02	−23.1	Stationary, Sep 23
Oct	25	19	04	−23.1	
Dec	4	19	10	−22.9	

1993

Jan	13	19	20	−22.6	Conjunction, Jan 8
Feb	22	19	30	−22.3	
Apr	3	19	35	−22.1	
May	13	19	36	−22.1	Stationary, Apr 26
Jun	22	19	31	−22.3	
Aug	1	19	25	−22.5	Opposition, Jul 12
Sep	10	19	20	−22.6	Stationary, Sep 27
Oct	20	19	20	−22.6	
Nov	29	19	26	−22.4	

1994

Jan	8	19	36	−22.1	Conjunction, Jan 12
Feb	17	19	46	−21.7	
Mar	29	19	52	−21.4	
May	8	19	54	−21.4	Stationary, May 1
Jun	17	19	50	−21.5	
Jul	27	19	44	−21.8	Opposition, Jul 17
Sep	5	19	38	−22.0	
Oct	15	19	38	−22.1	Stationary, Oct 2
Nov	24	19	42	−21.9	

1995

Jan	3	19	51	−21.5	Conjunction, Jan 17
Feb	12	20	01	−21.1	
Mar	24	20	08	−20.7	
May	3	20	11	−20.6	Stationary, May 5
Jun	12	20	09	−20.7	
Jul	22	20	03	−21.0	Opposition, Jul 21
Aug	31	19	57	−21.3	
Oct	10	19	55	−21.4	Stationary, Oct 6
Nov	19	19	58	−21.2	
Dec	29	20	06	−20.8	

1996

Feb	7	20	16	−20.3	Conjunction, Jan 21
Mar	18	20	24	−19.9	
Apr	27	20	28	−19.7	Stationary, May 9
Jun	6	20	27	−19.8	
Jul	16	20	22	−20.1	Opposition, Jul 25
Aug	25	20	15	−20.4	

Date	R.A.	Dec.	
Oct 4	20 12	−20.6	Stationary, Oct 10
Nov 13	20 14	−20.5	
Dec 23	20 21	−20.1	
1997			
Feb 1	20 30	−19.5	Conjunction, Jan 24
Mar 13	20 39	−19.0	
Apr 22	20 44	−18.7	
Jun 1	20 44	−18.7	Stationary, May 13
Jul 11	20 40	−19.0	Opposition, Jul 29
Aug 20	20 34	−19.4	
Sep 29	20 29	−19.7	Stationary, Oct 14
Nov 8	20 30	−19.6	
Dec 18	20 36	−19.2	
1998			
Jan 27	20 45	−18.7	Conjunction, Jan 28
Mar 8	20 54	−18.1	
Apr 17	21 00	−17.7	
May 27	21 02	−17.6	Stationary, May 17
Jul 6	20 58	−17.9	
Aug 15	20 52	−18.3	Opposition, Aug 3
Sep 24	20 47	−18.6	
Nov 3	20 46	−18.7	Stationary, Oct 19
Dec 13	20 51	−18.3	
1999			
Jan 22	20 59	−17.8	Conjunction, Feb 2
Mar 3	21 08	−17.1	
Apr 12	21 15	−16.6	
May 22	21 18	−16.4	Stationary, May 22
Jul 1	21 15	−16.7	
Aug 10	21 10	−17.1	Opposition, Aug 7
Sep 19	21 04	−17.5	
Oct 29	21 02	−17.6	Stationary, Oct 23
Dec 8	21 05	−17.3	
2000			
Jan 17	21 13	−16.8	
Feb 26	21 22	−16.1	Conjunction, Feb 6
Apr 6	21 30	−15.5	
May 16	21 34	−15.2	Stationary, May 25
Jun 25	21 32	−15.4	
Aug 4	21 27	−15.8	Opposition, Aug 11
Sep 13	21 21	−16.2	
Oct 23	21 18	−16.5	Stationary, Oct 26
Dec 2	21 20	−16.3	

Neptune

The right ascension and declination of Neptune, for epoch 2000, are given for every 40th day throughout the year. During the period 1988–2000, this planet will be passing slowly through Sagittarius and entering Capricornus, shining at about magnitude 7.8.

The planet follows its path around Zodiac in a series of very flat "loops," effectively backtracking in a westerly direction along its previous course for over five months, and then continuing in an easterly direction for the remainder of the year. The chart shows its approximate position in the sky at the time of each opposition; but for positive identification its accurate location should be plotted on a star map using the 40-day positions given here.

THE POSITION OF NEPTUNE AT OPPOSITION, 1988–2000

Date	R.A.	Dec.	
1988			
Jun 8	18^h 41^m	−22°.1	
Jul 18	18 37	−22.2	Opposition, June 30
Aug 27	18 33	−22.2	
Oct 6	18 33	−22.2	Stationary, Sep 18
Nov 15	18 37	−22.3	
Dec 25	18 43	−22.2	Conjunction, Dec 31
1989			
Feb 3	18 49	−22.1	
Mar 15	18 53	−22.0	
Apr 24	18 54	−21.9	Stationary, Apr 13
Jun 3	18 52	−22.0	
Jul 13	18 47	−22.1	Opposition, Jul 2
Aug 22	18 43	−22.2	
Oct 1	18 42	−22.2	Stationary, Sep 21
Nov 10	18 45	−22.2	
Dec 20	18 50	−22.1	

Date	R.A.		Dec.	
1990				
Jan 29	18	57	− 21.9	Conjunction, Jan 2
Mar 10	19	02	− 21.8	
Apr 19	19	04	− 21.8	Stationary, Apr 16
May 29	19	01	− 21.8	
Jul 8	18	57	− 21.9	Opposition, Jul 5
Aug 17	18	53	− 22.0	
Sep 26	18	52	− 22.1	Stationary, Sep 23
Nov 5	18	53	− 22.1	
Dec 15	18	59	− 22.0	
1991				
Jan 24	19	05	− 21.8	Conjunction, Jan 5
Mar 5	19	11	− 21.7	
Apr 14	19	13	− 21.6	Stationary, Apr 18
May 24	19	11	− 21.6	
Jul 3	19	07	− 21.7	Opposition, Jul 8
Aug 12	19	03	− 21.8	
Sep 21	19	01	− 21.9	Stationary, Sep 26
Oct 31	19	02	− 21.9	
Dec 10	19	07	− 21.8	
1992				
Jan 19	19	13	− 21.6	Conjunction, Jan 7
Feb 28	19	19	− 21.5	
Apr 8	19	22	− 21.4	Stationary, Apr 20
May 18	19	21	− 21.4	
Jun 27	19	17	− 21.5	Opposition, Jul 9
Aug 6	19	13	− 21.6	
Sep 15	19	10	− 21.7	Stationary, Sep 27
Oct 25	19	11	− 21.7	
Dec 4	19	15	− 21.6	
1993				
Jan 13	19	22	− 21.4	Conjunction, Jan 8
Feb 22	19	28	− 21.2	
Apr 3	19	31	− 21.1	Stationary, Apr 22
May 13	19	31	− 21.1	
Jun 22	19	28	− 21.2	Opposition, Jul 12
Aug 1	19	23	− 21.4	
Sep 10	19	20	− 21.5	Stationary, Sep 30
Oct 20	19	20	− 21.5	
Nov 29	19	23	− 21.4	
1994				
Jan 8	19	29	− 21.2	Conjunction, Jan 11
Feb 17	19	36	− 21.0	
Mar 29	19	40	− 20.8	
May 8	19	40	− 20.8	Stationary, Apr 25
Jun 17	19	38	− 20.9	
Jul 27	19	33	− 21.1	Opposition, Jul 14
Sep 5	19	30	− 21.2	

Date	R.A.		Dec.	
Oct 15	19	29	− 21.3	Stationary, Oct 2
Nov 24	19	32	− 21.2	
1995				
Jan 3	19	38	− 21.0	Conjunction, Jan 13
Feb 12	19	44	− 20.7	
Mar 24	19	49	− 20.5	
May 3	19	50	− 20.5	Stationary, Apr 27
Jun 12	19	48	− 20.6	
Jul 22	19	44	− 20.8	Opposition, Jul 17
Aug 31	19	40	− 20.9	
Oct 10	19	38	− 21.0	Stationary, Oct 5
Nov 19	19	41	− 20.9	
Dec 29	19	46	− 20.7	
1996				
Feb 7	19	52	− 20.4	Conjunction, Jan 16
Mar 18	19	57	− 20.2	
Apr 27	19	59	− 20.1	Stationary, Apr 29
Jun 6	19	58	− 20.2	
Jul 16	19	53	− 20.4	Opposition, Jul 18
Aug 25	19	49	− 20.6	
Oct 4	19	47	− 20.7	Stationary, Oct 6
Nov 13	19	49	− 20.6	
Dec 23	19	54	− 20.4	
1997				
Feb 1	20	00	− 20.1	Conjunction, Jan 17
Mar 13	20	05	− 19.9	
Apr 22	20	08	− 19.7	Stationary, May 1
Jun 1	20	07	− 19.8	
Jul 11	20	04	− 20.0	Opposition, Jul 21
Aug 20	19	59	− 20.2	
Sep 29	19	57	− 20.3	Stationary, Oct 8
Nov 8	19	58	− 20.3	
Dec 18	20	02	− 20.1	
1998				
Jan 27	20	08	− 19.8	Conjunction, Jan 19
Mar 8	20	14	− 19.5	
Apr 17	20	17	− 19.3	Stationary, May 4
May 27	20	17	− 19.4	
Jul 6	20	14	− 19.5	Opposition, Jul 23
Aug 15	20	09	− 19.8	
Sep 24	20	06	− 19.9	Stationary, Oct 11
Nov 3	20	07	− 19.9	
Dec 13	20	10	− 19.8	
1999				
Jan 22	20	16	− 19.5	Conjunction, Jan 22
Mar 3	20	21	− 19.1	
Apr 12	20	26	− 18.9	

Date	R.A.	Dec.	
May 22	20 26	− 18.9	Stationary, May 7
Jul 1	20 23	− 19.1	
Aug 10	20 19	− 19.3	Opposition, Jul 26
Sep 19	20 16	− 19.5	
Oct 29	20 16	− 19.5	Stationary, Oct 13
Dec 8	20 19	− 19.4	

2000

Date	R.A.	Dec.	
Jan 17	20 24	− 19.1	Conjunction, Jan 24
Feb 26	20 30	− 18.7	
Apr 6	20 34	− 18.5	
May 16	20 35	− 18.4	Stationary, May 8
Jun 25	20 33	− 18.6	
Aug 4	20 29	− 18.8	Opposition, Jul 27
Sep 13	20 25	− 19.0	
Oct 23	20 24	− 19.1	Stationary, Oct 15
Dec 2	20 27	− 19.0	

Index